RELIABILITY ANALYSIS AND ASSET MANAGEMENT OF ENGINEERING SYSTEMS

RELIABILITY ANALYSIS AND ASSET MANAGEMENT OF ENGINEERING SYSTEMS

Gilberto Francisco Martha de Souza
Department of Mechatronics and Mechanical Systems Engineering, Escola Politécnica – Universidade de São Paulo, São Paulo, Brazil

Adherbal Caminada Netto
Department of Mechanical Engineering, Escola Politécnica – Universidade de São Paulo, São Paulo, Brazil

Arthur Henrique de Andrade Melani
Department of Mechatronics and Mechanical Systems Engineering, Escola Politécnica – Universidade de São Paulo, São Paulo, Brazil

Miguel Angelo de Carvalho Michalski
Department of Mechatronics and Mechanical Systems Engineering, Escola Politécnica – Universidade de São Paulo, São Paulo, Brazil

Renan Favarão da Silva
Department of Mechatronics and Mechanical Systems Engineering, Escola Politécnica – Universidade de São Paulo, São Paulo, Brazil

ELSEVIER

Elsevier
Radarweg 29, PO Box 211, 1000 AE Amsterdam, Netherlands
The Boulevard, Langford Lane, Kidlington, Oxford OX5 1GB, United Kingdom
50 Hampshire Street, 5th Floor, Cambridge, MA 02139, United States

Library of Congress Cataloging-in-Publication Data
A catalog record for this book is available from the Library of Congress

British Library Cataloguing-in-Publication Data
A catalogue record for this book is available from the British Library

ISBN: 978-0-12-823521-8

Publisher: Matthew Deans
Acquisitions Editor: Brian Guerin
Editorial Project Manager: Gabriela D. Capille
Production Project Manager: Prasanna Kalyanaraman
Cover Designer: Mark Rogers

Typeset by STRAIVE, India

Dedication

To my wife Cleusa Cristina, our son Leonardo, our daughter Livia, and my mother Maria da Conceição.
Gilberto Francisco Martha de Souza

To the memory of my dear wife Lucia, to our children and grandchildren, to whom I have already dedicated my life.
Adherbal Caminada Netto

To my parents, Teresa Cristina and Marcos, and my sister, Amanda.
Arthur Henrique de Andrade Melani

To my wife Ranny and our boys, Natan and Otto.
Miguel Angelo de Carvalho Michalski

To my parents, Sueli and João.
Renan Favarão da Silva

Contents

Preface

At its heart engineering is about using science to find creative practical solutions. It's a noble profession.

Elizabeth Alexandra Mary Windsor (1926)

With the increasing advances in computational capacity, engineering design has come a long way from the times when slide rules and empirical manuals were the main resources at the disposal of designers and practitioners alike. Furthermore, materials science and engineering have experienced outstanding development in recent decades. Notwithstanding all the modern gadgetry and improved materials, however, the flawless engineering system remains elusive. In other words, faults will occur.

Modern engineering systems, which represent an enormous amount of invested capital in industries and sectors such as power, oil and gas, transportation, and processing plants, are incorporating increasing degrees of complexity, together with all its potential unpredictable emergent consequences and impact on humans, environment, and assets. This present state of affairs poses a formidable challenge for any competent asset management.

Therefore, terms like reliability, availability, health monitoring, risk analysis, and maintenance planning, once familiar only to limited circles of academia researchers and plant maintenance practitioners, are now under the spotlight and gaining increased attention from high-ranking personnel in the industry as well as in government.

Accordingly, this book deals with engineering systems asset management purporting that although faults may not be altogether avoidable, when timely detected and diagnosed by adequate methods they will not end up leading to catastrophic failures.

Numerous former and present students and colleagues of ours directly or indirectly contributed to the development of the material presented in this book. We would like to thank Prof. Bilal M. Ayyub at the University of Maryland, Prof. Silvio I. Nabeta at the University of São Paulo, Prof. Fernando J. G. Carazas at the National University of Colombia, Prof. Carmen E. P. Rodriguez at the University of Antioquia, Prof. Marjorie M. Belinello at the Federal University of Technology—Paraná, and Dr. Guilherme E. C. Barbosa, Dr. Marcos A. Coque Jr., Dr. Erick M. P. Hidalgo,

Dr. Carlos A. Murad, Mr. Andre G. Ramos, Mr. Italo S. de Melo, and Mr. Alécio J. Silva.

Some of the research that contributed to the development of this book had the financial support received from several different sources over several decades, including the Brazilian National Council for Scientific and Technological Development (CNPq), the Coordination for the Improvement of Higher Education Personnel (CAPES), the Brazilian Innovation Agency (FINEP), the Brazilian Navy, the University of São Paulo, Electricity of Portugal (EDP) and AES Tietê.

The authors would like to dedicate this book to all their colleagues who in every corner of the world strive to progress the vital and noble science and art of engineering.

São Paulo, summer of 2021.

CHAPTER

1

Introduction

Science cannot predict what is going to happen. Science can only predict the probability that something will happen (free translation).

César Lattes (1924–2005)

Modern engineering systems are designed based on design criteria whose hypotheses are based on information provided by basic sciences such as physics, mathematics, chemistry, biology, and others. The design criteria have a validity region that allows the design of engineering systems that will have a reduced probability of not achieving the desired operational performance. Despite the advances in basic sciences and design capacity, it is not possible to guarantee the design of a modern engineering system without assessing the likelihood of its failure.

1.1 Analysis of modern engineering systems

During recent years, the industrial sector that comprises value added in mining, manufacturing, construction, electricity, water, and oil and gas represents around 27% of the world's gross domestic product (The World Bank, 2020). To keep the productivity of this important economic industrial sector production is nowadays characterized by the use of modern pieces of equipment with complex communications between them. Developments in cloud computing technology lead manufacturing industries to face many opportunities in the field of production and maintenance.

In addition to the search for the efficiency of the production process, which includes achieving the product quality requirements, the industrial sector has also shown concerns with the safety of processes, in terms of not only preventing economic losses but also controlling the threat to human life and the environment. Accordingly, the problem of preventing accidents in a large industrial enterprise becomes especially relevant mainly

https://doi.org/10.1016/B978-0-12-823521-8.00001-3

in continuous production systems of asset-intensive industries such as oil and gas, chemical, and mining. For those industries, the culture of Health Safety, Security and Environment (HSSE) has become an organizational priority.

Aiming at avoiding production loss of performance or even interruption and accidents that may jeopardize operational safety, the industrial sites prioritize the use of modern machinery coupled to monitoring systems aiming at detecting any evidence of abnormal operation. Once this abnormal condition is detected, the system can be reconfigured to avoid loss of operational safety or even the production of low-quality items. So, the industrial assets may be considered complex engineering systems. One may define a complex engineering system as a system in which phenomena emerge as a consequence of multistate interaction among the system's components and their operational environments (Oliver, Kelliher, & Keegan, 1997).

The modern industrial assets are composed of pieces of machinery that can also be considered complex. Modern machinery architecture presents electromechanical components and sensing and controlling devices, including hardware and software. The design requirements should take into account very broad operational contexts associated not only with the production of a great variety of parts and products but also with the production management processes, including maintenance planning.

Eppinger and Browning define product or system architecture as "the arrangement of components interacting to perform specified functions," noting that architecture is represented by individual components, their relationships to one another and the environment, and principles guiding design (Eppinger & Browning, 2012). For complex engineering systems, there may be strong dependencies between the functioning of components. Those dependencies may cause difficulties in decomposing the product or system into smaller elements, such as subsystems, modules, and components that must be integrated to work together and achieve performance objectives.

In these circumstances, the reliability of an industrial asset is of particular importance. The potential damage caused by an accident at such an asset may lead to substantial material losses and, in some cases, can even cause a loss of human lives. That is why industrial asset reliability is immensely important. In terms of their reliability, industrial facilities are divided into simple and complex. Most of the industrial assets may be classified as complex engineered systems since they may be reconfigured aiming at avoiding a loss of performance or even an operational condition that could be risky to the plant (Lin, Li, & Kang, 2013).

Reliability, resilience, risk analysis, maintenance, and prognostics modeling and assessment for modern engineering systems have become very challenging due to factors such as complex dependency/interfaces, functional interactions among mechanical, electrical, software, and control subsystems, as well as the interaction with external environments. In view of the development of Industry 4.0 concepts, the risk analysis and maintenance planning tasks may take advantage of that concept considering the possibility of detecting in advance components' degradations that may affect system safety or performance (Zio, 2016).

The challenges of modeling the reliability as well as system performance prognosis of modern engineering systems are important research topics. The reliability modeling must consider the possibility of the systems adapting their configurations with different levels of autonomy as a function of data collected by several monitoring systems. Modern engineering systems often have multicomponents and the dynamic functional structure managed based on online data collection and analysis, with or without human supervision. The problem of failure interaction between components cannot be neglected since the degradation of a component may cause an overload in other components that affects their reliability. The differences in components' ages, due to maintenance substitution at different moments during system operational life, and different levels of technology (including technology readiness level) are also problems related to modern engineering systems' reliability modeling since they may be designed to operate for long periods of time and submitted to maintenance, updating, and even integration of new technologies.

In the age of Industry 4.0 and Industrial Internet of Things (IoT), the availability of data storage, processing, and analysis techniques such as data lake, data mining, big data, and cloud computing enables more informed decisions to be made in real time by using the database extracted from sensors in its equipment. During a system operational campaign, the Health Monitoring System (HMS) must identify component degradation and how this degradation is affecting the performance of the system. Through a structure supported by Artificial Intelligence (AI), data lake, and cloud computing, the HMS is the basis of the fault detection system that allows the indication of alarms to prevent failures. Integration of cloud platform with a standard condition monitoring technique reveals a new scope to develop technologies for modern complex engineering systems (Farsi & Zio, 2019).

The evolution of the HMS is the Prognostics and Health Management (PHM). The PHM is a field of reliability engineering that consists of the detection of the engineering component degradation, the fault-type diagnosis, the failure time prediction, and managing the maintenance planning aiming at avoiding critical failures. For modern engineering systems, the PHM can be used to predict the likelihood of occurrence of critical failures during an operational campaign based on determining the amount of the deviation or degradation of monitored process parameters from their expected normal situations. This estimate is supported by the evaluation of the residual life, also known as the Remaining Useful Life (RUL), of the piece of equipment after the detection of the abnormal monitoring signal.

Considering the advances in AI methods used in Industry 4.0 concept, the PHM system may be developed based on machine learning methods. The application of those methods requires the use of monitoring data to train the detection and diagnosis methods based on past failures. After the training processes, the AI methods are used not only to detect and diagnose faults but also to support maintenance planning decisions in asset management.

The use of the PHM process to evaluate the health and safety of modern engineering systems is important to guarantee any asset integrity. The asset integrity goal is to ensure that important equipment will be suitable for its intended application throughout an operational campaign. The systematic implementation of activities, such as inspections and tests, are basic pillars of asset integrity management. The use of risk-based and reliability techniques may support the definition of critical pieces of equipment that in case of failure may threaten system performance and safety. PHM techniques may be useful to avoid the failures of critical pieces of equipment and must be used to support asset integrity management.

Considering the discussion presented in the previous paragraphs, one can identify that the integrity of critical assets may be evaluated through the application of risk analysis and reliability techniques aiming at defining critical pieces of equipment and their failure modes that in case of occurrence may cause loss of performance. The main difficulty for the analysis is that most of the critical assets can be considered modern engineering systems and there are no consolidated methodologies to support that analysis.

1.2 Book proposal

The main goal of this book is to provide and discuss methods that can be used to evaluate the reliability and availability of modern engineering systems including simulation-based methods. Fault detection and diagnosis techniques based on AI are also discussed. The combination of reliability and AI techniques is proposed to compose a framework that can be used to define asset critical pieces of equipment and their failure modes, and for those failure modes, a PHM method is proposed to support asset integrity management.

The framework of the book development is presented in Fig. 1.1, considering the context of engineering systems. The reliability, availability, and risk analysis techniques may be considered as basic concepts supporting modern engineering systems' design in accordance with stakeholders'

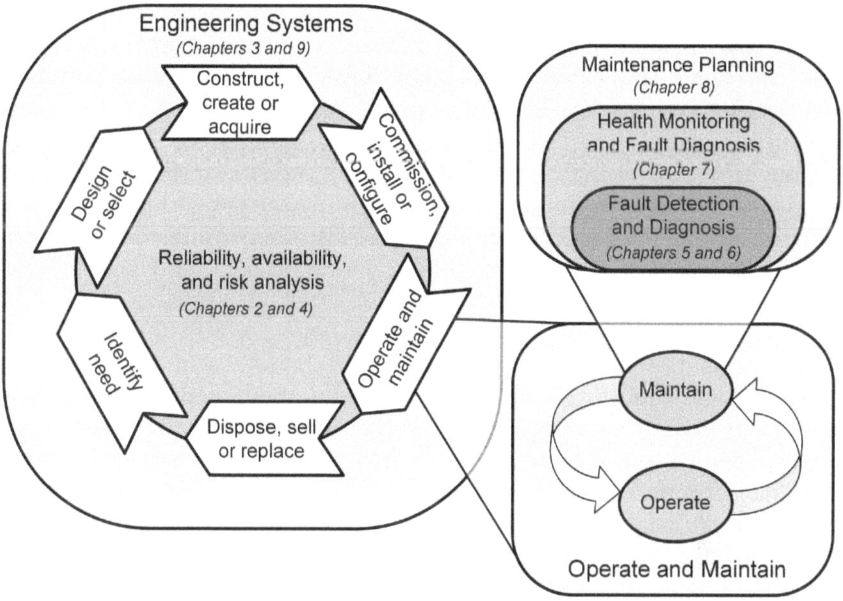

FIG. 1.1 Book development framework.

requirements. Based on the reliability and operational risk requirements, the engineering team must use the concepts of system engineering to synthesize the functional architecture of the system under design. Considering the functional needs, the designer must select the components and devices that will be part of the systems. Aiming at achieving design requirements the representation, modeling, and quantification of reliability must be executed. If necessary, redundancy allocation may be proposed. As part of reliability and risk analysis, the critical components must be identified and components manufacturing and assembly requirements must be defined in order to avoid imperfections and lack of quality that may affect the system reliability.

Although the system is designed in accordance with the requirements of operational reliability, its components must be subjected to maintenance actions to control the performance and availability of the system throughout its operational life. The maintenance planning of complex engineering systems affects the system life cycle costs. The maintenance planning not only has to control the degradation of critical components but also has to control system availability. Considering the evolution of monitoring systems and AI techniques, the designers must propose the use of PHM techniques to control system performance degradation and operational costs during an operational campaign. The use of PHM may be considered as one important pillar to support asset integrity management throughout the system's operational life.

To execute the analysis associated with the framework proposed in Fig. 1.1, the design team must have knowledge about reliability, risk analysis, and maintenance concepts including their relation with complex engineering systems' design philosophies. For these systems, some more advanced reliability and availability analysis techniques may be used aiming at modeling the possibility of system reconfiguration in the case of critical components' failure and representing the dependency between failures. Simulation-based methods may be also used to analyze the effects of system reconfiguration on production efficiency and operational safety and also to verify whether a given maintenance strategy is suitable for the system availability requirements that may include limits for operational costs.

The design teams must also have knowledge about methods that can be used to process monitoring databases in an integrated way to detect any degradation (health monitoring) of systems' performance that may support the definition of the maintenance schedule. Those methods may be based on artificial intelligence techniques including machine learning methods based on statistical techniques. Coupled with the detection method, diagnosis methods may be used to define the most probable set of pieces of equipment that may be causing the system's degradation.

The relations between those evaluations and the risk-based asset management goals proposed by ISO 55000 standard will be analyzed focusing on the effectiveness of maintenance planning to reduce the operational risk of modern systems. The presentation of those techniques applied to evaluate the reliability and availability of modern engineering systems are the main features of this book aiming at supporting the application of asset management tools.

1.3 Book organization

To satisfy the objectives of the book, the chapters gradually present several concepts, first addressing reliability, risk analysis, maintenance planning, and design of modern engineering systems and later presenting advanced concepts required to implement the prognostics and health management process of modern engineering systems, including the presentation of some practical problems.

After this brief introduction about the book's goals, Chapter 2 presents the basic concepts of reliability, risk analysis, and maintenance planning of engineering systems. Chapter 3 presents the basic concepts of modern engineering systems' design characteristics, including the fundamental of system engineering, concepts of robust design, and model-based engineering design, associating them with the basic features of modern engineering systems' architecture such as the composition of pieces of equipment with different nature—electronic, mechanical, and software, the possibility of reconfiguration, and interdependencies between pieces of equipment functions. Chapter 4 discusses the modern engineering systems' reliability and availability modeling including methods capable of dealing with system reconfiguration and dependencies between failures such as the Markov model, dynamic fault tree analysis, and simulation models such as Petri nets.

After presenting the concepts associated with reliability, risk analysis, and maintenance modeling and planning, Chapters 5–8 address the prognosis and health management process of modern engineering systems. Chapter 5 presents a brief overview of the fault detection methods used for health monitoring of engineering systems such as analytical (model-based), knowledge-based, signal-based, and data-driven methods. The main features, their relation to artificial intelligence requirements, and limitations of those methods are discussed. The main data sources for fault detection are addressed in this chapter. Chapter 6 presents the definition of fault diagnosis methods and their application to machinery fault root cause analysis. A brief overview of some of the following diagnosis methods is presented: rule-based methods, pattern recognition methods, and inference methods.

Chapter 7 presents a framework for Health Monitoring and Fault Diagnosis (HMFD) of modern engineering systems in view of the methods presented in Chapters 5 and 6. The main features and limitations of the framework are addressed. The relationship between the framework and artificial intelligence requirements is also discussed. Chapter 8 discusses the maintenance planning specificities associated with engineering systems aiming at supporting the application of the ISO 55000 standard for asset management. A brief description of the ISO 55000 series is presented and a maintenance planning framework based on ISO 55001 requirements is proposed considering the prognosis and health monitoring process associated with modern engineering systems.

Chapter 9 presents examples of reliability analysis and the application of the framework presented in Chapter 7. Two modern engineering systems are analyzed. The first application involves the analysis of a Combined Diesel or Gas (CODOG) naval propulsion that is a type of propulsion system where for every propeller shaft there are two diesel engines for cruising speed and one geared gas turbine for high-speed dashes. Both types of engines are connected to the shaft with clutches, but only one system is driving the ship at any time. The failure of some propulsion plant components causes its interruption and can leave the ship adrift. The PHM process for those systems may reduce the risk associated with their operation. The second application evaluates the power generation system of a Floating Production Storage and Offloading (FPSO) vessel used by the oil industry for exploration (production), storage of oil and/or natural gas, and offloading of production to tankers. The power generation system is critical since its failure causes the interruption of oil exploration and processing. Finally, Chapter 10 presents some reflections on the models proposed in the book, highlighting the advantages and limitations of their applications.

References

Eppinger, S. D., & Browning, T. R. (2012). *Design structure matrix methods and applications*: (p. 352). MIT Press.

Farsi, M. A., & Zio, E. (2019). Industry 4.0: Some challenges and opportunities for reliability engineering. *International Journal of Reliability, Risk and Safety: Theory and Application, 2*(1), 23–34. https://doi.org/10.30699/IJRRS.2.1.4.

Lin, Y., Li, D., & Kang, R. (2013). Reliability modelling and simulation of complex systems. *Chemical Engineering Transactions, 33*, 469–474. https://doi.org/10.3303/CET1333079.

Oliver, D. W., Kelliher, T. P., & Keegan, J. G., Jr. (1997). *Engineering complex systems with models and objects*: (p. 325) (1st ed.). McGraw-Hill.

The World Bank (2020). *Table 4.2 World development indicators: Structure of output*: (p. 5). (2020). http://wdi.worldbank.org/table/4.2.

Zio, E. (2016). Challenges in the vulnerability and risk analysis of critical infrastructures. *Reliability Engineering & System Safety, 152*, 137–150. https://doi.org/10.1016/j.ress.2016.02.009.

CHAPTER

2

Reliability and maintenance fundamentals

In God we trust, all others must bring data.

W. Edwards Deming (1900–1993)

All engineering design codes present recommendations consisting of equations that model a given failure mechanism, the external loads and material properties that must be considered in the analysis, and recommend safety factors that represent the uncertainties associated with the failure mechanism models and the magnitude of the parameters used in the analysis.

The uncertainty modeling in engineering design affects the safety associated with the product or system. The uncertainty analysis must be based on the evaluation of experimental data collected in the laboratory or based on in-field observations.

The probability of failure of a product or engineering system during an operational campaign has become the subject of studies since World War II, initially motivated by the development of electronic-assisted weapons such as the radar and the V1 and V2 bombs (Saleh & Marais, 2006). These studies have their interest increased in the last years due to the development of complex engineering systems, but they are still dependent on reliable data, as cited by Deming when developing statistical quality control studies.

The present chapter presents some basic concepts associated with equipment reliability and their application to modern engineering systems. For these systems, besides reliability requirements, the safety during its operation may be another design requirement. Safety analysis is usually executed based on risk analysis concepts whose basic aspects were proposed by the nuclear industry. The basic risk analysis concepts and techniques are also analyzed in this chapter. Finally, some maintenance

https://doi.org/10.1016/B978-0-12-823521-8.00005-0

concepts are also presented at the end of the chapter since modern engineering systems must rely on maintenance actions to support their performance throughout their entire operational life.

It is not the intention of this chapter to present a comprehensive review of the techniques used in reliability, risk analysis, and maintenance studies. Nevertheless, the concepts discussed in this chapter aim to present to the reader the necessary information to understand the methods used in the other chapters of the book.

2.1 Introduction

The design of engineering systems is based on requirements that express the needs of consumers or markets in terms of engineering language. Defining and expressing the design requireents is normally the initial step for an engineering system development project. Design requirement identification is an iterative process that coevolves with the product development process. Deficiencies in requirements could lead to the waste of time and money and even the failure of the project (Li, Zhang, & Ahmed-Kristensen, 2014).

The development and introduction into the use of an engineering system inherently require increasingly large commitments of resources as it progresses from concept through engineering, production, and operational use. The better the requirements are established, the less the need for design updates during the entire system life cycle. The translation of these needs into design requirements allows the design team to execute functional allocation and to select the components responsible for performing them. According to Kossiakoff, Sweet, Seymour, and Biemer (2011), the design requirements can be classified into four categories, which are:

i. *Operational requirements*, which refer to the mission and purpose of the system. The set of operational requirements will describe and communicate the user's objectives with the use of the system after the system is deployed and operated and may be named capability requisites;
ii. *Functional requirements*, which refer largely to what the system should do. These requirements should be action-oriented and should describe the tasks or activities that the system performs during its operation. Those requirements must be preferentially quantitative in order to be measured during system commissioning and operational phases;
iii. *Performance requirements*, which refer to how well the system should perform its requirements and affect its environment and must be also defined quantitatively; and

iv. *Physical requirements*. These refer to the characteristics and attributes of the physical system and the physical constraints placed upon the system design and may include appearance, general characteristics, such as volume and weight.

Modern engineering systems such as electric power plants, offshore oil and gas exploring facilities, and other infrastructure assets must achieve high performance with minimum operational life cycle costs. To achieve these goals, the designers must consider as performance requirements those that deal with the asset operational safety and maintenance costs. Reliability, availability, and safety are characteristics that will be considered as important operational requirements for complex engineering systems since they are related to the prediction of systems performance in the future.

2.2 Reliability concepts

Reliability of a device can be defined as the possibility of performing its required function, for a specified length of time, related to the required operational campaign, under given conditions, including loading and environmental aspects, in a specified operational context defined by the operational requirements.

According to MIL-HBDK-338 (DoD, 1998), reliability is defined as:

i. the duration or probability of failure-free performance under stated conditions or
ii. the probability that an item can perform its intended function for a specified interval under stated conditions.

The second definition is more suitable to be used for systems in as much as it is equivalent to the definition of mission reliability.

The definition of reliability according to ISO/IEC 2382-14 (ISO, 1997) is the ability of a functional unit to perform a required function under given conditions for a given time interval. It is worth mentioning that there are definitions proposed by other authors in many books and papers, but all of them consider three basic pillars: time interval (or mission duration), required performance, and operational context.

Usually, the specified time interval is considered as a parameter t, and the probability $P(T > t)$ that the random variable time to failure T will be greater than t is given by the reliability function $R(t)$, also referred to as the survival function, as shown in Eq. (2.1):

$$R(t) = P(T > t) \tag{2.1}$$

The reliability function expresses the probability that the device operates without failure in a time period t, therefore, considering that a system that does not fail for $T \leq t$ must fail at some $T > t$, a direct relationship must be built between reliability and probability of failure, as presented in Eq. (2.2).

$$R(t) = P(T > t) = 1 - P(T \leq t) = 1 - F(t) \tag{2.2}$$

where $F(t)$ is the Cumulative Distribution Function (CDF) of the time to failure that gives the probability, $P(T \leq t)$, that the time to failure T will be smaller or equal than the specified time t, or, the probability that the device will fail before time t.

The databases used for reliability analysis are based on experimental analysis and present time to failure of samples that are tested according to stated conditions listed in reliability requirement. So, Eq. (2.2) is usually used to define any device reliability. Failure is defined as the event, or inoperable state, in which any component does not perform as previously specified. Based on the reliability definition, the way in which the component or device can fail to perform its intended function is named failure mode.

The cumulative distribution function $F(t)$ and the reliability function $R(t)$ are related to the Probability Density Function (PDF) $f(t)$ of the time to failure by Eqs. (2.3) and (2.4), where $f(t)$ describes how the failure probability is spread over time.

$$f(t) = \frac{dF(t)}{dt} \tag{2.3}$$

$$f(t) = -\frac{dR(t)}{dt} \tag{2.4}$$

The basic properties of the probability density of the time to failure are as follows:

i. $f(t)$ is always nonnegative and
ii. the total area beneath $f(t)$ is always equal to one.

Fig. 2.1 illustrates the link between the reliability function $R(t)$, the cumulative distribution function $F(t)$, and the probability density function $f(t)$ of the random variable time to failure T.

Once the probability density distribution that represents a component's reliability is defined, it may be interesting to compute the moments of the distribution. Moments contain less information than the probability distribution, but they are useful to summarize some characteristics of the distribution. The most common characteristic associated with the reliability distribution is the mean, or first raw moment, named Mean Time To Failure (MTTF), which is the expected or mean value of the time to failure T. This value is calculated as presented in Eq. (2.5):

$$MTTF = \int_0^{\infty} t \cdot f(t) dt \tag{2.5}$$

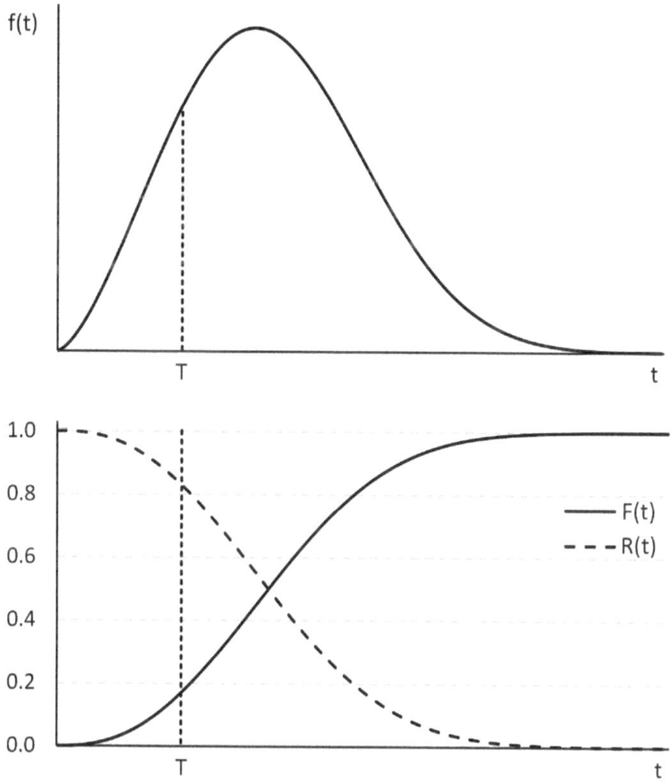

FIG. 2.1 Reliability function, cumulative distribution function of the time to failure, and failure probability density function.

It is important to note that although MTTF is an important metric, it does not indicate the characteristics of the reliability distribution of the device under analysis. Attempting to use a single number to describe an entire lifetime distribution can be misleading and may lead to poor business decisions. The reliability of a device should be specified as a percentage value with an associated time.

The next section presents the definition of component failure rate which is an important measure that can be defined from the reliability distribution. The failure rate represents the frequency with which a component fails. The knowledge about component failure rate allows the reliability analyst to gain additional insight into the nature of the component failure.

2.2.1 Failure rate function

The instantaneous hazard function, or hazard rate, often denoted by instantaneous failure rate $\lambda(t)$, measures, for nonrepairable components, the probability that a surviving component will fail in the next infinitesimal time interval Δt. The failure rate is the limit of the probability that a failure occurs per unit time interval Δt given that no failure has occurred before time t. It can be written as shown in Eq. (2.6):

$$\lambda(t) = \lim_{\Delta t \to 0} \frac{P[(t < T < t + \Delta t) | T > t)]}{\Delta t} = \frac{f(t)}{R(t)} \tag{2.6}$$

Replacing $f(t)$ given by Eq. (2.4) into Eq. (2.6) leads to Eq. (2.7):

$$\lambda(t) = \frac{1}{R(t)}\left(-\frac{dR(t)}{dt}\right) \tag{2.7}$$

The hazard or failure rate function of nonrepairable components and devices over operating time is usually represented by a curve with a bathtub shape, characterized by three distinct regions, as illustrated in Fig. 2.2. The first part is a decreasing failure rate, known as early failures or infant mortality; the second part is a constant failure rate, known as random failures; the third part is an increasing failure rate, known as wear-out failures. The bathtub curve, displayed in Fig. 2.2, does not depict the failure rate of a single item but describes the idealized relative failure rate of an entire population of components and devices over time.

The period of time on the left side of Fig. 2.2 is a region of high but decreasing failure rates. This is referred to as the period of infant mortality, or early failures. Most of the failures in the infant mortality region are lack of quality and result from inherent defects due to poor manufacturing and assembly, including human error during installation and operation. It is expected that the number of devices presenting low quality represents a very minor portion of the entire population so the high failure rate tends to decrease rapidly. The statistical quality control techniques used in manufacturing processes aim at reducing variability and consequently minimize the occurrence of infant mortality failures.

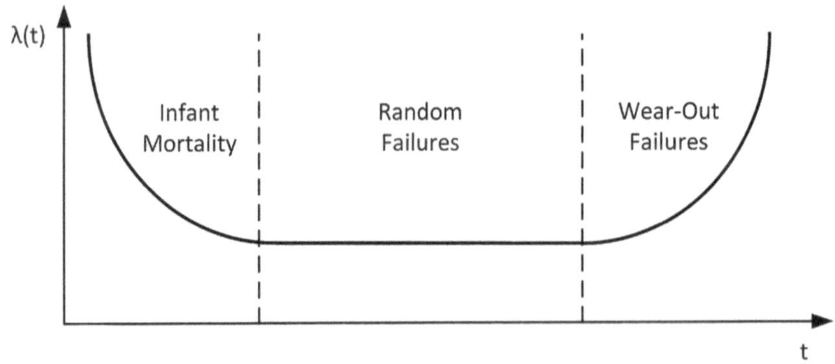

FIG. 2.2 Typical bathtub curve.

The middle section of the bathtub curve contains the smallest and almost nearly constant failure rates and is referred to as the useful life. This behavior is characteristic of failures caused by random events and hence referred to as random failures. During this period of time, failures do not follow a predictable pattern and occur randomly due to the unexpected changes in operational loading rather than from any inherent defect in the device. The design criteria of the device are used to define the useful life because they aim at controlling cumulative damage physical phenomena during this period.

In the wear-out region, the failure rate increases with time as a result of irreversible aging or cumulative damage effects. The failures are attributed to degradation or wear-out, which accumulates and accelerates over time. A design goal for high-reliability components is to extend the life of components prior to the onset of wear-out.

The bathtub curve can be considered generic because not all types of devices always present all described regions since the failure rate curve is dependent on the nature of the device. For electronic devices, for example, on the one hand, the failure rate curve presents typically the infant mortality and useful life regions, that is, there is the presence of early failures and random failures, without wear characteristics. On the other hand, mechanical devices' failure rate curves present infant mortality and wear-out, in which early failure is associated with problems of manufacturing, assembly, or materials that may induce them. During use, mechanical equipment usually fails due to wear, corrosion, or fatigue. For those failure mechanisms, the damage is cumulative over the time of operation, until it reaches a value critical, precipitating failure.

The device reliability is modeled using some typical probability functions. That selection is based on the analysis of test data characterized by a specific pattern. The methods used to select the probability distribution, such as the maximum likelihood method and probability plotting, are detailedly presented in many books (Ayyub & McCuen, 2016; Lewis, 1994) and will not be discussed in this one.

Some characteristics of those distributions are briefly presented to support complex engineering systems' analysis. More details about those distributions may be found in literature (Ayyub & McCuen, 2016; Lewis, 1994; Modarres, Kaminskiy, & Krivtsov, 2016; O'Connor & Kleyner, 2011).

Four probability distributions are usually used to model devices' reliability: Weibull, normal, lognormal, and exponential. The probability density functions of those distributions are presented in Table 2.1.

The most commonly used probability function is the Weibull distribution that has the exponential distribution as one of the special cases (Prabhakar Murthy, Xie, & Jiang, 2003). The Weibull distribution is a three-parameter function widely used in reliability and life data analysis due to its versatility. Depending on the values of the parameters, the Weibull distribution can be used to model a variety of life behaviors. An

TABLE 2.1 Most usual probability density functions used to model reliability.

Distribution	Parameters	Probability density function equation
Two parameter Weibull distribution	η: scale factor β: shape parameter	$f(t) = \frac{\beta}{\eta}\left(\frac{t}{\eta}\right)^{\beta-1} e^{\left[-\left(\frac{t}{\eta}\right)^{\beta}\right]}$
Lognormal distribution	μ_{ln}: mean in the logarithm domain σ_{ln}: standard deviation in the logarithm domain	$f(t) = \frac{1}{t\sigma_{ln}\sqrt{2\pi}} e^{\left\{-\frac{[ln(t)-\mu_{ln}]^2}{2\sigma_{ln}^2}\right\}}$
Normal distribution	μ: mean σ: standard deviation	$f(t) = \frac{1}{\sigma\sqrt{2\pi}} e^{\left[-\frac{(t-\mu)^2}{2\sigma^2}\right]}$
Exponential distribution	λ: rate parameter	$f(t) = \lambda e^{(-\lambda t)}$

important aspect of the Weibull distribution is how the values of the shape parameter (β) affect its characteristics like the shape of the PDF, reliability, and failure rate curves. For $\beta = 1$, the Weibull distribution represents the exponential distribution characterized by a constant failure rate in time (e.g., the useful life region of the bathtub curve). The Weibull functions can model a changing failure rate in time and are widely used to model infant mortality ($\beta < 1$) and wear-out ($\beta > 1$).

The lognormal distribution is a two-parameter function that models a time to failure dataset whose logarithms are normally distributed. The lognormal distribution is commonly used to model the lives of devices whose failure modes are of a fatigue-stress nature. Since this includes most, if not all, mechanical systems, the lognormal distribution may have widespread application. The failure rate curve is very sensitive to variations in the standard deviation that is a distribution parameter and can model infant mortality and wear-out failures.

The normal distribution, although widely used for quality control problems, has minor application in reliability modeling mainly because the left-hand limit of the distribution extends to negative infinity that may be inappropriate for life data analysis. However, that model may be considered in the reliability analysis of mechanical devices when the failure rate represents the wear-out phase of the bathtub curve.

The previous analysis is applied to devices that may be tested in laboratory facilities under controlled conditions. The greater the device complexity, the greater the difficulty of performing reliability tests, as the failure delineation may be inaccurate given the diversity of failure modes of the device components and the variability of their effects on the device performance.

It is very difficult to perform reliability test for engineering systems given the diversity of failure modes that it can present. The reliability analysis of engineered systems is dependent on the component reliability. The time dependence of system failures can be viewed in terms of component reliability and failure mechanisms.

In the sequence of this section, the systems' reliability analysis is discussed based on the presentation of the techniques usually used to execute that investigation.

2.2.2 Systems' reliability

To extend reliability analysis to a system level, a definition of "system" shall be initially presented. An engineering system may be defined as a set of connected components or devices that operate together aiming at achieving an operational performance under a given operational context. Considering the definition of reliability, the system's operational performance is associated not only with the number of components but also with their functional arrangement, including the allocation of redundant components or devices for the most critical functions. The criticality of a function is related not only to the effects on system performance in case of loss or degradation of that function but also to the effects on the environment or threats to human beings.

It is very difficult to perform reliability tests with engineering systems due to the difficulty in simulating the operational context in a test facility. Some tests may be executed with prototypes although the failure database is used for not only reliability predictions but also design performance improvement.

Sometimes a failure database may be built on information gathered during the field operation of a group of units of the system. The database supports reliability analysis that will represent the actual system performance. In case of nonacceptable reliability performance, the system must suffer a design review that is usually costly.

The design team must apply systems reliability analysis techniques to predict the system reliability during the design process aiming at achieving the reliability target defined for that system. Due to the increasing complexity of engineering systems, the identification and evaluation of criticality with the failure of individual components is usually the starting point for efficient reliability analysis. For this purpose, several techniques such as Fault Tree Analysis (FTA), Failure Mode and Effects Analysis (FMEA), and Reliability Block Diagram (RBD) are traditionally used during the design process (Ayyub & McCuen, 2016; Lewis, 1994; Modarres et al., 2016; O'Connor & Kleyner, 2011). Those techniques are presented in the next sections. In addition to these techniques, there are more advanced methods that will be presented in Chapter 4.

To develop a system reliability analysis, the analyst must represent the functional links between the components and devices that constitute the system. That analysis is commonly executed using the functional tree, which is a diagram showing the dependencies between the functions of a system as shown in Fig. 2.3. The functions of all components and devices are linked together in a hierarchical way and the model resulting from this diagram illustrates what will be done by the systems and by each of its components.

2.2.2.1 *Reliability block diagram*

Based on the functional tree, the analyst can develop the system RBD, which is a diagrammatic method for showing how component reliability contributes to the success or failure of a system operation. An RBD is drawn as a series of blocks connected in a specific configuration aiming at defining system reliability based on the reliability of its components. The system is breaking down in terms of reliability analysis chosen by the analyst but each block must have its reliability defined in order to estimate system reliability.

The most used configuration types for system reliability are as follows:

- **i.** Series configuration,
- **ii.** Simple parallel configuration,
- **iii.** Combined (series and parallel) configuration,
- **iv.** Standby redundancy, and
- **v.** K-out-of-N parallel configuration.

These configurations are represented in Fig. 2.4.

In a series configuration, a failure of any component results in the failure of the entire system, i.e., all components must be performing their functions so the system will perform its intended function. If the system is composed of n components with statistically independent failure

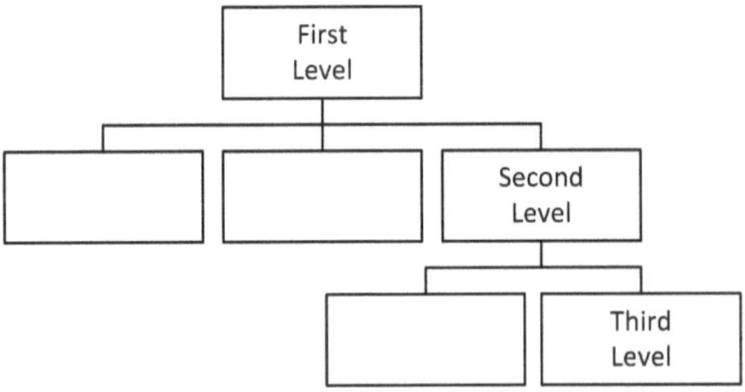

FIG. 2.3 Functional tree.

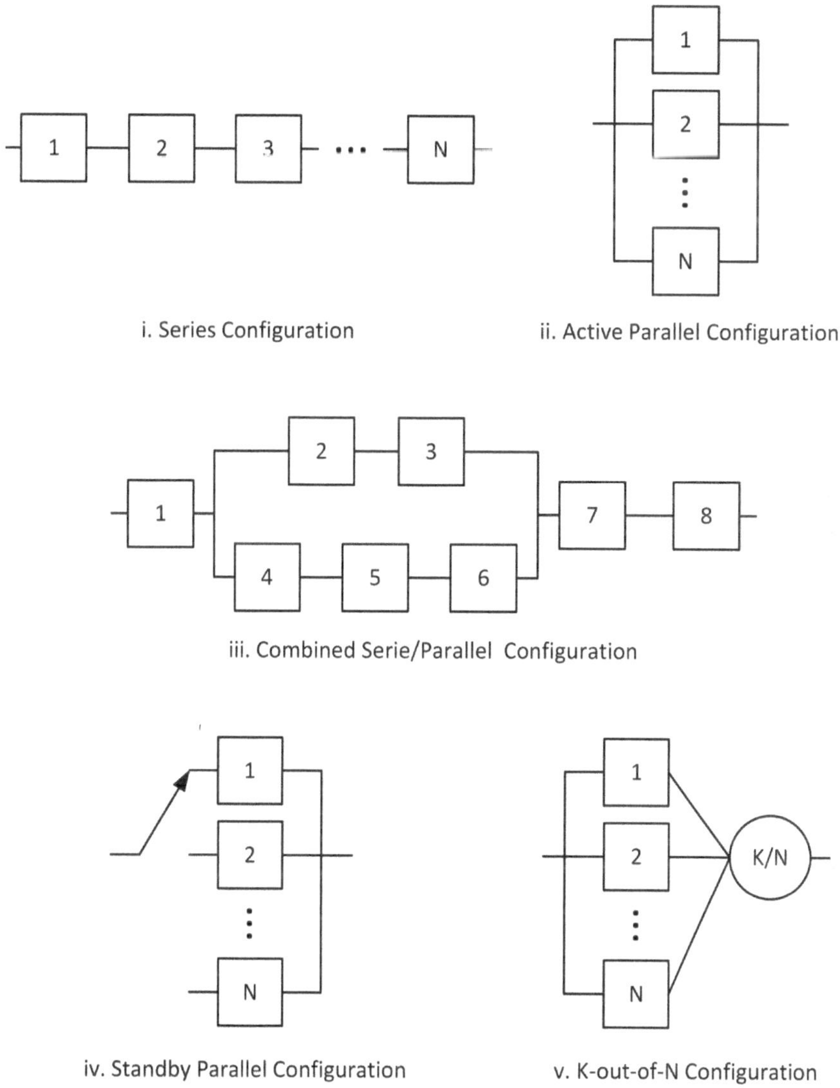

FIG. 2.4 Some reliability block diagram configurations.

modes, the system reliability is expressed by the multiplication of component reliability, expressed by Eq. (2.8) or Eq. (2.9):

$$R_{system}(t) = R_1(t) \cdot R_2(t), \ldots, R_n(t) \tag{2.8}$$

$$R_{system}(t) = \prod_{i=1}^{n} R_i(t) \tag{2.9}$$

Based on Eq. (2.9), it is noticeable the great deterioration of system reliability with an increasing number of components. Typical engineering systems such as automobiles, locomotives, ships, and telecommunications networks may be represented by series systems. To improve their reliability, a great improvement in component reliability is required, which can increase system purchase price and may not be technically feasible due to limitations in components' design and manufacturing technologies. Accordingly, an alternative to the requirements for increased component reliability is to provide redundancy in part or all of a system.

Redundancy is represented by the use of parallel arrangements in some components or part of a system aiming at increasing the number of components capable of performing a specific function considered critical for the system operation. The redundant system may be represented by active parallel components, also known as simple parallel configuration, where all units operate during mission time. In the case of n-redundant components, all units must fail to cut the information path and thus cause a system failure. The unreliability of the system, for statistically independent components, is expressed as shown in Eq. (2.10):

$$F_{system}(t) = F_1(t) \cdot F_2(t), \ldots, F_N(t) \tag{2.10}$$

Given the definition presented in Eq. (2.2), the system reliability is defined as presented in Eqs. (2.11) or (2.12).

$$R_{system}(t) = 1 - [(1 - R_1(t)) \cdot (1 - R_2(t)), \ldots, (1 - R_N(t))] \tag{2.11}$$

$$R_{system}(t) = 1 - \prod_{i=1}^{N} (1 - R_i(t)) \tag{2.12}$$

The K-out-of-N configuration is a special case of parallel redundancy. This type of configuration requires at least K operating components out of the total N parallel components to guarantee system operation according to the required performance. Usually, in K-out-of-N configuration systems all components are identical and are active. Considering that the components are statistically independent and identical and with the same reliability distribution, $R(t)$, the system reliability is expressed by Eq. (2.13), which is a probabilistic analysis based on the binomial distribution.

$$R_{system}(t) = \sum_{i=K}^{N} \binom{N}{K} (R(t))^i (1 - R(t))^{N-i} \tag{2.13}$$

In the case where the K-out-of-N components are not identical, the system reliability must be calculated based on the event space method where all possible operational combinations are considered in order to obtain the system's reliability. This analysis is detailedly presented in some references (Lewis, 1994; Modarres et al., 2016).

Another commonly used redundancy in system reliability improvement is the standby redundancy. Standby redundancy configurations consist of items that are inactive and available to be called into service when an active item fails. Considering a system with two parallel units, in an active redundancy, both units are operating, whereas in standby parallel, the second unit is not brought into operation until the first fails. The reliability of a redundant standby system is the reliability of unit 1 over the mission time t (i.e., the probability that it succeeds the whole mission time) plus the probability that unit 1 fails at time t', prior to t, and the probability that standby unit 2 successfully functions for the remainder of the mission in an active operation mode. The reliability function can be defined as presented in Eq. (2.14):

$$R_{system}(t) = R_1(t) + \int_0^t R_2(t-t')f_1(t')dt' \qquad (2.14)$$

In a standby arrangement besides the use of the inactive component, there must be sensoring and switching devices that must recognize the failure of the active component and switch to the second unit that must turn on immediately. The sensoring and switching system must have very high reliability in order to guarantee the system's reliability. The standby component must not deteriorate in an inactive time. Those hypotheses are used to derive Eq. (2.14). Detailed analyses of standby redundancy may be found in Modarres et al. (2016).

Finally, an engineering system is a combination of series and parallel components. The system reliability is calculated based on the combination of those components' arrangements. In some cases, the system may present complex block arrangements where the components are linked in such a way that they cannot be defined as basic series or parallel configurations. Several methods exist for obtaining the reliability of a complex system including the decomposition method, the event space method, and the path-tracing method. Classic reliability analysis books present those methods that are not discussed within this chapter since it aims at presenting an overview on the various topics and not providing an extensive and detailed description of the subject matter.

Although the use of RBD allows the evaluation of the system reliability, for each component failure the system may achieve different degraded operational performances. Depending on the degree of these degradations, the mission can be finished and, in some cases, the system may present serious damage or even threaten human life. When the system is operating in a degraded condition, but its performance is better than the value used to define reliability, it is said that the system is in a faulty condition or is presenting a fault. The evolution of a fault will cause a failure. Thus, the unreliability space must be analyzed in order to define critical degradation caused by component failures. The analysis that relates components' potential failures with system operational performance degradation is associated with the failure mode and effects analysis technique usually cited as FMEA.

2.2.2.2 *Failure Mode and Effects Analysis (FMEA)*

The FMEA is a simple qualitative method to reveal possible component failures and to predict the failure effects on the system. The method was developed to support the design of spaceships in the 1950s (Dhillon, 1992). The analysis starts with a component to identify possible failure modes and then investigate what will happen if this component fails. After the completion of the analysis, one can reveal the significant potential failure modes and important effects on system performance. The analysis is centered on the system's components and maps the system performance degradation in case of failure modes occurrences. Based on FMEA analysis, it is possible to define critical components as for system operational degradation or threats to human life or environment (DoD, 1984; Palady, 1995; Stamatis, 1995).

If FMEA analysis is used as a system design technique, it provides important information regarding improvements in systems' architecture, redundancy allocation, and component reliability requirements. If the system is already operating, it allows operating and maintenance teams to define the most suitable operational and maintenance planning aiming at avoiding critical components' failures.

The FMEA analysis involves defining answers to the following questions:

i. How could each component of the product fail or which are its failure modes?
ii. What mechanisms could produce these failure modes?
iii. What would be the effects of these failures occurred?
iv. How is this failure detected? and
v. What will be planned during the design or operational phase to compensate for the failure?

Aiming at organizing and standardizing FMEA presentation, the analysis is presented in a table where the answer to the previous questions is presented. There are many variants for that table but one of the most used formats is the one proposed by SAE J1739 standard (SAE, 2009) that is reproduced in Table 2.2.

The analysis starts with the definition of the components' functions that are presented in the functional tree. The failure modes for each component must be listed. A failure mode is the manner in which a component potentially fails to meet or deliver the intended function and associated requirements. Failure modes may include the failure to perform the function within defined limits, inadequate or poor performance of the function,

TABLE 2.2 FMEA table.

Device function	Potential failure mode	Potential effects of failure	Severity (S)	Potential causes/ mechanisms of failure	Occurrence (O)	Current control (prevention/ detection)	Detection (D)	RPN
What are the functions, features, or requirements?	What can go wrong? • No function • Partial/ overfunction/ degraded overtime • Intermittent function • Unintended function	What are the effects?	How bad are the effects?	What are the causes?	How often do they happen?	How the failure modes can be prevented/ detected	How good are those methods at preventing/ detecting the failure modes	SxOxD

the intermittent performance of the function, and/or performing an unintended or undesired function. There may be many failure modes for each function. For each failure mode, there must be defined the potential failure mechanisms that are associated with physical mechanisms such as wear, fatigue, corrosion, or other degradation processes that may trigger the failure mode development. For each failure mode, the immediate consequences of its occurrence on system operational performance may be evaluated.

The Severity (S) of a failure mode, associated with the degradation degree of the system performance, is usually described by a numerical scale varying between 1 and 10, where 1 is insignificant and 10 is very damaging and may result in loss of equipment and human and/or environmental harm. To the potential cause is also assigned a numerical score varying from 1 to 10 related to how often it occurs, named Occurrence (O), where 1 represents never and 10 represents very frequent. The method of detecting the failure mode development (preferentially before it becomes a failure) is also assigned a score between 1 and 10, named Detection (D), where 1 means certain and 10 means impossible to detect.

Works published by SAE, Ford, and Palady (Ford, 2011; Palady, 1995. SAE, 2009) present numerical scales that can be used in defining S, O, and D. The Risk Priority Number (RPN) is an overall danger score calculated by multiplying S, O, and D. Failure modes with higher RPNs are design or operational priorities. Tables 2.3–2.5 present suggestions for those classifications, adapted from SAE and Ford (Ford, 2011; SAE, 2009). Although those scales, mainly the severity one, are developed based on automotive industry requirements, the substitution of the term "vehicle" by equipment, product, or device allows their application in other industrial sectors.

The analyst must provide recommended actions to be taken to reduce the higher RPN. These actions can focus on one or more of the severities, occurrence, and detection scores. In some cases, the analysis must be done for higher severity or occurrence scores that may harm system safety or operational performance. The effect of those actions may be estimated through a new analysis of RPN.

As the FMEA method looks at a single potential failure mode of each component and its consequence on the system performance, it is named "bottom-up" analysis. This analysis does not consider the effects associated with combinations of failure modes from two or more components. Such combinations may potentiate the detrimental effects on system performance. Despite this weakness, FMEA analysis provides valuable engineering results to improve the design and operational performance such as follows:

i. identify redundant and fail-safe design requirements,
ii. identify single item failure modes,

TABLE 2.3 FMEA severity ranking.

Classification of effect	Effect	Severity of effect	Ranking
Failure to meet safety and/or regulatory requirements	Hazardous—without warning	Potential failure mode affects safe vehicle operation and/or involves noncompliance with government regulation without warning	10
	Hazardous—with warning	Potential failure mode affects safe vehicle operation and/or involves noncompliance with government regulation with warning	9
Loss or degradation of the primary function	Very high	Loss of primary function (vehicle inoperable, does not affect safe vehicle operation)	8
	High	Degradation of primary function (vehicle operable, but at a reduced level of performance)	7
Loss or degradation of secondary function	Moderate	Loss of secondary function (vehicle operable, but comfort/ convenience functions inoperable)	6
	Low	Degradation of secondary function (vehicle operable, but comfort/ convenience functions at a reduced level of performance)	5
Annoyance	Very low	Appearance problem or audible noise, vehicle operable, item does not confirm and noticed by most customers (> 75%)	4
	Minor	Appearance Problem or Audible Noise, vehicle operable, item does not confirm and noticed by many customers (50%).	3
	Very minor	Appearance Problem or Audible Noise, vehicle operable, item does not confirm and noticed by discriminating customers (<25%)	2
No effect	None	No discernible effect	1

iii. identify most suitable inspection and maintenance requirements and planning,

iv. identify manufacturing and assembling requirements to improve system performance, and

v. identify components for a redesign.

TABLE 2.4 FMEA occurrence ranking.

Probability of failure	Criteria: occurrence of cause (design life/reliability of item/vehicle)	Criteria: occurrence of cause (incidents per items/vehicles)	Ranking
Very high	New technology/new design with no history	≥100 per thousand ≥1 in 10	10
High	Failure is inevitable with new design, new application, or change in duty cycle/operating conditions	50 per thousand 1 in 20	9
	Failure is likely with new design, new application, or change in duty cycle/operating conditions	20 per thousand 1 in 50	8
	Failure is uncertain with new design, new application, or change in duty cycle/operating conditions	10 per thousand 1 in 100	7
Moderate	Frequent failures associated with similar designs or in design simulation and testing	2 per thousand 1 in 500	6
	Occasional failures associated with similar designs or in design simulation and testing	0.5 per thousand 1 in 2000	5
	Isolated failures associated with similar designs or in design simulation and testing	0.1 per thousand 1 in 10,000	4
Low	Only isolated failures associated with almost identical design or in design simulation and testing	0.01 per thousand 1 in 100,000	3
	No observed failures associated with almost identical design or in design simulation and testing	≤0.001 per thousand 1 in 1,000,000	2
Very low	Failure is eliminated through preventive control	Failure is eliminated through preventive control	1

In some cases, the FMEA analysis is named FMECA where C stands for criticality. The development of FMECA analysis follows the same steps previously presented, but the occurrence analysis is expressed based on failure rate estimate associated with each failure mode and it considers the conditional probability that the failure effect will result in the identified severity classification, given that the failure mode occurs. The

TABLE 2.5 FMEA detection ranking.

Opportunity for detection	Likelihood of detection by design control	Likelihood of detection	Ranking
No detection opportunity	No known control available to detect the cause/mechanism of failure or failure mode	Almost impossible	10
Not likely to detect at any stage	Design analysis/detection controls have a weak detection capability of cause/mechanism of failure or failure mode	Very remote	9
Post-design freeze and prior to launch	Product verification/validation after design freeze and prior to launch with pass/fail to test (remote likelihood current control will detect cause/ mechanism of failure or failure mode	Remote	8
	Product verification/validation after design freeze and prior to launch with a test to failure testing (very low likelihood current control will detect cause/mechanism of failure or failure mode)	Very low	7
	Product verification/validation after design freeze and prior to launch with degradation testing (low likelihood current control will detect cause/ mechanism of failure or failure mode)	Low	6
Prior to design freeze	Product validation (reliability testing, development, or validation tests) prior to design freeze using pass/fail to test (moderate likelihood current control will detect cause/mechanism of failure or failure mode)	Moderate	5
	Product validation (reliability testing, development, or validation tests) prior to design freeze using a test to failure (moderately high likelihood current control will detect cause/mechanism of failure or failure mode)	Moderately high	4
	Product validation (reliability testing, development, or validation tests) prior to design freeze using degradation testing (high likelihood current control will detect cause/mechanism of failure or failure mode)	High	3

Continued

TABLE 2.5 FMEA detection ranking—cont'd

Opportunity for detection	Likelihood of detection by design control	Likelihood of detection	Ranking
Virtual analysis—correlated	Design analysis/detection controls have a strong detection capability (very high likelihood current control will detect cause/mechanism of failure or failure mode)	Very high	2
Detection not applicable; failure prevention	Failure cause or failure mode cannot occur because it is fully prevented through design solutions (e.g., proven design standard, best practice, or common material) (current control almost certain to detect cause/mechanism of failure or failure mode)	Almost certain	1

FMECA procedure, according to the MIL-STD-1629A (DoD, 1984) document, uses the quantitative criticality analysis method instead of RPN analysis, consisting of:

i. defining the reliability/unreliability for each item, at a given operating time;
ii. identifying the portion of the items unreliability that can be attributed to each potential failure mode;
iii. rating the probability of loss (or severity) that will result from each failure mode that may occur; and
iv. calculating the criticality for each potential failure mode by obtaining the product of the three factors—Item unreliability × mode ratio of unreliability × probability of loss.

The criticality for each item is obtained by the sum of the criticalities for each failure mode that has been identified for the item. Although more laborious to execute, the FMECA analysis uses more information about the component reliability to support the design and operational improvement.

As previously noted, despite being very useful for systems design and operational analysis, the most important FMEA weakness is to disregard the occurrence of more than one failure mode simultaneously, which can take the system to a critical operating condition. In that case, the reliability analyst must use a technique that should relate a critical operating condition to components' failure combinations.

2.2.2.3 *Fault Tree Analysis (FTA)*

Aiming at considering the combination of components' failure modes and their effects on system performance, the method named FTA was proposed in 1962 at the Bell Laboratories by H.A. Watson, under a US Air

Force Ballistics Systems Division contract to evaluate the Minuteman I Intercontinental Ballistic Missile (ICBM) Launch Control System (Beresh, Ciufo, & Anders, 2007). FTA is based on a logical decomposition of the causes of an undesired performance to a lower level defined by the component's failure. Besides its use in reliability analysis, this method is also used in the field of safety engineering as initially proposed by the US Nuclear Regulatory Commission document NUREG 0492 (Vesely, Goldberg, Roberts, & Haasl, 1981) developed for probabilistic safety analysis of nuclear power stations.

FTA is a deductive, top-down method aimed at analyzing the effects of components' failures on a complex system. The first step of a fault tree analysis consists of the definition of the top event to be analyzed and of the boundary conditions for the analysis. Then, failure events that are the necessary and sufficient causes that result in the top event are identified and connected to the top event through logic gates. Logic symbols graphically represent the gates used to interconnect the low-level events that contribute to the top-level event according to their causal relations. Fig. 2.5 lists the most commonly used logic symbols together with a brief description of their interpretation, IEC 61025 (IEC, 2006), while Fig. 2.6 presents an example of a typical FTA.

Name of the Symbol	**Description**	**Graphic Representation**
Top or mediator event	Fault event or deterioration event associated with a logic gate	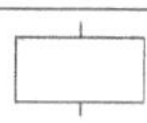
Primary basic fault events	Basin initiating fault requiring no further development.	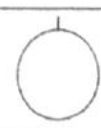
Secondary basic fault event	Fault event that is not developed up to its own causes because of lack of required information	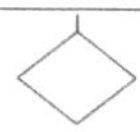
AND gate	Logic gate. Output event occurs if all input events occur simultaneously.	
OR gate	Logic gate. Output event occurs if any one of the input events occurs.	
INHIBIT gate	Input produces output when a conditional event occurs.	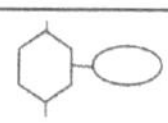
PRIORITY AND	Output event occurs only if the input events occur in a specific sequence	
EXCLUSIVE OR gate	Output event occurs if only one of the input events occurs.	
VOTING gate	Output event occurs if at least *k* of *n* input events occur.	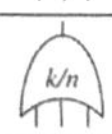

FIG. 2.5 FTA symbols.

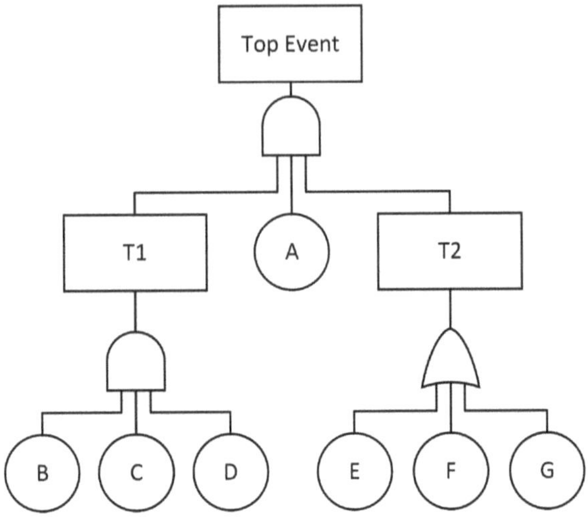

FIG. 2.6 FTA example.

The main objective of FTA is to define the minimum cut sets or, in other words, to assess the combination of events that triggers the top event. To define the minimum cut sets of a fault tree, the analyst may use logical expressions based on probability concepts.

Considering that the logic gates used in fault tree are directly related to Boolean algebra operator's disjunction (OR gate) and conjunction (AND gate), it is possible to translate a fault tree into an entirely equivalent Boolean function by using the rules of Boolean algebra. More information about Boolean algebra and its application on fault tree analysis may be found in Modarres et al. (Modarres et al., 2016).

The minimum cut sets may be classified according to the number of components' failures that may happen simultaneously to cause the top event. First-order minimum cut sets are those where the failure of one component causes the top event. Second-order minimum cut sets require that two components fail to cause the top event. So, the higher the number of simultaneous components' failure necessary to cause the top event, the higher the cut set order. For a given system's undesired performance analysis, the higher the order of the minimum cut sets, the more resistant is the system to multiple components' failure.

The fault tree analysis provides a logical expression that combines the components' failures that cause the top event. If the probability of occurrence of the components failures is known, a quantitative analysis may be done considering the basic rules of probability analysis. The probability of the top event is given by the probability of the union of the minimum cut sets. If the components' failures are independent, conditional probability rules must be used to define top event probability. The dependency between components' failures usually arises from sharing the same power source, effects of the operational environment, functional dependencies, and even human actions.

As shown in Fig. 2.6, FTA's graphical representation is hierarchical and named according to its branches. It has strong readability and is easy to understand, which makes FTA a useful tool to conduct reliability analysis by industry and certification authorities. However, given the fact that FTA attempts to build a mathematical model to evaluate the occurrence of a physical operational condition, for complex engineering systems, the tree construction is time-consuming and is very dependent on the judgment of the analyst.

System reliability analysis must be executed using at least two of the previously presented techniques aiming at reducing the disadvantages of each technique. Usually, a reliability block diagram is developed since it provides numerical reliability evaluation and allows estimating the effects of redundancy allocation. FMEA and FTA analysis may be executed as complementary analysis aiming at defining some specific relations among components' failures and undesirable system operational performance.

2.3 Risk analysis

The reliability analysis of an engineering system defines the probability that the system will execute its mission according to an operational context. The reliability analysis may also present a relation between components' failures and systems loss of performance, including harm to the system owner or even to human life and environment.

The concept of risk was introduced in engineering system design aiming at defining the probability of occurrence of a given failure scenario and the consequences associated with that scenario. The failure scenario is associated with the occurrence of an initial event, usually components' failures or even operating errors, which propagate their effects through other system's devices and subsystems causing loss of performance or even an accident that could affect the system integrity and cause personnel injuries or even deaths. The safety of the system is related to the acceptance of risks associated with a system operation (Aven, 2003; Aven, 2012; Ayyub, 2014; Kumamoto & Henley, 1996).

Risk analysis for modern engineering systems was introduced during the development of nuclear power plants for electrical energy generation. The aim of risk analysis was to identify critical failure scenarios that could cause loss of integrity of the power plant containment with leakage of radioactive material (Frankel, 1983). The concept of risk analysis was later expanded to chemical process plants after some critical accidents such as the one involving Union Carbide in Bhopal, India, which caused a great number of fatalities and severe environmental pollution (AIChe, 1989).

Risk can be viewed to be a multidimensional quantity that includes failure scenario-occurrence probability, scenario-occurrence consequences, and consequence significance associated with population and environment at risk. To facilitate risk analysis representation, data are presented as a pair of the probability of occurrence of a failure scenario ($p_{scenario}$) and the consequences associated with the scenario's occurrence (C_{scenario}), as presented in Eq. 2.15, (Ayyub, 2014):

$$\text{Risk} = \left\{ \left(\mathbf{p}_{\text{scenario1}}, \mathbf{C}_{\text{scenario1}}\right), \left(\mathbf{p}_{\text{scenario2}}, \mathbf{C}_{\text{scenario2}}\right), \ldots, \left(\mathbf{p}_{\text{scenarioN}}, \mathbf{C}_{\text{scenarioN}}\right) \right\} \tag{2.15}$$

For a complex engineering system, the group of pairs is used to define the system's risk profile. That profile must be compared to risk acceptance criteria aiming at evaluating if the risk associated with the system is acceptable or some changes are necessary for system design or operation aiming at reducing system operational risk.

The mentioned procedure to evaluate the risk associated with a system is named quantitative risk analysis since the probability of occurrence of the failure scenarios must be evaluated. To define risk quantitatively, a method must be used to evaluate the failure scenarios associated with a given initiating event that propagates through the system and causes its deterioration.

The initiating event is associated with the presence of hazards. A hazard is any source of potential damage or harm to something or someone. Risk is the chance or probability that someone or something will be harmed or experience an adverse effect if exposed to a hazard. During the operation of complex engineering systems, the exposition to hazards is usually caused by components' failures that may be associated with external events, operational errors, or degradation associated with physical failure mechanisms.

The risk analysis procedure for a complex engineering system involves the process of identifying hazards, event-probability assessment, and consequence assessment. The risk analysis process must give answers to three basic questions: (1) What can go wrong? (2) What is the likelihood that it will go wrong? (3) What are the consequences if it does go wrong? Based on the answer to the previous questions, the analyst must compare the risk profile associated with the complex engineering system with established risk acceptance criteria to determine whether the risk level is acceptable or not. If the estimated risk level is too high compared to the acceptance criteria, risk-reducing measures need to be identified and implemented.

Once the risk analysis is executed, the system's operator must control the risk profile according to accepted levels during the entire operational life. The risk control techniques involve the prevention of failures and/or mitigation of consequences in case of occurrence of some failure scenarios. The merge of risk analysis and risk control constitutes what is named risk management.

Many authors (Aven, 2011; Ayyub, 2014; Kumamoto & Henley, 1996) discuss risk management concepts. This chapter aims to present some basic risk aspects that are used in complex engineering systems to support the development of diagnosis and prognosis methods aiming at avoiding failures that can cause unacceptable risks.

2.3.1 Risk management

The risk management process is composed basically of two steps: risk analysis and risk control.

The risk analysis aims at defining the risk profile associated with the operation of a complex engineering system. Risk analysis must be based on the understanding of how systems' components may fail and expose the system to a given hazard and what are the consequences of those failures. The risk analysis process models the propagation of a failure through the system aiming at defining the barriers for that propagation and the combination of failures that define a given failure scenario. For each failure scenario, the consequences associated with system operational degradation and threats to human life and the environment must be listed.

The risk analysis procedure can be executed by some of the following basic steps:

- Definition of analysis objectives and systems,
- Hazard analysis and definition of failure scenarios,
- Collection of data in a life cycle framework, and
- Qualitative and/or quantitative risk assessment.

The definition of systems boundaries is important to develop the analysis. This definition is partially based on what aspects of the system's performance are of concern. For large systems, not only the physical boundaries must be defined but also the system's geographical location must be considered to define the population and environment features that may be affected by system abnormal operation. The analysis objectives must also be defined as a primary step since it defines the extension of the analysis and if quantitative analysis will be necessary.

Besides the system boundary, it is also important to establish a resolution limit for the system. The depth of the system model needs to be sufficient for the specific problem under analysis and may consider information availability. If the analysis is narrow, some important details may not be considered. On the contrary, if the analysis is too broad, the need for information will increase and the analysis will probably be more costly and difficult. Considering the risk analysis of operating complex systems, the resolution is limited to components' levels since their failures will probably be the initiating event of failure scenarios.

After defining the risk analysis objectives and system boundaries, the system breakdown structure must be developed to support failure scenarios' definition. That system breakdown structure shall use the functional tree already presented in this chapter since it identifies the function(s) that must be performed for the operation of the system, respectively, in the subsystems, devices, and components' levels.

Based on the functional tree, the hazard analysis shall be defined. Usually, for complex engineering systems, the hazardous events are associated with components' failures that may put into danger not only the system but also the operating personnel, environment, and population surrounding the system. Considering as an initiating event, the failure of a component, the risk analysis must define the failure scenario associated with the initiating event and must define the consequences associated with that scenario.

Several tools can be used to develop risk analysis studies and some of them are presented in Table 2.6. Those tools may be suitable to define the initiating event and the system operational scenario associated with the initiating event. For each scenario, those tools will provide the failure propagation scenario and an evaluation of varying consequence categories including items such as economic loss, loss of life, or injuries and environmental impacts.

Some other techniques can be used in the early design phases of the sytems but are not suitable for complex engineering systems' risk analysis because they are mainly dependent on nonstructured interviews made with managers, operators, and maintainers that have experience with similar systems, such as experience-based identification and brainstorming. Those techniques identify hazards that may be the root cause of system risk scenarios but do not support quantitative analysis. More detailed characteristics, disadvantages, and advantages of each of those methods can be obtained in specific literature regarding risk analysis (Ayyub, 2014; Kumamoto & Henley, 1996; Wang & Roush, 2000).

Once the risk scenarios are defined, a qualitative or quantitative analysis shall be executed. Some methods presented in Table 2.6 only allow qualitative analysis such as what-if, Preliminary Hazard Analysis (PrHA), FMEA, and Hazard and Operability (HAZOP) study. The likelihood of occurrence of each identified initiating event and the consequences estimates associated with that occurrence may be defined based on expert opinion or risk analysis team judgment. This approach may be sufficient to assess the risk of a complex engineering system, depending on the available resources and information available for the analysis.

TABLE 2.6 Risk analysis techniques.

Method	Scope
What-if	Identifies hazards, hazardous situations, or specific accident events that could result in undesirable consequences. The method does not provide a structured methodology to identify the failure scenarios
Hazard and Operability (HAZOP) study	Identifies system operating parameters' deviations and their causes that can lead to undesirable consequences and determine recommended actions to reduce the frequency and/or consequences of the deviations. The analysis is dependent on the monitoring variables (online and offline) used to control system operation. The method provides a structured methodology to identify the failure scenarios. This is an inductive modeling approach
Preliminary Hazard Analysis (PrHA)	Identifies and prioritizes hazards and their causes leading to undesirable consequences. It determines recommended actions to reduce the frequency and/or consequences of the prioritized hazards. Preliminary, because it is used as the first approach for engineering systems risk analysis. This is an inductive modeling approach
Failure Modes and Effects Analysis (FMEA)	Identifies the components' (equipment) failure modes and the impacts on other devices, subsystems, and the system. This is an inductive modeling approach
Fault Tree Analysis (FTA)	Identifies combinations of equipment or components' failures and human errors that can result in an undesirable event. This event may be an initiating event or any other event that is listed in the scenario analysis. If the basic event failure probabilities are known, FTA provides the undesirable event probability of occurrence. This is a deductive modeling approach
Event Tree Analysis (ETA)	Identifies various sequences of events, both failures, and successes that can lead to an accident. It is used to model the failure scenario associated with an initial event. This is an inductive modeling approach
Cause consequence diagram	Represents all system outcomes, given an initial event, on a diagram that contains a full textual description of the system's behavior, both failures and success of devices' operation that can lead to an accident. Each device operation is analyzed based on an FTA diagram. If the failure probabilities of the devices are known, the diagram provides the probability of occurrence of each scenario. The diagram is a combination of FTA and ETA being a mix of inductive and deductive modeling approaches
Probabilistic Risk Analysis (PRA)	Methodology for quantitative risk assessment developed by the nuclear engineering community for risk assessment.

Continued

TABLE 2.6 Risk analysis techniques—cont'd

Method	Scope
	This comprehensive process may use a combination of risk assessment methods including PrHA, FMEA, FTA, and ETA. The method is based on the evaluation of the probability of occurrence of each event that is part of a failure scenario. The consequences can be expressed in monetary value or loss of lives. It is a mix of inductive and deductive modeling approaches

The consequences and the likelihood of an initiating event are usually evaluated based on categories such as those proposed in Tables 2.7 and 2.8. A matrix may be constructed that assigns risk ratings (usually varying from very low to very high) to risk or conditions based on combining likelihood and consequence scales, as shown in Fig. 2.7. Risks with high probability and high impact are likely to require further analysis, including quantification, and require very stringent risk management.

Quantitative risk analysis assigns numerical values to the amount of damage that would occur in case of occurrence of a given failure scenario including calculating the probability of a given failure scenario originated from an initiating event. This quantitative analysis, usually named Probabilistic Risk Analysis (PRA), uses tools such as FTA, ETA, and cause-consequence diagram aiming at defining the probability of occurrence of failure scenarios that threaten the system operation. The consequences are usually expressed in terms of human, environmental, and economic losses related to relevant scenarios. The risk is quantified in a risk number (or index) or a graph that relates probabilities and consequences.

A quantitative risk analysis, besides the identification of all the possible system failures with related probabilities of occurrence, requires the evaluation of probability distributions of variables of interest to estimate the damage to equipment and communities surrounding the failed component. This evaluation is usually time-consuming and requires numerical data that may be scarce when dealing with systems designed with very stringent operational safety requirements.

2.3.2 Risk evaluation

Based on the qualitative or quantitative analysis, the risk analyst must evaluate the risk on grounds of the results of the former analyses. This phase is named risk evaluation. In this phase, the decision is made as to whether or not the risk is tolerable. In the cases of qualitative analysis, the analysis is based on the risk matrix, and intolerable risks may be reduced by the installation of barriers to reduce the likelihood of

TABLE 2.7 Example of consequences categories.

	Critical factors			
Category	**People safety**	**Operations**	**Environment**	**Reputation**
Insignificant	May cause minor injuries that would be treated at the site with no long-term effects	Light damage to pieces equipment without commitment of operation	Insignificant damage	External or internal reputation not affected. No effort or expense is required to recover
Minor	May cause several minor casualties that require medical attention off-site with no long-term effects	Light damage to systems/ pieces of equipment with minor impact on the operation	Minor environmental damage	External and internal reputation minimally affected. Little effort or expense to recover
Moderate	May cause several casualties (in-site or off-site) that require hospitalization with no long-term effects	Moderate damage to systems/ pieces of equipment with moderate impact on the operation	Moderate damage	External and internal reputation damaged. Some efforts and expenses required to recover
Major	May cause off-site serious casualties resulting in long-term physical impairment. May cause in-site loss of life	Severe damage to systems (slow repair) with great impact on the operation	Cause extensive local environmental damage	External and internal reputation severely damaged. Considerable efforts and expenses required to recover
Catastrophic	My cause in-site and off-site loss of life	Catastrophic damage leading to loss of industrial installation	Cause severe damage in local sensitive areas or extending to other locations	External and internal reputation irrevocably damaged

TABLE 2.8 Example of initiating event frequency categories.

Frequency score	Category	Frequency
A	Extremely rare	Conceptually possible, but without references in the industrial sector under analysis
B	Rare	Not expected to occur, despite occurrence listed in historical references related to similar facilities in the industrial sector
C	Unlikely	Unlikely to occur during the useful life of a set of similar units
D	Possible	Possible to occur once during the life of the installation
E	Likely	Possible to occur many times during the life of the installation

Likelihood ↑ / Consequence →	Insignificant	Minor	Moderate	Major	Catastrophic
E	MODERATE	HIGH	EXTREME	EXTREME	EXTREME
D	MODERATE	HIGH	HIGH	EXTREME	EXTREME
C	LOW	MODERATE	HIGH	HIGH	EXTREME
B	LOW	MODERATE	MODERATE	HIGH	HIGH
A	LOW	LOW	LOW	MODERATE	MODERATE

FIG. 2.7 Risk evaluation matrix.

undesirable scenarios (causation barriers) and to limit the consequences of such events if they should occur (Vinnem, 2007).

In cases of operational systems, the maintenance planning may also be improved to avoid the occurrence of initial events associated with components, devices, or equipment failures. This improvement can be achieved through the application of modern maintenance philosophies that will be later analyzed in the present chapter. Those measures are considered risk control techniques. To select the most suitable measure to eliminate intolerable risks the analyst must employ decision-making techniques that aim to find an equilibrium between the consequences of a risk scenario and the cost of adoption of the available measures to avoid that risk scenario. Many decision-making techniques may be used for this task, such as cost-benefit analysis, decision trees, and a great number of multicriteria decision-making methods (MCDM). More details about those methods can be found in Velasquez and Hester (2013) and Ceballos, Lamata, and Pelta (2016).

In cases of quantitative analysis, the same actions mentioned in the previous paragraph may be taken, but the risk assessment may be based on the comparison of calculated risk with risk indexes provided by professional societies or standards related to the industrial sector where the system operates.

Common measures for probabilistic analysis of risks include the "individual risk" and the "societal risk" assessments. "individual risk" is the risk of a person suffering injuries in a location close to the system under analysis, including the likelihood of the injury occurring and the time period over which the injury might occur, usually one year. Details are reported in some references such as AIChe (1989), Mannan (2005), and Bottelberghs (2000).

Individual risk does not, however, completely describe situations where a single accident could kill or injure large numbers of people. For those situations, the "societal risk," which is expressed as the relation between the probability of occurrence of a given risk scenario and its consequences in terms of deaths, is used in the analysis. The probability is usually expressed as the average frequency with which the risk scenario is expected to occur on an annual basis. The societal risk is expressed according to the F-N curve as presented in Fig. 2.8 and represents the frequency of exceedance of a given number of fatalities due to the occurrence of a risk scenario. The diagram is divided into three regions: tolerable risk, intolerable risk, and a region where the risk must be kept "As Low As Reasonably Practicable" (ALARP).

In addition to people being exposed to risk the system's infrastructure or even surrounding buildings may be affected by the occurrence of a given failure scenario. In cases of system infrastructure, the effect is not

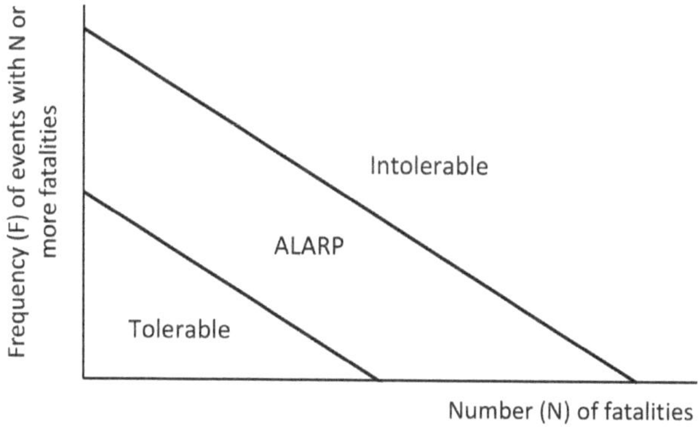

FIG. 2.8 F-N curve.

only the damage caused on the buildings and pieces of equipment but also the loss due to the reduction of production performance. Those losses are usually named economic losses. The economic loss may be represented by an F-D curve which relates the probability of a risk scenario to the expected economic loss expressed in monetary values proposed by Jansen (1988), similar to F-N curve. According to Jonkman, van Gelder, and Vrijling (2003), this model has been used in natural disaster modeling, such as flooding.

Also associated with an economic loss, there may be environmental loss linked to a risk scenario. The problem with a risk assessment for environmental accidents is that the environment is extremely complex since the effects of a risk scenario may include both short-term and long-term effects on land, water, and ecosystems. Wilday, Ali, and Wu (1998) present a comprehensive analysis of environmental risk indexes used to evaluate the tolerability of environmental damage. The indexes are based on toxic releases associated with a given risk scenario. The indexes are related to the following quantities:

- The toxicity factor gives a measure of the level of toxicity in the environment caused by the particular release;
- The recovery factor gives a measure of the time that the environment would take to recover from the release.

The authors propose curves similar to F-N and F-D diagrams to analyze environmental risk.

Based on the comparison between risks associated with a complex engineering system and risk acceptance criteria, if a system has a risk value above the risk acceptance level, the design and operational teams must take actions to address safety concerns and improve the system through risk reduction measures. Risk managers make decisions based on risk assessment and other considerations including economic, political, environmental, legal, reliability, producibility, and safety, using decision-making techniques such as those previously presented.

Finally, it is important to remind that the probabilities of risk scenario occurrences may vary during the system operational time. That variability must be considered in risk analysis through constant risk assessment and evaluation review (Zio, 2018).

2.4 Maintenance practices and planning

Although an engineering system is designed according to reliability requirements, its operational life may be much longer than an operational campaign for which those requirements are defined. In order to control system reliability degradation, maintenance actions must be taken to avoid its failure during an operational campaign. The development of maintenance plans for a complex engineering system is a laborious task that requires a lot of knowledge about the failure modes and their probability of occurrence for the devices that make up the system.

The maintenance plan can be defined as a group of tasks according to a maintenance policy that are executed in a system in order to assist the system in maintaining its functionality during operation. It shall be based on the combination of the basic maintenance practices—corrective, preventive, and predictive—with the mechanisms that are the root causes of the device's failure modes. The maintenance practices must control the development of the root causes reducing the devices, and consequently, the system failure probability, controlling reliability degradation (Waeyenbergh & Pintelon, 2004).

The maintenance practices can be classified into two categories: Corrective Maintenance (CM) and Preventive Maintenance (PM). Corrective maintenance is a maintenance task that is performed after the failure of an item of equipment or system. Preventive maintenance includes time-directed tasks that are performed according to a given schedule or after a given amount of use.

Condition-based tasks are performed after identification and diagnosis reflecting the actual status of the degradation of the devices that constitute the equipment or system at the present time. Predictive maintenance can be defined as continuous or periodic monitoring and diagnosis in order to give the condition or status of a device or component degradation evolution prior to equipment or system failure.

The selection of the most suitable set of maintenance practices for an engineering system is dependent on the failure modes presented by that system. Some philosophies help the maintenance teams to select the most suitable maintenance plan (or concept or strategy as usually cited in literature). The main goal of those philosophies is to keep the system life cycle cost down and ensure proper operation during an operational campaign.

The requirements for a maintenance plan have changed a lot since the end of the Second World War. Until the mid-1950s, the requirements of products' quality and production demands allowed the acceptance of long periods of machine downtime associated with the application of corrective maintenance. Furthermore, until that time, studies on the reliability of mechanical systems had not been systematically developed which is a basic requirement to apply preventive maintenance practice.

Between the early 1960s and the end of the 1970s, there was an increase in the perception of the quality demanded of products by consumers, requiring greater control of the machines' performance. Competition between companies had also increased due to the increase in the consumer market. The need for machinery availability had increased, and supported by the evolution of mechanical systems' reliability studies, the maintenance planners were capable of defining time intervals for submitting pieces of machinery to preventive maintenance. During the maintenance intervention, some parts were substituted to avoid failure during the next operational campaign. The maintenance intervention could be planned in advance supporting the production plan.

From the early 1980s, until the end of the twentieth century, there was an increase in the use of electronic devices as part of machinery subsystems and other engineering systems. The electronic components were (and still are) used as actuators, sensors, and control devices. The development of portable computers and microprocessors also allowed the development of numerical controlled machines and later of robots. The monitoring of parameters that can be associated with the evolution of the degradation of a given device that affects system operational performance supports the application of predictive maintenance practice. The main objective of predictive maintenance is to perform the intervention on the system only when the degradation is detected, reducing its downtime. During this period, the aeronautical industry adopted the reliability-centered maintenance philosophy to support the development of maintenance plans aiming at improving operational safety and availability. In general, predictive maintenance is associated with reliability analysis techniques aiming at avoiding failures that can jeopardize system operational safety (Nowlan & Heap, 1978).

From the early 2000s until 2015, the perception of safety requirements for industrial assets operation supports the development of the risk-based maintenance philosophy, aiming at supporting maintenance planning for assets whose failures have not only operational consequences but also consequences to the environment and surrounding communities. The predictive practice also had its application consolidated based on the evolution of monitoring systems and methodologies, including the application of artificial intelligence concepts, used to characterize degradation based on monitored signals.

From 2015 until now, with the proposition of ISO 55000 series of standards (ISO, 2014), the maintenance activity has broadened its focus to include actions aimed at managing physical asset operations in alignment with business strategy. The control of risks associated with asset operation is the basic goal of maintenance. The maintenance activity is also influenced by the concepts of Industry 4.0. The maintenance focus, using modern smart technology, is not only the degradation detection but also the use of artificial intelligence concepts to diagnose the root cause of a given degradation and to prognosis the operational time until failure to support maintenance intervention. The main goal of the maintenance planning is to support asset integrity with the minimum operational cost taking advantage of distributed monitoring systems and concepts of the Internet of Things.

In the sequence of the chapter, two maintenance planning philosophies that can be associated with the concept of asset management will be described, which are reliability-centered maintenance and risk-based maintenance.

2.4.1 Reliability-centered maintenance

Reliability-Centered Maintenance (RCM) is a method for maintenance planning developed within the aircraft industry aiming at solving problems of maintenance and operational safety of the wide-body aircraft (Nowlan & Heap, 1978) and later adapted to several other industries (Moubray, 1997). The main goal of this philosophy is to keep maintenance attention on critical equipment whose failure may affect system operational performance and safety. The methodology is strongly supported by reliability analysis techniques and aims at keeping system reliability during an operational campaign. On a generic level, IEC 60300-3-11 (IEC, 2009) defines RCM as a *systematic approach for identifying effective and efficient preventive maintenance tasks for items in accordance with a specific set of procedures and for establishing intervals between maintenance tasks.*

The RCM presents a well-established method for reliability analysis of complex systems, whose objective is to identify the critical components, defined as those components that in case of failure cause, the immediate shutdown of the system (Nowlan & Heap, 1978), as well as to reduce maintenance downtime and costs associated with them. This goal is achieved with an adequate selection of maintenance activities and policies (Moubray, 1997).

The RCM focuses efforts on the preservation of the functions of the critical components of the system through maintenance planning, avoiding, or eliminating maintenance tasks that are not strictly necessary (IEC, 2009). Various research papers and specialized books describe the RCM concept and its relationships with reliability engineering (Bloom, 2005; Rausand, 1998; SAE, 2009; Smith & Hinchcliffe, 2003).

But the most important contribution of these works is the analysis methodology, where the following steps can be highlighted:

1. System selection and information gathering;
2. Selection of the system limit;
3. Description of the functional system, the definition of the functions, and functional failures;
4. Analysis of modes and effects of FMEA failures;
5. Application of decision diagrams;
6. Selection of the most suitable tasks;
7. Consolidation of maintenance tasks; and
8. Permanent feedback.

Steps 1 to 4 are typical reliability analysis tasks and were already discussed in the previous items of the present chapter. The most important novelty of the RCM methodology is the decision diagram used to select the most suitable maintenance practice to avoid the occurrence of the critical component failure modes. The selection of the critical components is based on FMEA analysis through the evaluation of RPN index. The RCM decision-making diagram is presented in Fig. 2.9, adapted from Nowlan and Heap (1978).

In this diagram, it is possible to identify the selection process of maintenance tasks considering the consequence of the critical component failure on the system operation. The selection process initially classifies the component failure as evident or hidden for the operation team during a given operational condition. For evident failures, the diagram also considers the effects on environmental and system safety and system operational performance.

Based on that diagram, it is noticed that the method prioritizes the application of predictive maintenance practices aiming at controlling component degradation and maximizing the system operational life before the execution of preventive maintenance intervention. The method also emphasizes the necessity of applying not only technical but also economic analysis to verify the feasibility of predictive practice application.

In case of impossibility to apply predictive practices, the method recommends time-based preventive maintenance and it is noteworthy that as a last alternative, the method recommends the application of corrective practices, except for environmental or systems' safety consequences. Again, the decision-making process is guided by economic and technical aspects.

The RCM method is strongly dependent on the operation and maintenance teams' expertise on the failure modes and reliability of the critical components or pieces of equipment in the system operational context. Furthermore, the RCM decision-making diagram requires the use of a detailed time to failure and time to repair and maintenance costs database, aiming at supporting the application of economic engineering concepts to support the selection of the most suitable maintenance practice. Those requirements are normally met by industrial sectors whose operation requires high-value assets whose downtime causes a loss in production volume and/or quality. Accordingly, RCM is usually used by mining, steel processing, chemical processing, power generation, and water and sewage treatment industries. The metallurgical sector, including the automotive industry, is still consolidating the use of that philosophy.

The RCM methodology also allows the use of many decision-making criteria aiming at optimizing the preventive maintenance and predictive inspection interval based on cost-reliability balance. Thus, its application contributes to keep the life cycle cost down and ensures proper operations.

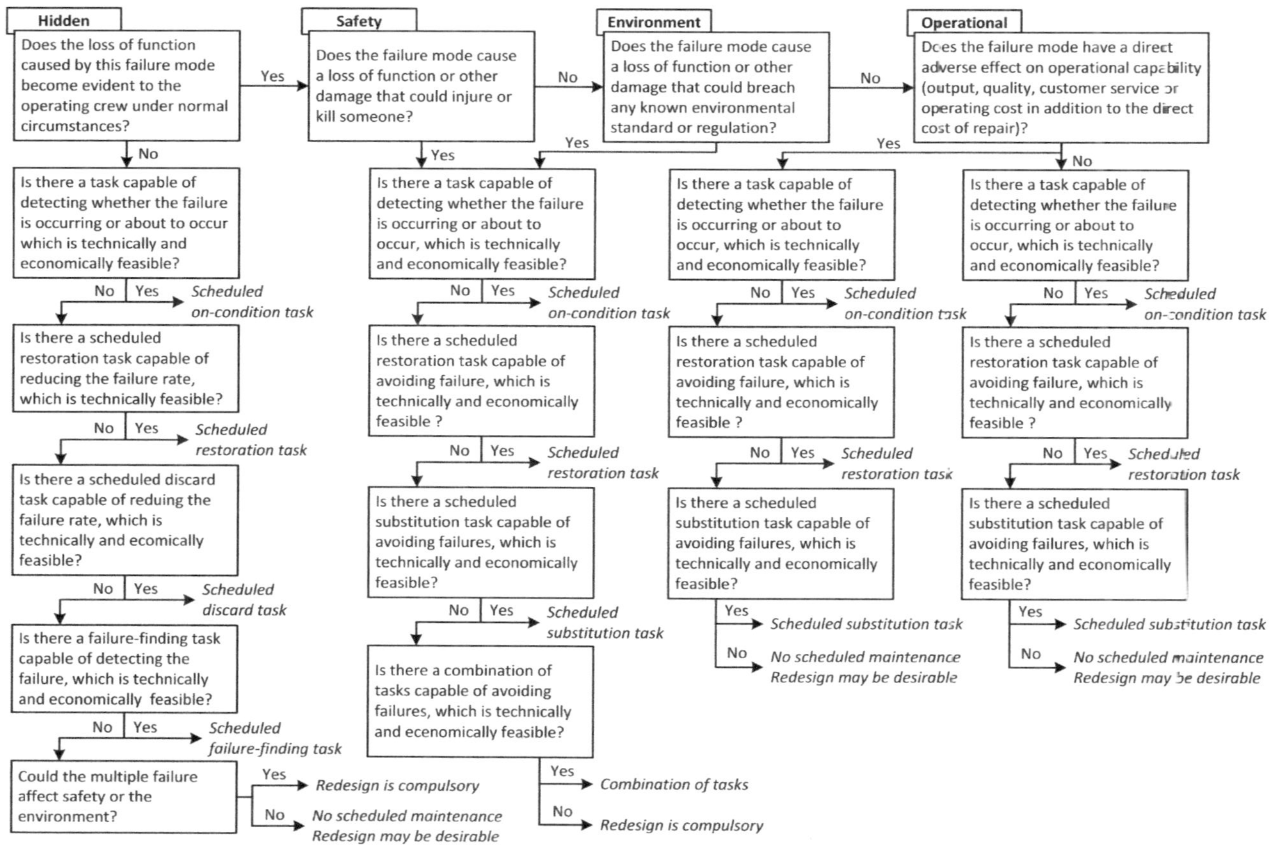

FIG. 2.9 RCM decision-making diagram.

Although the RCM is very useful to define the most suitable maintenance policy for a piece of equipment considered critical for the system operation, in cases of complex engineering systems composed of various assets that operate in an interconnected manner in order to perform the system's function, the selection of the critical assets is an important task.

2.4.2 Risk-based maintenance

The Risk-Based Maintenance (RBM) philosophy was proposed in the 1990s aiming at using risk concepts to select the most critical pieces of equipment of a complex engineering system (Khan & Haddara, 2003, 2004). The criticality of the piece of equipment is associated with its risk profile. The equipment risk profile, based on risk analysis concepts, represents the hazards, both to humans and to the environment, caused by the unexpected failure of the equipment. The pieces of equipment failure modes represent the basic events of the risk analysis, and their effects on the system operational performance represent the consequences. Based on quantitative analysis, the probability of occurrence of a given failure scenario must be estimated and the consequences to the environment, humans, and system operation must be quantitatively expressed. The pieces of equipment with the most critical risk profile are the main focus of maintenance planning. The resulting maintenance program minimizes the risk of the system and the maintenance cost.

The application of RBM consists of the following steps that were presented in Fig. 2.10 (Arunraj & Maiti, 2007; Selvik & Aven, 2011):

i. Identifying the pieces of equipment and their components, subsystems, system, and their relationships, using the functional tree analysis;
ii. Collecting failure data, failure model, and failure rate: the database,
iii. Risk assessment and evaluation, preferably using the quantitative techniques previously presented in this chapter.

Once the risk profiles of the pieces of equipment of an asset are determined, the maintenance planners will select the most critical ones. The selection emphasizes equipment whose failure affects operational safety or endangers human life or the environment. For those pieces of equipment, predictive and preventive maintenance techniques are the recommended practices in order to avoid critical risks. The selection of the most suitable maintenance practices for a piece of equipment may be done using the guidelines of the RCM philosophy indicating a complementary application of those methods (Kiran, Prajeeth Kumar, Sreejith, & Muralidharan, 2016).

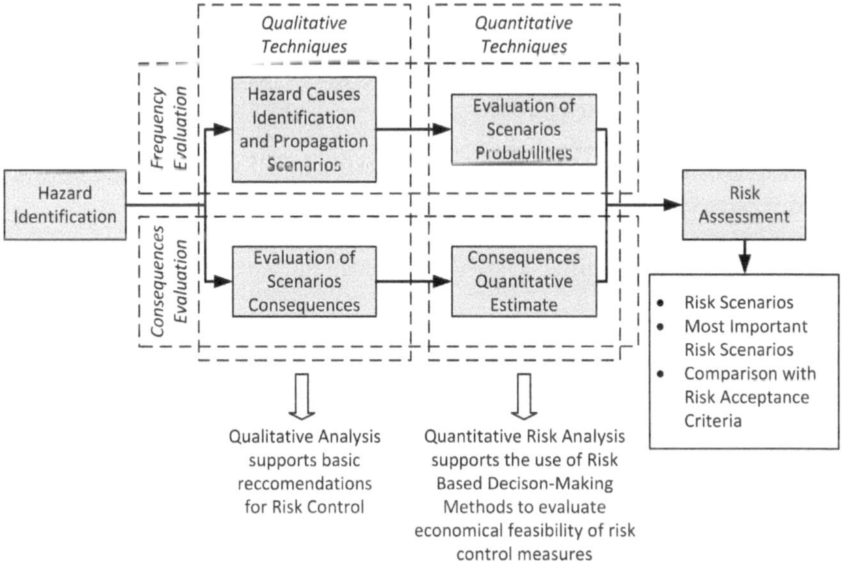

FIG. 2.10 Steps for risk-based maintenance.

The maintenance planning should be assigned to lower the risk to meet the acceptance criterion and to reduce the probability of failure. The definition of the optimal maintenance strategy must also be based on a decision-making method that seeks the balance between costs (including maintenance costs and costs incurred as a result of failure) and the minimization or elimination of risks.

The effect of maintenance planning on the system performance may be measured on different levels, for example through resulting production capacity, quality, availability, and reliability, and through the occurrence of accidents and other undesirable events.

In general, the basic principles for the risk-based maintenance program are:

- Assets with greater risk are maintained and monitored more frequently to achieve tolerable risk criteria and
- Assets with lower risk have a less stringent maintenance program.

The benefits of the risk-based maintenance philosophy are:

- Reduces the risk of asset failures to an acceptably low level;
- Supports decision-making about how best to allocate limited maintenance budgets; and
- Provides opportunities to identify and eliminate low-value maintenance tasks.

Considering the characteristics of the RBM philosophy, this method is usually applied to safety-critical systems such as chemical and petrochemical industries, transportation systems, and nuclear industry. More recently, due to some accidents, offshore oil and gas operations have adopted the application of RBM. In the future, the RBM will probably be adopted by industrial sectors where risk control is part of the organization's strategy if the magnitude of consequence far outweighs the cost of equipment. A limitation is that a highly sophisticated analysis team is needed to quantify the risk reduction with the application of different maintenance tasks.

An additional driver for RBM application is the publication of the ISO 55000 series of standards on asset management (ISO, 2014). These standards require companies to develop a set of structured documents that cover the maintenance of assets including data, information, and costs throughout their life cycle. The standard guidelines use risk as a basis for decision-making and continuous process improvement. More details about that standard are presented in Chapter 8.

2.4.3 Availability

For modern engineering systems submitted to maintenance actions, the reliability is not only the index that shall be used to characterize its performance. The reliability may be associated with the number of failures during an operational campaign. In case of failure, the time required to make corrective repairs is important because during the repair action the equipment must be paralyzed although there are demands for its use (Lewis, 1994).

The predictive and preventive maintenance actions are executed to avoid or eliminate the occurrence of failures so their influence is considered on the devices' reliability analysis. Furthermore, the predictive actions are established in advance on a usually annual calendar, making it possible not to schedule production during the periods of maintenance intervention. In cases of predictive maintenance, the intervention can usually be planned in advance and the production planning department can rearrange the production plan and open a time interval for maintenance intervention. Broadly, it can be said that predictive and preventive actions can be minimally planned to cause a minimum disturbance for production. On the contrary, in case of a device failure, it must be submitted to a corrective intervention, causing an unexpected disturbance in production.

The time to repair is also a random variable since it is dependent on the nature of failure, on the ability to diagnose the cause of failure, and on the availability of equipment and skilled personnel to carry out the repair procedure. The probability that equipment will be repaired in a given period of time is defined as maintainability and described by a probability distribution. If T_{repair} is a random variable that characterizes the time required to carry out a repair, then the repair time distribution function or simply the maintainability, $M(t)$, is given by Eq. (2.16) (Lewis, 1994):

$$M(t) = P(T_{repair} \leq t). \tag{2.16}$$

And, the repair time probability density function is expressed by Eq. (2.17):

$$m(t) = \frac{dM(t)}{dt}. \tag{2.17}$$

The mean time to repair (MTTR) is defined by Eq. (2.18)

$$MTTR = \int_0^\infty t \cdot m(t) dt. \tag{2.18}$$

Similarly, to the reliability analysis, a repair rate $\nu(t)$ can be defined. It represents the probability that the component is repaired between t and $t + \Delta t$, given that it failed in time instant t and expressed by Eq. (2.19):

$$\nu(t)\Delta t = \frac{P(t \leq T_{repair} \leq t + \Delta t)}{P(T_{repair} > t)}. \tag{2.19}$$

Considering Eq. (2.20), the repair rate is defined by Eq. (2.21).

$$m(t)\Delta t = P(t \leq T_{repair} \leq t + \Delta t) \tag{2.20}$$

$$\nu(t) = \frac{m(t)}{1 - M(t)} \tag{2.21}$$

Once the reliability and maintainability parameters are calculated, the system availability can be estimated. Availability is a measure of the percentage of time during which a piece of equipment is capable of producing its end product at some specified acceptable level. The most common definitions for availability are (Lewis, 1994):

i. Instantaneous availability of a device at time t, $A(t)$, is the probability that the device is up at time t;
ii. Average or interval availability, defined for a fixed time interval, T, as presented in Eq. (2.22), is the value of the instantaneous availability averaged over T. This time interval may be the mission time required for the device under analysis.

$$\overline{A}(T) = \frac{1}{T} \int_0^T A(t) dt. \tag{2.22}$$

iii. The steady-state, or asymptotic, availability is defined by Eq. (2.23):

$$A_{\infty} = \lim_{T \to \infty} \frac{1}{T} \int_{0}^{T} A(t)dt \tag{2.23}$$

For a single device, the steady-state availability can be estimated as presented in Eq. (2.24):

$$A_{\infty} = \frac{MTTF}{MTTF + MTTR} \tag{2.24}$$

In a simple way, the device availability is controlled by two parameters:

- Mean Time To Failure (MTTF), previously defined in "Reliability concepts" section.
- Mean Time To Repair (MTTR), which is a measure of how long, on average, it will take to bring the device back to normal serviceability when it does fail.

Equipment availability analysis may be defined as a combination of the availability of devices that constitute the equipment considering its functional architecture. Equipment availability analysis can be executed using reliability block diagrams or more sophisticated techniques as presented in Chapter 4.

2.5 Concluding remarks

To conclude this chapter, it must be reminded that the reliability and maintenance of modern engineering systems are important aspects that must be considered during the system design. The system reliability is defined during the design phase. The basic system functional architecture and component reliability affect system operational behavior. During the operational phase, the most suitable maintenance program must be defined aiming at controlling risks associated with components' failures. The maintenance plans shall avoid the most critical failures that severely affect system operational behavior or may endanger the environment or human life. These plans may balance the maintenance costs and system reliability and risk profile.

Finally, it is important to highlight that modern engineering systems' reliability, maintainability, and risk analysis require specialized technical teams with experience in the design and operation of similar systems and also great knowledge about the fundamental concepts and tools of reliability and risk analysis.

References

AIChe. (1989). *Guidelines for chemical process quantitative risk analysis*. American Institute of Chemical Engineers AICHE/CCPS.

Arunraj, N. S., & Maiti, J. (2007). Risk-based maintenance—Techniques and applications. *Journal of Hazardous Materials*, *142*(3), 653–661. https://doi.org/10.1016/j.jhazmat.2006.06.069.

Aven, T. (2003). *Foundations of risk analysis: A knowledge and decision-oriented perspective*. John Wiley & Sons.

Aven, T. (2011). A risk concept applicable for both probabilistic and non-probabilistic perspectives. *Safety Science*, *49*(8–9), 1080–1086. https://doi.org/10.1016/j.ssci.2011.04.017.

Aven, T. (2012). The risk concept—Historical and recent development trends. *Reliability Engineering & System Safety*, *99*, 33–44. https://doi.org/10.1016/j.ress.2011.11.006.

Ayyub, B. M. (2014). *Risk analysis in engineering and economics*. Chapman and Hall/CRC. https://doi.org/10.1201/b16663.

Ayyub, B. M., & McCuen, R. H. (2016). *Probability, statistics, and reliability for engineers and scientists*. CRC Press. https://doi.org/10.1201/b12161.

Beresh, R., Ciufo, J., & Anders, G. (2007). Basic fault tree analysis for use in protection reliability. In *2007 Power systems conference: Advanced metering, protection, control, communication, and distributed resources* (pp. 1–7). IEEE. https://doi.org/10.1109/PSAMP.2007.4740894.

Bloom, N. (2005). *Reliability Centered maintenance: Implementation made simple*. McGraw-Hill Education.

Bottelberghs, P. H. (2000). Risk analysis and safety policy developments in the Netherlands. *Journal of Hazardous Materials*, *71*(1–3), 59–84. https://doi.org/10.1016/S0304-3894(99)00072-2.

Ceballos, B., Lamata, M. T., & Pelta, D. A. (2016). A comparative analysis of multi-criteria decision-making methods. *Progress in Artificial Intelligence*, *5*(4), 315–322. https://doi.org/10.1007/s13748-016-0093-1.

Dhillon, B. S. (1992). Failure modes and effects analysis—Bibliography. *Microelectronics Reliability*, *32*(5), 719–731. https://doi.org/10.1016/0026-2714(92)90630-4.

DoD. (1984). *Procedures for performing a failure mode, effects and criticality analysis. US MIL-STD-1629 (1974), US MIL-STD 1629A (1980), US MIL-STD 1629A/Notice 2 (1984)*. United States Department of Defense.

DoD. (1998). *Military handbook MIL-HDBK-338 electronic reliability design handbook*. United States Department of Defense.

Ford. (2011). *Failure mode and effects analysis handbook version 4.2*. Ford Motor Company.

Frankel, E. G. (1983). *Systems reliability and risk analysis. Vol. 1*. Springer Netherlands. https://doi.org/10.1007/978-94-009-6920-9.

IEC. (2006). *IEC 61025 fault tree analysis (FTA)*. International Electrotechnical Commission.

IEC. (2009). *IEC 60300-3-11 dependability management—Part 3-11: Application guide—Reliability centred maintenance*. International Electrotechnical Commission.

ISO. (1997). *ISO/IEC 2382-14 information technology—Vocabulary—Part 14: Reliability, maintainability and availability*. International Organization for Standardization.

ISO. (2014). *ISO 55000 asset management—Overview, principles and terminology*. International Organization for Standardization.

Jansen, H. M. A. (1988). Financieel-economische risicolimiet. In *Ministerie van Volkshuisvesting*. Ruimtelijke: Ordening en Milieubeheer.

Jonkman, S. N., van Gelder, P. H. A. J. M., & Vrijling, J. K. (2003). An overview of quantitative risk measures for loss of life and economic damage. *Journal of Hazardous Materials*, *99*(1), 1–30. https://doi.org/10.1016/S0304-3894(02)00283-2.

Khan, F. I., & Haddara, M. (2004). Risk-based maintenance (RBM): A new approach for process plant inspection and maintenance. *Process Safety Progress, 23*(4), 252–265. https://doi.org/10.1002/prs.10010.

Khan, F. I., & Haddara, M. M. (2003). Risk-based maintenance (RBM): A quantitative approach for maintenance/inspection scheduling and planning. *Journal of Loss Prevention in the Process Industries, 16*(6), 561–573. https://doi.org/10.1016/j.jlp.2003.08.011.

Kiran, S., Prajeeth Kumar, K. P., Sreejith, B., & Muralidharan, M. (2016). Reliability evaluation and risk based maintenance in a process plant. *Procedia Technology, 24*, 576–583. https://doi.org/10.1016/j.protcy.2016.05.117.

Kossiakoff, A., Sweet, W. N., Seymour, S. J., & Biemer, S. M. (2011). *Systems engineering principles and practice*. John Wiley & Sons, Inc. https://doi.org/10.1002/9781118001028.

Kumamoto, H., & Henley, E. J. (1996). *Probabilistic risk assessment and Management for Engineers and Scientists*. IEEE Press.

Lewis, E. E. (1994). *Introduction to reliability engineering* (2nd ed.). John Wiley & Sons.

Li, X., Zhang, Z., & Ahmed-Kristensen, S. (2014). The sources and methods of engineering design requirement. In *Moving integrated product development to service clouds in the global economy. Proceedings* (pp. 112–121). IOS Press. https://doi.org/10.3233/978-1-61499-440-4-112.

Mannan, S. (2005). *Lees' loss prevention in the process industries*. Elsevier. https://doi.org/10.1016/B978-0-7506-7555-0.X5081-6.

Modarres, M., Kaminskiy, M. P., & Krivtsov, V. (2016). *Reliability engineering and risk analysis*. CRC Press. https://doi.org/10.1201/9781315382425.

Moubray, J. (1997). *Reliability-centered maintenance* (2nd ed., p. 448). Industrial Press, Inc.

Nowlan, F. S., & Heap, H. F. (1978). *Reliability-centered-maintenance*.

O'Connor, P. D. T., & Kleyner, A. (2011). *Practical reliability engineering*. John Wiley & Sons, Ltd. https://doi.org/10.1002/9781119961260.

Palady, P. (1995). *Failure modes and effects analysis: Predicting & preventing problems before they occur*. PT Publications.

Prabhakar Murthy, D. N., Xie, M., & Jiang, R. (2003). *Weibull models*. John Wiley & Sons, Inc. https://doi.org/10.1002/047147326X.

Rausand, M. (1998). Reliability centered maintenance. *Reliability Engineering & System Safety, 60*(2), 121–132. https://doi.org/10.1016/S0951-8320(98)83005-6.

SAE. (2009). *SAE J1739 potential failure mode and effects analysis in design (design FMEA), potential failure mode and effects analysis in manufacturing and assembly processes (process FMEA)*. Society of Automotive Engineers.

Saleh, J. H., & Marais, K. (2006). Highlights from the early (and pre-) history of reliability engineering. *Reliability Engineering & System Safety, 91*(2), 249–256. https://doi.org/10.1016/j.ress.2005.01.003.

Selvik, J. T., & Aven, T. (2011). A framework for reliability and risk centered maintenance. *Reliability Engineering & System Safety, 96*(2), 324–331. https://doi.org/10.1016/j.ress.2010.08.001.

Smith, A. M., & Hinchcliffe, G. (2003). *RCM—Gateway to world class maintenance* (1st ed., p. 337). Butterworth-Heinemann.

Stamatis, D. H. (1995). *Failure mode and effect analysis: FMEA from theory to execution*. ASQC Quality Press.

Velasquez, M., & Hester, P. T. (2013). An analysis of multi-criteria decision making methods. *International Journal of Operations Research, 10*(2), 56–66.

Vesely, W. E., Goldberg, F. F., Roberts, N. H., & Haasl, D. F. (1981). *Fault tree handbook (NUREG-0492)*. https://www.nrc.gov/docs/ML1007/ML100780465.pdf.

Vinnem, J. E. (2007). *Offshore risk assessment*. London: Springer. https://doi.org/10.1007/978-1-84628-717-6.

Waeyenbergh, G., & Pintelon, L. (2004). Maintenance concept development: A case study. *International Journal of Production Economics, 89*(3), 395–405. https://doi.org/10.1016/j.ijpe.2003.09.008.

Wang, J. X., & Roush, M. L. (2000). *What every engineer should know about risk engineering and management*. CRC Press.

Wilday, A. J., Ali, M. W., & Wu, Y. (1998). Index method for cost-effective assessment of risk to the environment from accidental releases. In *IChemE symposium series no. 144*.

Zio, E. (2018). The future of risk assessment. *Reliability Engineering & System Safety, 177*, 176–190. https://doi.org/10.1016/j.ress.2018.04.020.

CHAPTER

3

Engineering systems fundamentals

The whole is not simply the sum of its parts.

Aristotle (384–322 BCE) In: Metaphysics

The reader will probably find the above rendering of the famous phrase attributed to Aristotle a bit unusual. In his well-known translation, however, Ross (1924) provides a text even more apart from the standard misquote: "the totality is not, as it were, a mere heap, but the whole is something beside the parts." Well, but the interested reader may also ask why is it relevant to revisit those philosophical definitions when dealing with matters in scientific fields such as engineering systems. The answer lies in the fact that modern thought, particularly modern scientific thought, has been so strongly influenced by René Descartes' philosophy that it is still difficult for many scientists and engineers, no matter how learned or well informed, to depart from the deep-rooted idea of breaking down systems in as many smaller parts as possible and then proceed gradually from the simple to the complex. As a matter of fact, Descartes (1637) was so convinced of his reasoning that he begins his brilliant Discourse by saying that "If this discourse seems too long to be read at a sitting you may divide it into six parts." No doubt history has shown that he succeed in being very convincing too.

We are all acquainted with the notion that as scientific investigation progresses, new theories, although may not invalidate, will bring out the limitations of older ones. Perhaps one of the most notorious cases is that of Albert Einstein's and Isaac Newton's respective theories relating to energy and matter. The latter will always be regarded as one of the most excellent realizations of the human mind; nonetheless, its limitations are now well understood.

https://doi.org/10.1016/B978-0-12-823521-8.00012-8

In everyday practice, however, the *status quo* is likely to be protracted as much as possible. By the end of the 19th century, the first automobiles were simply coaches without horses. It takes some time for the human mind to get used to new ideas and concepts, or, for that matter, to be aware of the need for change when the familiar and simpler world of the past is no longer present.

3.1 Systems theory

Around the middle of the 20th century, systems theory was developed by researchers from fields as diverse as biology, psychiatry, and philosophy with the aim of considering solutions for complex systems where division into subsystems leads to the distortion of results due to the fact that some properties arise from relationships and interactions among subsystems. Such properties, therefore, can only be understood when they are pondered in their entirety by an approach that considers not only the technical but also the managerial and social aspects of the problem.

Ashby (1956), a psychiatrist and pioneer in cybernetics, realized that in the study of some systems, such as the cerebral cortex of the free-living organism, the anthill as a functioning society, and the human economic system, complexity could not be wholly evaded; besides, despite being outstanding in their practical importance, those systems were intractable by what he called the older methods. So, he added, "*today we see psychoses untreated, societies declining, and economic systems faltering, the scientist being able to do little more than to appreciate the full complexity of the subject he is studying.*" His hopeful conclusion, however, was that modern science was already at that time taking the first steps toward studying "complexity" as a subject in its own right.

In his quest for the unification of science, i.e., for a General System Theory, von Bertalanffy (1969, 1972), an outstanding biologist, contended that there are general aspects and viewpoints alike in different sciences and that formally identical or isomorphic laws in different fields can be frequently found. In many cases, isomorphic laws hold for certain classes or subclasses of "systems," irrespective of the nature of the entities involved. Therefore, considering that apparently general system laws exist that apply to any system of a certain type, irrespective of the particular properties of the system and of the elements involved, the grounds for a general science of "wholeness" can be stated as follows:

1. *"There is a general tendency towards integration in the various sciences, natural and social.*
2. *Such integration seems to be centered in a general theory of systems.*

3. *Such theory may be an important means of aiming at exact theory in the nonphysical fields of science.*
4. *Developing unifying principles running 'vertically' through the universe of the individual sciences, this theory brings us nearer to the goal of the unity of science.*
5. *This can lead to a much-needed integration in scientific education."*

As a consequence of his holistic approach to the study of scientific phenomena, Bertalanffy put forth the idea that in the past science tried to explain observable phenomena by reducing them to an interaction of elementary units that would be investigated independently of each other. In contrast, contemporary science is concerned with what is somewhat vaguely termed "wholeness," i.e., problems of organization, phenomena not resolvable into local events, and dynamic interactions manifest in difference of behavior of parts when isolated or in a higher configuration. In other words, this new way of thinking was concerned with systems not understandable by investigation of their respective parts in isolation. He added that conceptions and problems of this nature have appeared in all branches of science, irrespective of whether inanimate things, living organisms, or social phenomena are the object of study. Accordingly, he defined a system as a set of elements standing in interrelation among them and with the environment.

As a philosopher concerned with management, and sharing ideas with his colleagues at the University of California at Berkeley during the first days of mainframe computers, Churchman (1968) started by reminding us that a system can be defined in many ways, but that all "definers" will agree that it is a set of parts coordinated to achieve a set of goals. Then, he proposed that a system should be considered according to the model shown in Fig. 3.1, although he acknowledged that there are other ways of thinking about systems.

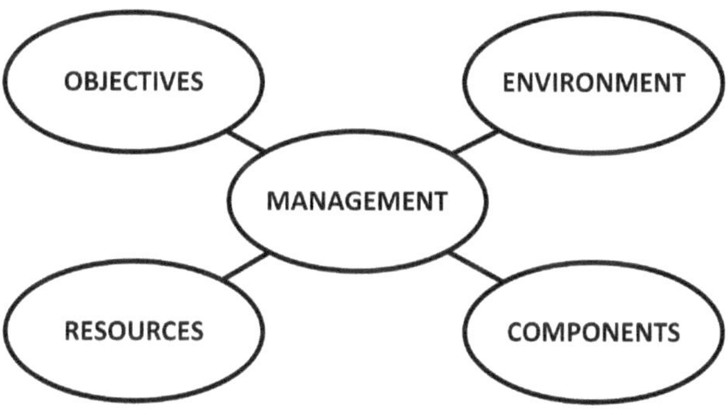

FIG. 3.1 Model of a management system.

In the model of Fig. 3.1, "objectives" means the objectives of the whole system, not those of its parts. An important point is that formulated objectives that do not relate to the performance of the system are not true objectives of the system. A simple test can be carried out: will the system sacrifice other goals in order to attain an objective? If the answer to this simple question is yes, then that objective is a true objective of the system otherwise it is not. Therefore, it is necessary to have adequate means of assessing the performance of the whole system, that is, to establish precise and specific metrics for the system's overall performance.

The environment is composed of whatever is "outside" the system, meaning all things that represent restrictions imposed upon the system. Besides being outside the system, the environment will also influence in part the functioning of the system. Once more a simple test can be applied: Is it possible to do something about "that"? No. Is "that" important regarding the system's objectives? Yes. Then, "that" is in the environment.

The resources are the general reservoir that provides what is necessary for the components or processes to accomplish the basic missions, activities, or tasks that the system should perform. Finally, the system management is concerned with creating plans for the system while considering the other elements: global objectives, the environment, resources, and components. In other words, management establishes the components' objectives, allocates resources, and controls system performance.

Ackoff (1990) who often worked in close collaboration with Churchman maintained that "if each part of a system, considered separately, is made to operate as efficiently as possible, the system as a whole will not operate as efficiently as possible," implying that the performance of a system is more influenced by the way its elements interact, than by their functioning independently from one another. He proposed that a system is a set of elements that satisfies the following three conditions:

1. *"The behavior of each element has an effect on the behavior of the whole;*
2. *The behaviors of the elements and their effects on the whole are interdependent;*
3. *However subgroups of the elements are formed, each has an effect on the behavior of the whole and none has an independent effect on it."*

This way of viewing systems leads to the obvious conclusion that the "whole" cannot simply be divided into components for analysis as if they were independent. In addition, it seems clear that due to the interactions between components the system will possess properties that its parts do not have. In other words, if a system is taken apart it will lose its essential properties.

Systems theory provides the theoretical foundation for tackling engineering systems as an integrated whole, despite their many specialized subsystems and elements, so as to achieve the overall objectives.

3.2 Engineering systems

Traditionally, engineering problems are dealt with using analytic reduction; i.e., inherently complex systems are divided into simpler parts for individual consideration. Then, different models and techniques are applied in the analysis of such separate components as if each subsystem operated independently and not interacting with others. Results are combined under the assumption that each component has the same behavior whether acting alone or in the whole. In so doing, analysts frequently overlook important nonlinear interactions and feedback loops.

Having emerged out of needs and concerns in the fields of aeronautics, aerospace, and defense industries, engineering systems is a relatively new academic field—at the crossroads where engineering intersects the management and the social sciences—that promotes the development of new approaches, frameworks, and theories to analyze, design, deploy, and manage complex technological systems which do not cease to augment in size, scope, and complexity due to rapid technological development. Having to cope with problems and consequences that are essentially dynamic, difficult to predict, and multidimensional, involving collections of interconnected relationships and parts, this new field seeks to take into consideration the fact that designing and operating modern very complex systems require not only traditional engineering skills, but also inter alia knowledge of public policy issues and awareness of societal norms and preferences.

In addition to being a mix of electronic, electrical, and mechanical components, due to the rapid development of technology in recent decades engineering systems have incorporated large amounts of software. As a consequence, they have become complex systems where intricate interactions occur between software, subsystems, equipment, instruments, and people. These are aspects that now greatly influence the operation and maintenance of modern engineering systems, particularly in regard to their safety and related effects such as human and environmental hazards.

It should be said that software as such poses no threat to anything or anyone. However, it can be potentially hazardous when situated in a complex system context. Since software does not degrade like hardware items nor fail randomly, complex system hazards related to software will be caused by software flaws derived from either specification, design or implementation errors, or a combination thereof, as well as from uncontrolled interactions among different elements comprised by the system (Abdulkhaleq, Wagner, & Leveson, 2015). This fact must be thoroughly understood and carefully taken into consideration when designing complex engineering systems, so as to allow for adequate control through all levels of the system hierarchy.

At this point, before proceeding, it is hoped that the attentive reader will have realized that the term "complex" should not be confused with the term "complicated."

3.2.1 Properties

One aspect that makes the Cartesian approach appear limited is the concept of emergence, which is nowadays often considered in connection with the behavior of complex engineering systems. It was first conceived by Bivona (2019) and Lewes (1875) while considering the difference between the nature of effects more than a century ago: "*... there is a co-operation of things of unlike kinds. The emergent is unlike its components ... and it cannot be reduced to their sum or their difference.*" On the one hand, he contended, resultant effects are obtained by the mere addition or subtraction of the causes that produce them. A simple physical example would be the total weight of an object calculated merely by adding the weights of the parts that make it up. On the other hand, emergent effects could not be thus obtained because they are novel compared to the causes from which they emerge. The example then proposed by Lewes referred to the mental properties that emerge from neural processes and yet are not properties of the parts of the neural processes from which they emerge.

Although there has been some argument in the literature as to what emergence precisely means (Damper, 2000; Goldstein, 2001; Johnson, 2006), it is also now accepted by many authors in diverse scientific fields that complex systems may possess what came to be called emergent properties (Mitchell, 2009; Nakamura, 2019; Pons, Pons, & Pons, 2013). Therefore, instead of engaging in any argument, this section simply intends to provide both researchers and practitioners with a concept that may be useful when pondering about engineering systems, at least until future progress in science might indicate more rigorous ways of doing so. Nevertheless, in way of reassurance to skeptic readers it must be said that here emergence at least is not considered as something "magical," but just as something that cannot be entirely explained by reduction and causality.

Perhaps the simplest definition that can be given for an emergent property is as follows: a property that emerges from the relationships between system components. In other words, emergent properties are collective properties that necessarily arise from the interactions of the parts of the larger system, the "whole" as it were. Take life for instance, it can be seen in terms of interactions, relationships, and collective properties of many molecular systems and their constituents. Therefore, it seems appropriate to consider that "*emergence and complexity refer to the appearance of higher-level properties and behaviours of a system that obviously come from the collective dynamics of that system's components*" (Aziz-Alaoui & Bertelle, 2009).

In systems engineering terms, it can be said that since an emergent property is a resultant of the interactions of individual components, it can only be perceived and measured, or at least assessed, after the integration of said components into a system. This is clearly the case with the safety of modern engineering systems where socio-technical interactions have become of the essence. A power plant safety, for instance, cannot simply be determined by assessing "the safety" of individual components such as generators without reference to their respective situations within the "whole," and this includes the way in which they are operated and maintained. Thus, since individual components may behave differently according to circumstances, it should be realized that plant safety can only be evaluated by considering all possible interactions between components, operators, maintainers, and the environment.

On the other hand, individual component properties can be combined in order to produce important resultant properties. Such is the case of reliability. As seen in Chapter 2, it is perfectly meaningful to consider the reliability of an individual generator and then use its value when building a block diagram and assessing the reliability of the entire power plant. Therefore, although they may be sometimes wrongly confused, reliability and safety are essentially different system properties.

3.2.2 Hierarchy and control

When modeling and designing complex engineering systems, it is convenient to consider a hierarchy of levels of organization. From bottom up, each level will be more complex than the previous level. Besides, upper levels will possess emergent properties that cannot be perceived at lower levels. By highlighting the fundamental differences between complexity levels, bringing out what separates them, and what links them, hierarchical models provide a convenient way of identifying control requirements and supplying the necessary actions. Existing complex systems can also be analyzed in this way.

Fig. 3.2 shows the situation of a Power Plant (PP) in the hierarchy of a country's energy sector, where there is a National Energy Agency (NEA) linked to the Ministry of Mines and Energy (MME) and responsible for regulating and controlling both technically and economically the generation, transmission, and distribution of electric energy, while a National Operator of the Energy System (NOS) is responsible for operating the national grid.

Whether performed by humans or by automated controllers, controls need a model of the process being controlled. This model should include the required relationship among the system variables, the current values of the system variables, and the ways the process can change state in order

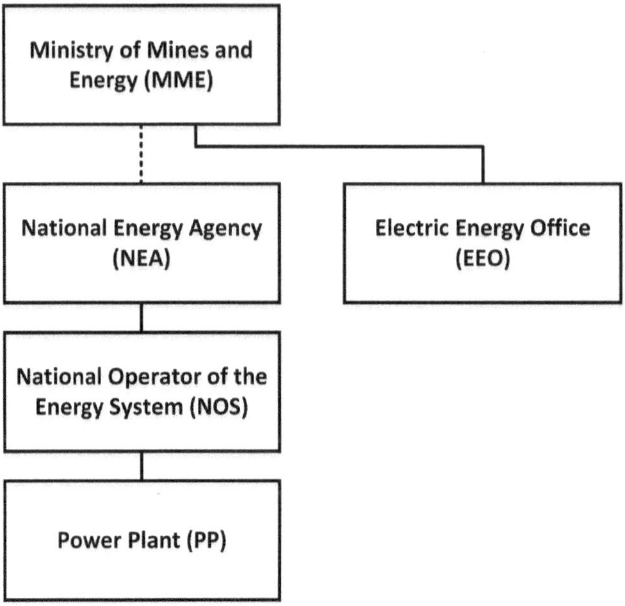

FIG. 3.2 Hierarchy of a country's electrical sector.

to determine what control actions are needed. In addition, it should be updated through various forms of feedback. When the process model embedded in the controller does not match the process, incorrect control commands are given, required control actions are not provided, potentially correct control commands are provided too early or too late, or control is stopped too soon or applied too long. Component interaction failures can usually be explained in terms of incorrect process models.

The type of high-level control provided by the electrical sector hierarchy of Fig. 3.2 is shown in Fig. 3.3. Moving down along the complex system hierarchy, the control structure at plant level is shown in Fig. 3.4.

The most basic concept in complex engineering system safety is a constraint (Marais, 2005) in the way that events leading to losses occur only because necessary constraints were not successfully enforced. Failures can then be understood in terms of which safety constraints were violated at each level, and why the constraints were inadequate or not adequately enforced. This concept allows the incorporation of software, human decision-making, adaptation, and social and managerial factors into the analysis.

When failures are described in terms of a structured hierarchy of control based on adaptive feedback mechanisms, adaptation plays a central role in the understanding and prevention of failures. At each level of the hierarchical structure, inadequate control may result from causes such

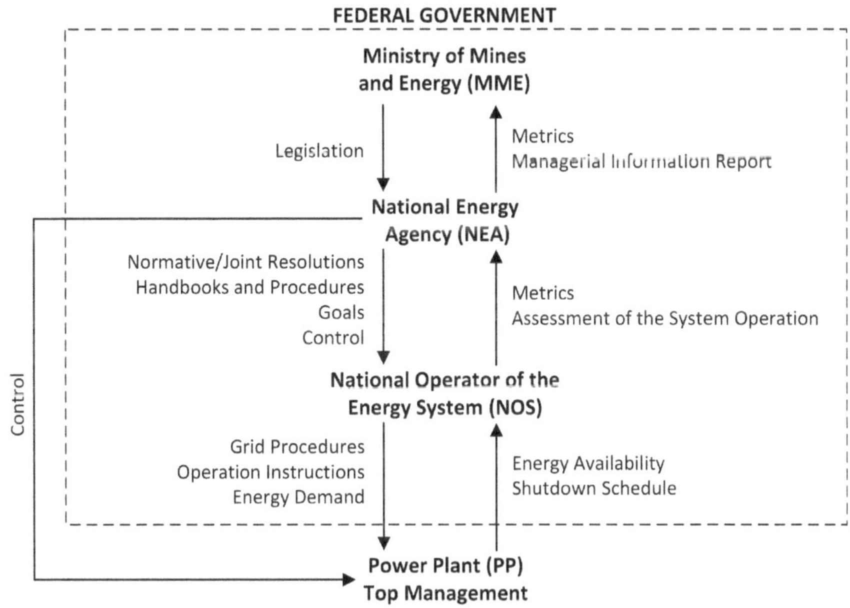

FIG. 3.3 High-level control structure.

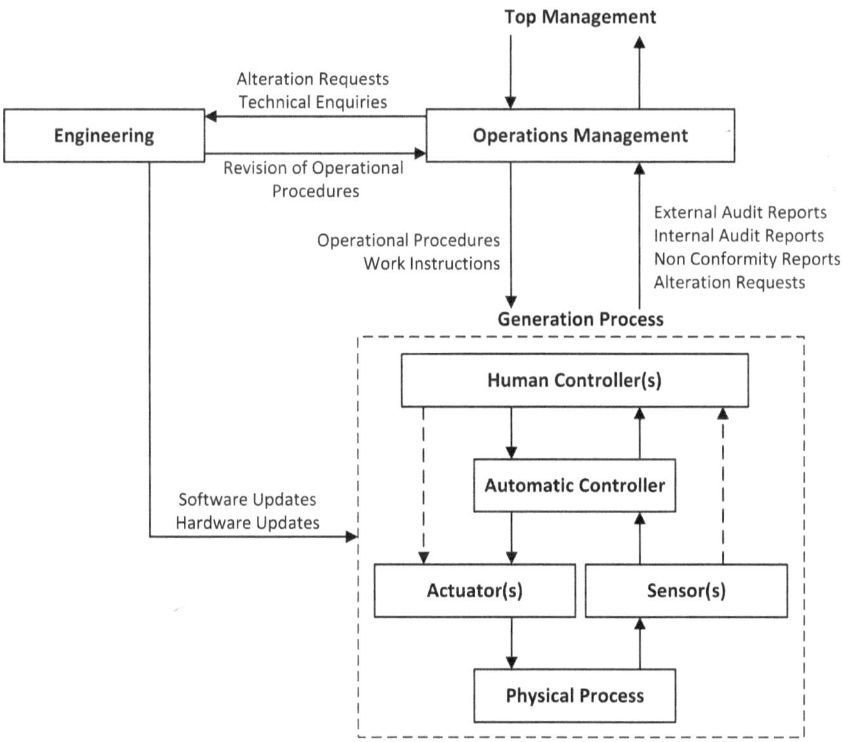

FIG. 3.4 Plant-level control structure.

as ineffective commands, incorrect execution at a lower level, or deficient communication or processing of feedback about constraint enforcement. In each control structure and at each level, the proper constraints will vary but in general may include design and process constraints, management constraints, manufacturing constraints, and operational constraints.

3.3 Complex engineering system hazard analysis

The identification of potential hazards is fundamental to establishing safety and reliability along the design, production, and operational stages of complex engineering systems life cycles. In the last decades, under the impact of serious plant disasters, accidents have increasingly been considered as emergent phenomena. However, emergent properties may be overlooked by well-established reliability and risk analysis techniques, such as Hazard and Operability (HAZOP) study, Fault Tree Analysis (FTA), Failure Modes and Effects Criticality Analysis (FMECA), and Operation and Maintenance Causal Tree Analysis (OMCTA). Those methods, although suitable for individually pinpointing critical components and therefore assuring the continuity of plant operation, come short of providing the same degree of assurance as to the safe operation of safety-critical software-dependent systems. Accordingly, new methods have been proposed for hazard analysis, that is, the process of identifying hazards and their potential causal factors in plants viewed as complex engineering systems.

3.3.1 Systems-Theoretic Process Analysis (STPA)

One such hazard analysis method is STPA based on Levenson's (2011) Systems-Theoretic Accident Model and Processes (STAMP) that has already been successfully used in many domains including aerospace, defense, energy, chemical, healthcare, and transportation systems (Fleming & Levenson, 2015). Here, hazard is understood as a system state or set of conditions which may or may not be compounded by a particular set of unfavorable environment circumstances, and that will lead to an undesired system behavior or loss.

STPA was conceived as a top-down system engineering analysis technique comprising two main steps:

1. Identify reliability constraints—system and component requirements—as well as conditions that may lead to the violation of such constraints, i.e., conditions brought about when the process model embedded in the controller does not match the process being controlled.

2. Determine how each potentially hazardous condition identified in Step 1 could occur. Examine the levels of the control loop to see what could cause them. Evaluate existing measures. Where there are multiple controllers of the same component—safety constraint—identify conflicts and potential coordination problems. Consider how the designed controls could degrade over time and build in protection.

The information obtained while identifying the hazardous control actions (Step 1) is required to perform the identification of the causes of such inadequate control actions (Step 2).

A vital feature of a hydropower plant is the adequate control of the speed governor (SG) for the required power generation. An application of STPA is exemplified in Fig. 3.5 and Tables 3.1 and 3.2 at the SG level.

When using STPA, the analyst will be considering the occurrence of hazardous system states as a dynamic control problem, as opposed to the result from a chain of failure events. This attitude implies a fundamental change in emphasis: from preventing failures to enforcing constraints on system behavior.

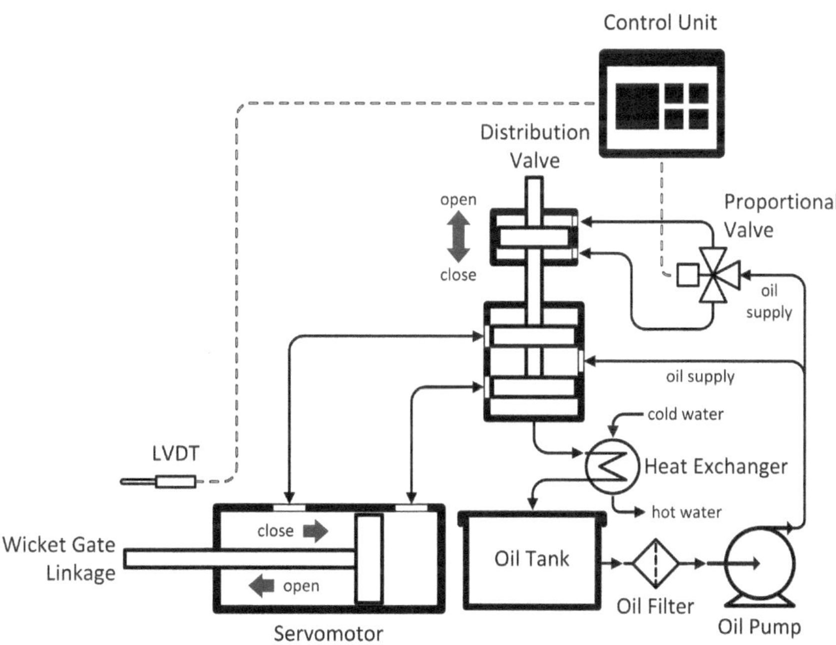

FIG. 3.5 SG illustrative diagram.

TABLE 3.1 Hazardous actions.

Control action	Agent	Hazard	Consequence
Vary and keep the opening angle of wicket gates according to power generation demand and available water flow	Control System	Not provided	Prevents synchronism of generating unit. No generation.
		Provided too late	Loss of synchronism. Trip of generating unit.
		Stopped too soon	Loss of synchronism. Trip of generating unit.

TABLE 3.2 Definition of safety restrictions.

Unsafe control action	Safety restriction
Control system does not vary the opening angle of wicket gates	Alternative shutdown capability
Control system does not keep the opening angle of wicket gates	Alternative shutdown capability

3.3.2 Functional Resonance Analysis Method (FRAM)

This method proposed by Hollnagel (2012) primarily for accident investigation purposes in the context of resilience engineering, but also for risk assessment of complex socio-technical systems, combines adequate hazard identification and analysis capabilities. It has also been successfully used in many diverse fields such as aerospace, high-risk industrial nuclear and oil plants, the healthcare industry, railway network safety, and ferry accidents at sea. The method aims at describing outcomes using the idea of resonance arising from the variability of system performance, and is based on four basic principles, to wit: equivalence of successes and failures; approximate adjustments; emergence; and functional resonance.

By "equivalence of successes and failures" is meant that failures and successes can be considered equivalent because they have the same origins. In other words, it is purported that if an action leads to the expected outcome, it will be considered correct, and if it produces an unexpected outcome, it will be considered an error as illustrated in Fig. 3.6.

Here, systems are divided into tractable and intractable. In the latter, of which socio-technical systems constitute an example, existing conditions cannot be completely described or specified. Besides, resources—money, time, materials, etc.—are inevitably limited and may sometimes be insufficient. As a consequence, people either individually or in groups and

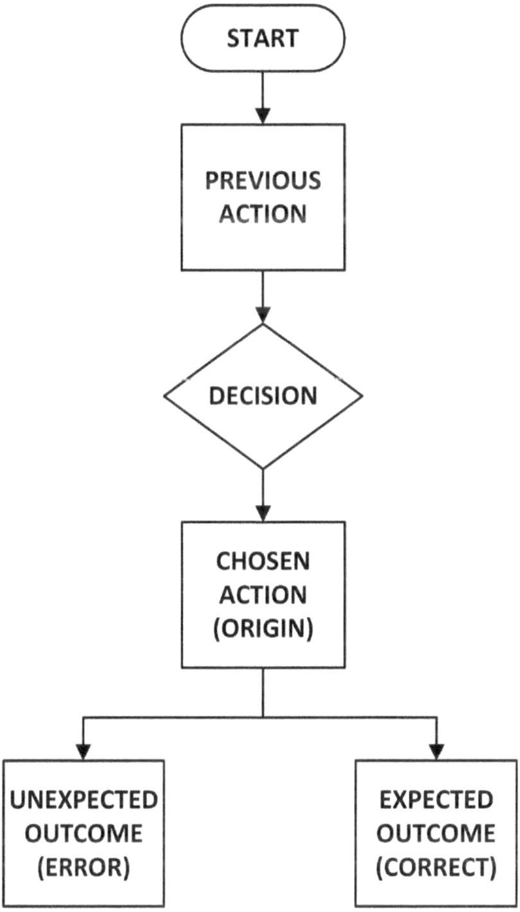

FIG. 3.6 Equivalence of successes and failures.

organizations for that matter will have to adjust their performance according to varying actual conditions, and such adjustments will not be exact, but rather approximate. The variability introduced by such approximate adjustments is at the same time cause for both desired and undesired outcomes.

As already extensively discussed previously, a system overall performance, either normal or faulty, is an emergent rather than a resultant phenomenon that cannot be entirely explained using the principles of decomposition and causality, that is, simply by referring to its specific components or parts.

The concept of functional resonance, derived from the theory of stochastic resonance, is defined as a detectable result emerging out of nonintentional interactions of the normal variability of multiple signals. In other words, failures may occur when the variability of many functions causes a reinforcement effect through tight couplings not identifiable as cause-effect chains, and resulting in the variability of some functions exceeding normal limits.

Traditional risk assessment methods are based on linear models and on failure probabilities calculation. Complex socio-technical systems, however, are essentially nonlinear; i.e., risks are emergent rather than resultant phenomena and can be represented by nonlinear combinations of performance variability. Thus, instead of focusing on individual failures, risk assessment should consider how undesirable outcomes can result from performance variability. The concept of functional resonance can be employed to identify those combinations of variability that may be in the root of such unwanted results.

In order to develop a FRAM model, the first step consists in identifying the functions required to accomplish the basic missions, activities, or tasks that the system must perform. And this should be done in accordance with the breadth-first principle: It is more important to identify all the required functions than to describe some or even all functions in great depth. Then, each function should be described in terms of the aspects that are essential for the function to be carried out, as shown in Fig. 3.7 and described in Table 3.3. The completeness and consistency of the thus obtained model should be checked, for instance, through feedback by interested parties.

The next three steps in a FRAM analysis are identifying and describing variabilities; defining how variabilities may be combined, that is, the aggregation of variability; and finally considering ways of mitigating those variations capable of producing unwanted effects, or possibly favoring those variations that might lead to successful outcomes.

The FRAM model, such as generically illustrated in Fig. 3.8, represents both the set of functions involved in the activity being analyzed and the potential couplings among them. On the other hand, an "instantiation" describes the couplings among functions for a given scenario, i.e., a concrete instance of the model for specific conditions.

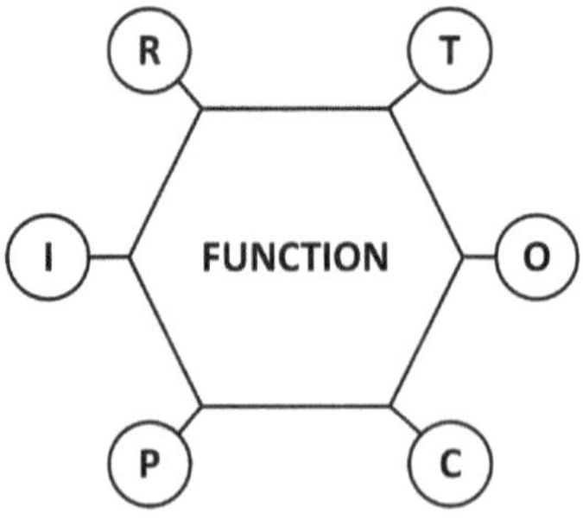

FIG. 3.7 System function and aspects.

TABLE 3.3 Description of aspects.

	Aspect	Description
I	Input	That which activates the function or is either used or transformed to produce the output.
P	Precondition	System conditions that must be present before the function can be carried out. Should correspond to an Output from another function.
R	Resource	That which is necessary or consumed—matter, energy, competence, software, tools, work force, etc.—when the function is activated.
C	Control	That which supervises or regulates the function: plans, procedures, instructions, etc.
T	Time	Temporal constraints—start, duration, finish—affecting how the function is carried out.
O	Output	That which results from the function: matter, energy or information. May be the signal for the activation of a subsequent function.

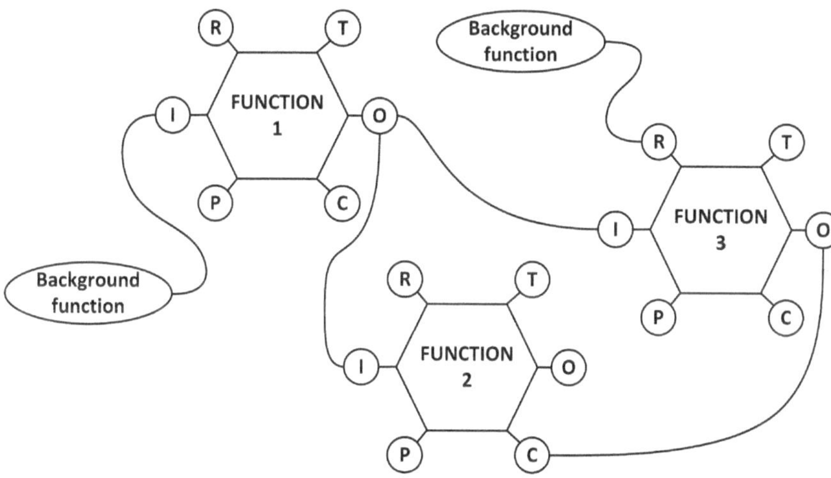

FIG. 3.8 FRAM model illustrative example.

The FRAM proposes that activities be described in terms of the functions involved. However, it does not specify any functions or any specific way of disposing them. It can thus be said that the FRAM uses a method to produce a model and not the other way around. Besides, since the processes that originate phenomena are the same either on a small or a large scale, i.e., there are no levels in a FRAM model, from the perspective of failures there is no need to differentiate between high- and low-level failures, or to have separate theories for organizational and human failures.

3.4 Systems engineering concepts applied to engineering systems design

3.4.1 Fundamentals of systems engineering

The simplest way to define systems engineering is to say that it is a total system approach for the development of intended systems. Here, the term total means that such an approach comprises the following important aspects:

1. Viewing the system as a whole;
2. Having a life cycle perspective;
3. Promoting interdisciplinary collaboration;
4. Implementing a multilevel management plan; and
5. Carrying out all that in a highly disciplined way.

Systems engineering is composed of two significant disciplines, to wit: the technical knowledge domain and the engineering management domain (U.S.A., 2001). The objective of systems engineering is to produce what might be called engineered systems, that is, technical systems capable of achieving specified operational objectives over a life cycle, from the identification of stakeholders' needs until phase-out and subsequent adequate disposal. Such human-made or human-modified systems will be produced employing multiple human and material resources; be composed of interacting subsystems and components; provide responses and exhibit behaviors; be influenced by external factors; and interact with their environment.

3.4.2 Functional analysis and allocation

Firstly, it is important to distinguish between requirements and constraints. The former are established according to the stakeholders' expectations in connection with the mission to be accomplished or purpose to be fulfilled, including operational goals and criteria for defining effectiveness or success. The latter, on the other hand, are related to the acceptable ways the system can achieve the mission objectives notwithstanding limitations, particularly those associated with budget, schedule, and safety.

After having identified and analyzed the top-level requirements and constraints, the next step in the systems engineering process illustrated in Fig. 3.9 consists in the logical decomposition of such requirements, creating a top-down system architecture that breaks down parent requirements and deploys them to the lower design levels. This allows individual roles, relationships, dependencies, and interfaces between elements to be plainly established and understood (NASA, 2016).

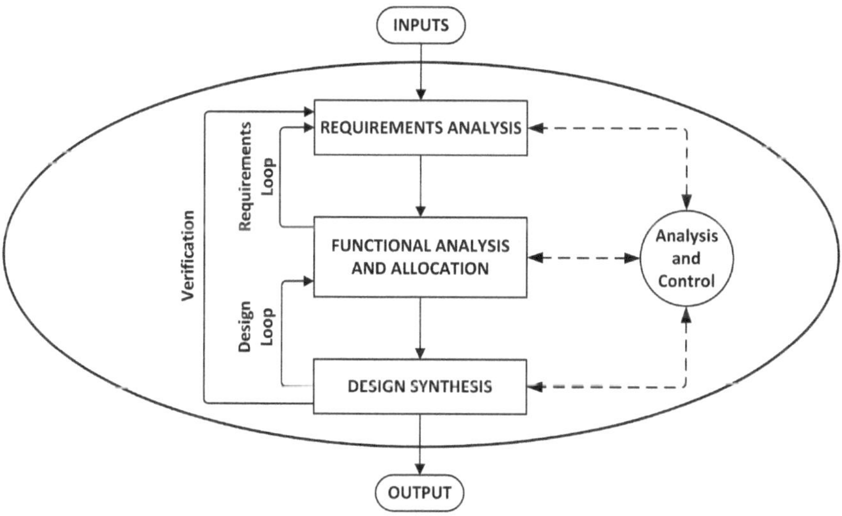

FIG. 3.9 The systems engineering process.

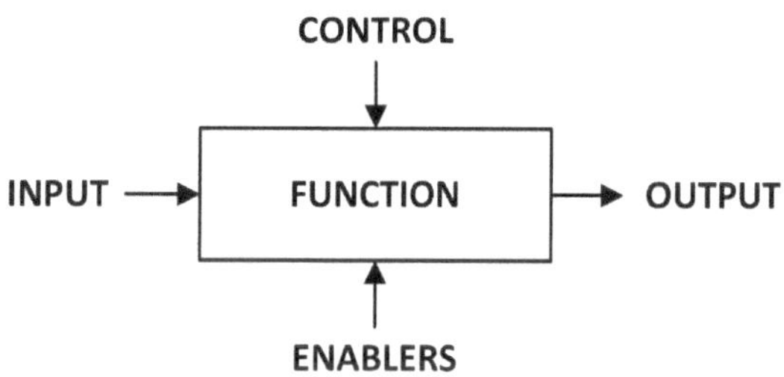

FIG. 3.10 System function.

Functions, which can be represented by a rectangular box in a functional flow block diagram such as the one in Fig. 3.10, transform a set of inputs into a set of outputs and derive their denomination from mathematical functions which, as is well known, act on input values and produce different output values.

Function names will typically be made up of an action verb and a noun summarizing the task performed, e.g., "close gate." Enablers are the entities that render the action possible and are not themselves transformed. Controls are the means employed to command, direct, limit, or check the function's operation.

The purpose of functional analysis is to transform the requirements resulting from the first step in the systems engineering process into a consistent description of system functions, which will provide the designer with the information needed about what the system must do, how well it must perform, and what are the limits imposed by constraints. This top-down process used to develop system structure and decompose functional and performance requirements, that is, to identify, describe, and relate the functions a system should perform to fulfill its goals and objectives, comprises the following stages:

- Decompose top-level functions identifying at successively lower levels what actions should be carried out;
- Identify how well functions should be performed;
- Identify and define all internal and external functional interfaces;
- Identify logical groupings of functions that facilitate the use of modular components and open-system solutions to reduce functional interfaces;
- Incorporate functional characteristics of existing components in the analysis and allocation;
- Examine life cycle functions as applicable;
- Perform trade studies to investigate alternative solutions to meet requirements; and
- Return to the requirement analysis step whenever necessary for re-examination of higher-level requirements.

The output of the Functional Analysis and Allocation step is the functional architecture, i.e., a logical hierarchical decomposition and sequencing of system functions and their associated performance requirements. Tools can be used to support the development of the functional architecture, such as:

- Functional flow block diagram (FFBD), a multi-tier, time-sequenced, step-by-step flow diagram used to describe system requirements in functional terms, as illustrated in Fig. 3.11.
- Integration definition for function modeling diagram (IDEF0), a modeling tool for the analysis, development, re-engineering, and integration of information systems.
- Timeline analysis sheet (TLS) that defines concurrency, overlapping, and sequential relationships of functions and tasks.

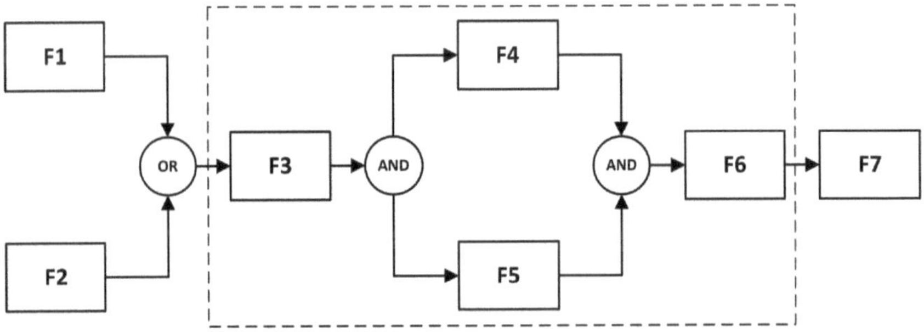

FIG. 3.11 FFBD illustrative example.

- Requirements allocation sheet (RAS), a tool that provides traceability and maintains consistency between functional architectures and designs based thereupon.

It should be noted that although the process of functional analysis and allocation identifies, describes, and relates the functions that a system must perform as well as their performance requirements, it does not prescribe how such functions will be performed.

3.4.3 Reliability allocation

For modern complex engineering systems, reliability is critical for many reasons. It is not only the key to the system's operational effectiveness, but also a major factor as regards logistics infrastructure and costs associated with failures. Thus, a certain level of basic reliability must be achieved for the system to be useful according to its intended purpose. The primary objective is the capacity to achieve the expected probability of success and the minimization of the risk of failure within established availability, cost, power, weight, and volume constraints.

The reliability allocation or apportionment process consists in assigning reliability requirements to various subassemblies and individual components within a system to attain the specified overall system reliability. The process will be carried out along the different stages of system design as information about components and the operating environment becomes available. After system reliability requirements have been specified, the allocation of reliability to subsystems and related components is facilitated by the generation of a reliability block diagram (RBD), such as the one shown in Fig. 3.12.

Essentially, reliability allocation involves solving the inequality in Eq. (3.1).

$$f(R_1(t), R_2(t), \ldots, R_{n-1}(t), R_n(t)) \geq R_s(t) \tag{3.1}$$

where:

R_s (t) is the system reliability goal at time t;

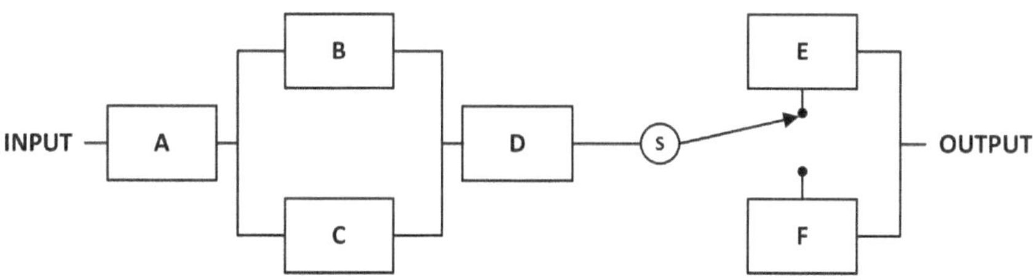

FIG. 3.12 RBD illustrative example.

$R_i(t)$ is the reliability of the i-th unit or subsystem at time t; and
f is the function that relates components' reliabilities to system reliability.

For a simple series system, Eq. (3.1) becomes Eq. (3.2).

$$R_1(t) \cdot R_2(t) \ldots R_{n-1}(t) \cdot R_n(t) \geq R_s(t) \tag{3.2}$$

Being a resultant property, the reliability of a system will entirely depend on the reliability of its component parts. However, since system type and complexity may vary considerably there is no single method for reliability apportionment. For system elements of known design where field reliability data are readily available, such as, for instance, the so-called commercial off-the-shelf items, reliability assessment can be made based on experience. On the other hand, for elements that are new and where design information is not available complexity or weighing factors should be assigned, that are estimated based on mission scenarios involving aspects such as anticipated environmental conditions, use factors, duty cycles, and stresses. Reliability allocation methods used to deploy reliability requirements to system components are usually of two types: weighing factors and optimal reliability allocation.

The weighting factor methods take into consideration factors that are meaningful for system performance in order to assign reliability requirements to system components. Then, subsystems' reliabilities are calculated as the product of an allocation factor and the system's required reliability. As an example, in the method, designed by ARINC Research Corporation, a subsidiary of Aeronautical Radio, Inc., the allocation procedure applies to subsystems that are independent, in series, and have constant failure rates. In addition, it is assumed that the subsystems' mission times and the system mission time are the same. Calculations for the weighting factor, w_i and component failure rate, λ_i, respectively, are presented in Eqs. (3.3) and (3.4).

$$w_i = \frac{\lambda_i}{\sum_i^n \lambda_i} \tag{3.3}$$

$$\lambda_i = w_i . \lambda_s \tag{3.4}$$

where:

$1 \leq i \leq n$;
n is the number of units or subsystems;
λ_s is the system's failure rate.

When expert knowledge about the system is available, ARINC provides reasonable reliability allocation values. Several other weighting factor methods exist, and new ones are regularly proposed and discussed by researchers such as Cruz (2016), Yadav and Zhuang (2014), and Kim, Yang, and Zuo (2013).

Optimal reliability allocation methods involve the application of mathematical programming techniques in order to deal with the challenge of looking for the right combination of individual reliabilities that will maximize system reliability and minimize system cost, subject to constraints such as weight and volume. This generally involves the minimization of a reliability cost function, such as that in Eq. (3.5) that refers to a series configuration for the sake of simplicity:

$$C_s = \sum_{i=1}^{n} C_i(R_i) \tag{3.5}$$

where:

$R_s \geq R_r$;
$R_{i,min} \leq R_i \leq R_{i,max}$;
$1 \leq i \leq n$;
C_s is the total cost of the system;
C_i is the cost of ith stage or subsystem;
R_r is the system reliability requirement;
R_s is the system reliability goal;
R_i is the reliability of ith stage or subsystem;
$R_{i,max}$ is the maximum achievable reliability of the ith component; and
$R_{i,min}$ is the minimum achievable reliability of the ith component.

The reliability cost function, that is, the relationship between cost and component reliability, could be established using empirical data or, as generally but not exclusively done, using an exponential increasing function. The solution will be obtained through the application of several techniques such as heuristic, nonlinear, dynamic, or mixed integer programming. Just as for the weighting factor methods, new proposals for solving reliability optimization problems are also frequently being presented by researchers such as Wang et al. (2020), Li et al. (2016), Li, Liu, and Huang (2014), Wattanapongsakorn and Levitan (2004), and Elegbede, Chu, Adjallah, and Yalaoui (2003).

As a consequence of the increasing complexity of modern engineering systems, that has led to a considerable rise in their susceptibility to faults, despite the possible increase in cost, weight, volume, and power consumption, whenever the cost of failure is significant enough redundancy may become an appealing or even the only option to improve their reliability and availability.

One way of implementing redundancy is to have multiple units operating in parallel such as the dual modular redundancy represented by subsystems B and C in Fig. 3.12. When the cost of failure is extremely high, as in aerospace applications for instance, triple or even quadruple redundancy may be considered. As seen in Chapter 2, the parallel operation of subsystems will increase overall system reliability.

Instead of having all subsystems operational all the time, however, in practice it is more common to use what is termed standby or backup redundancy as provided by the switched alternation of subsystems E and F in Fig. 3.12, where the former operates full time while the latter only takes over in case of failure. It can be easily shown that the reliability of a system with standby redundancy is higher than the reliability of the system when using operating redundancy (Blanchard, 2014).

New methods of redundancy allocation, that is, for an optimal balance between redundancy and the aforementioned cost, weight, volume, and power consumption constraints are once more regularly being sought by researchers through the use of diverse techniques and optimization algorithms such as those proposed by Ashraf et al. (2019), Gholinezhad and Hamadani (2017), Kim and Kim (2017), Ardakan and Hamadani (2014), and Tian and Zuo (2008).

Of special concern here are subsystems intended to perform safety-related functions, which are vital for the protection of people, property, and the environment. Frequently, they are low-demand passive subsystems that stay dormant until required to operate and thus represent a challenge to system reliability and availability due to the fact that dormant failures will generally go undetected. In this case, the usual solution adopted for detecting such hidden faults in the passive state is to require periodic functional tests. If such tests are carried out at regular intervals and the system is considered to be as-good-as-new afterward, the long-term average Probability of Failure on Demand (PFD) of a subsystem operating in a low-demand mode, that is, the probability of not being able to perform its specified safety function if a demand occur is given by Eq. (3.6) (Rausand, 2014; Rausand & Høyland, 2004).

$$PFD_{avg} = \frac{1}{\tau}\int_0^{\tau} PFD(t)dt \tag{3.6}$$

where:

τ is the regular test interval;
$PFD(t)$ is the time-dependent probability of failure in a test interval, i.e., probability that $T < \tau$.

Demands may be triggered by either dangerous detected or undetected faults. Full proof tests should be designed and planned in order to be able to detect the latter, as well as all elements' faults that can contribute to their occurrence. After the detection and repair of such previously dangerous undetected faults, it is assumed that the safety instrumented subsystem will be as close as practicable to the as-good-as-new condition.

The test interval should be adequately determined based on the dangerous failure rate of the subsystem so as to make PFD_{avg} as low as possible. In other words, a dangerous undetected fault may be revealed and corrected before the occurrence of a demand, thus avoiding a hazardous event, with an as high as possible average probability given by Eq. (3.7).

$$P(duf\ revealed \mid duf\ present)_{avg} = e^{-\lambda_{de}\tau/2} \tag{3.7}$$

where:

duf is short for dangerous undetected fault;
λ_{de} is the demand rate or frequency of demands.

Safety-related functions may be performed using different subsystem architectures: single, series, parallel, and more generally *k-out-of-n* (*koon*) dormant redundant subsystems, i.e., n component subsystems that fail if and only if at least k of the n components fail. Li (2016) showed that for *koon* subsystems periodically inspected/maintained the reliability function $R(t)$ will be expressed as in Eq. (3.8).

$$R(t) = \sum_{k}^{n} \frac{n!}{k!(n-k)!} \left(e^{-\lambda t}\right)^{k} \left(1 - e^{-\lambda t}\right)^{n-k} \tag{3.8}$$

where λ is the failure rate.

If T is considered to be the unattended period of operation or predetermined test/maintenance interval and $R(T) = e^{-\lambda T}$ for the exponential distribution, the mean time between failures (MTBF) can be represented as in Eq. (3.9).

$$MTBF = \frac{T}{1 - \sum_{k}^{n} \frac{n!}{k!(n-k)!} R(T)^{k} (1 - R(T))^{(n-k)}} \tag{3.9}$$

Eq. (3.9) establishes one of the possible mathematical relationships between MTBF, or failure rate for that matter, and the test/maintenance interval for safety-related *koon* redundant subsystems with potential dormant failures.

3.5 Robust design

Considering that a product or process is said to be robust when it is insensitive to the effects of sources of variability, even when those sources themselves have not been eliminated, Taguchi (1986), combining statistical methods with a deep understanding of engineering problems, introduced the robust product design methodology that focus on reducing the influence of noise factors that degrade the performance of systems, either products or processes. It comprises three steps: system design, parameter design, and tolerance design.

System design is the traditional conceptual design phase where ideas and specialized knowledge pertinent to the development at hand are put together by the design team in order to envisage the adequate system configuration that will satisfy functional and cost requirements. At this phase, improvement involves scientific and technological innovation and breakthrough.

After the system has been decided on, the parameter design phase is dedicated to selecting the optimum levels for the system parameters that will make it robust, that is, less sensitive to variations due to noise factors. System parameters such as material properties, element dimensions, and process setting parameters whose range of values can be selected by designers can also be designated as controllable or control factors. On the other hand, error or noise factors are difficult, even impossible, or too expensive to control and constitute sources of uncertainty. External noise factors are induced by variations in environmental conditions such as vibration, humidity, shock, storage, and usage. Internal noise factors are represented by wear and tear, as well as by the aging of materials and consequent deterioration of their properties. Finally, piece-to-piece characteristic deviations from target values, for instance, give rise to between-product noises.

Even though designers strive to obtain the optimum combination of control factors in the preceding parameter design phase, their effort may not always be sufficient to reduce the effect of internal and external noises. Therefore, the variation of the major noise factors must be restricted to narrower ranges even though it may result in higher costs. This is done in the tolerance design phase where the allowable deviations in the parameter values are specified, tightening tolerances if necessary.

For Taguchi, parameter design was the most important part of the design methodology because it could both improve quality and lower cost. Although, as he contended, many designers skipped directly from system design to tolerance design and ignored the critically important parameter design stage. Since the effect of error factors can often be considerably reduced by parameter design, he advised that it should always be done first and tolerance design second.

Thus, instead of measuring quality by means of tolerance ranges to one side and the other of the target value m for parameter y, he proposed, as illustrated in Fig. 3.13, a continuous quality loss function $L(y)$ that can be expanded in a Taylor series around m as given by Eq. (3.10).

$$L(y) = L(m + y - m) = L(m) + \frac{L'(m)}{1!}(y - m) + \frac{L''(m)}{2!}(y - m)^2 + \dots \quad (3.10)$$

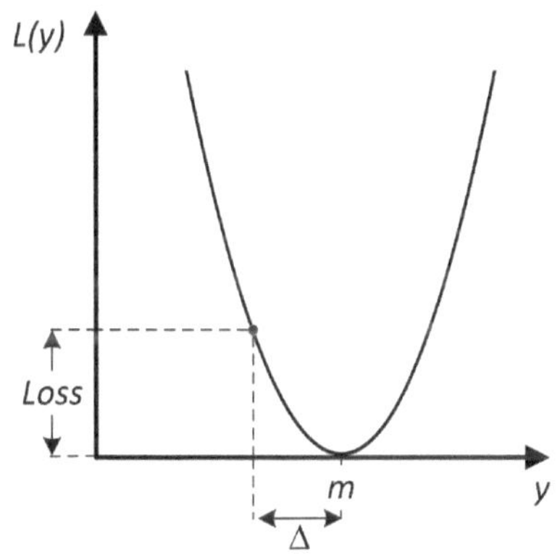

FIG. 3.13 Quality loss function.

TABLE 3.4 Orthogonal array L_8.

No.	Col.						
	1	2	3	4	5	6	7
1	1	1	1	1	1	1	1
2	1	1	1	2	2	2	2
3	1	2	2	1	1	2	2
4	1	2	2	2	2	1	1
5	2	1	2	1	2	1	2
6	2	1	2	2	1	2	1
7	2	2	1	1	2	2	1
8	2	2	1	2	1	1	2

By definition $L(m) = 0$, and since $L(y)$ is minimal at $y = m$: $L'(m) = 0$. Therefore, the loss function can be approximated as in Eq. (3.11), where Δ is the distance from the target value.

$$L(y) \cong k(y - m)^2 = k\Delta^2 \tag{3.11}$$

High values of the proportionality factor k mean elevated sensitivity to change, while low values indicate robust behavior.

Taguchi stated that the efficiency of both parameter design and tolerance design can be raised by means of experimental designs and advocated the use of orthogonal arrays (Taguchi, 1987) such as the L_8 (2^7) exemplified in Table 3.4, where the column to the left indicates experiment numbers and the upper row column numbers. It can be seen that for each experiment in any two columns there are four possible combinations, to wit: (1, 1), (1, 2), (2, 1), and (2, 2). Since these combinations appear the same number of times in a pair of columns, the columns are said to be balanced or orthogonal. Seven factors being studied at two levels, for instance, temperature at 21 and 120°C, could be assigned to each column, and eight experiments could be carried out in any order. The 1 and 2 coding for the low and high levels of the factors may be called the orthogonal or the effect coding (Montgomery, 2012).

A complete parameter design layout consists of an inner array of control factor settings and an outer array of noise factor settings. The combination of the inner and outer arrays constitutes the product array that is used to systematically test the several combinations of control factor settings over all combinations of noise factors. That is, the entire L_j outer array is applied to each of the n experiments in the inner array, so there are $j \times n$ combinations in all for which to calculate the value of y. The "optimal" parameter settings will be those that maximize the appropriate signal-to-noise ratio (S/N). Eqs. (3.12)–(3.14) show three standard types of S/N for different desired performance responses or ways of specifying tolerances for parameters, respectively, "nominal-is-best" for reducing variability around the target value; "smaller-is-better" for making the system response as small as possible; and "larger-is-better" for making the system response as large as possible.

$$(S/N)_m = 10\log\left(\frac{\overline{y}^2}{s^2}\right) \tag{3.12}$$

Where $\overline{y}$ is the mean response and s the standard deviation.

$$(S/N)_S = -10\ \log\left(\frac{1}{n}\sum_{i=1}^{n} y_i^2\right) \tag{3.13}$$

$$(S/N)_l = -10\ \log\left(\frac{1}{n}\sum_{i=1}^{n}\frac{1}{y_i^2}\right) \tag{3.14}$$

The signal-to-loss ratios are derived from the loss function $L(y)$ and are expressed in decibels. Once their values have been calculated for each run of an experiment, it is possible to analyze results and select the factor levels that maximize S/N and bring the mean to the desired value, i.e., target, maximum or minimum. Accordingly, controllable or control factors that affect both product variability and average, as well as those that affect variability only, can be used to reduce variations in the system. Factors that affect average only can then be used to bring it to the desired value. Finally, factors that do not affect either variability or average should be set to the least cost level.

When designing complex engineering systems, certainly an important factor to be considered is uncertainty resulting particularly from natural randomness, imperfect knowledge, and limited data. Therefore, robust design methodologies are being increasingly considered in systems engineering applications and several approaches and algorithms are constantly being proposed in order to extend the possibilities of use of robust design, with an aim to enhancing product reliability (Massein, Daney, & Papegay, 2016; Otto, Wang, & Uyan, 2019; Salazar & Rocco, 2007; Shahraki & Noorossana, 2013, 2014; Wang et al., 2015; Wang, Li, & Savage, 2015).

3.6 Model-based engineering fundamentals

Model-based engineering (MBE) can be described as "*an approach to engineering that uses models as an integral part of the technical baseline that includes the requirements, analysis, design, implementation, and verification of a capability, system, and/or product throughout the acquisition life cycle*" (NDIA, 2011).

But what is a model? It can be said that a model is a simplified representation of a system, such as car, ship, or airplane models used as toys or hobby objects. Models can, however, fulfill many needs other than recreational or decorative, and the modeling of systems has become an essential process of modern engineering. More formally, model is "*a physical, mathematical, or otherwise logical representation of a system, entity, phenomenon, or process*" (U.S.A., 1998).

The use of models as well as of simulation in the design of modern complex engineering systems provides insights for:

- Recording and analyzing requirements;
- Allocating functions;
- Allocating reliability;
- Evaluating trade-offs, including costs;
- Enhancing quality;
- Evaluating and reducing risks; and
- Assessing and improving performance.

Models may be used at many levels from components to systems of systems, at many design stages from specification to verification, and throughout the life cycle of the system.

As systems engineering practices become more formalized, structured, and rigorous, Model-Based Systems Engineering (MBSE) constitutes an approach that can help scientists and engineers manage system complexity. As defined by The International Council of Systems Engineering (INCOSE), MBSE is "*the formalized application of modelling to support system requirements, design, analysis, verification, and validation activities beginning in the conceptual design phase and continuing throughout development and later life cycle phases*" (SEBoK Editorial Board, 2020).

Models can be formal or informal. However, since the primary objective of system modeling is to use models supported by a well-defined modeling language, although less formal representations such as written descriptions of systems can still be used, a model must now meet certain standards for it to be considered within the scope of MBSE. In addition, models can be physical, i.e., a concrete representation of the system, or abstract. The latter can be analytical, similar to mathematical, or descriptive, similar to logical. Descriptive models may typically describe the functional architecture of a system. Analytical models can be further classified into time-invariant models that may represent, for instance, the reliability prediction of a system or component, and time-variant models that may represent a ship's fuel consumption over time, for example. It is also possible to have hybrid descriptive and analytical system models.

The Systems Modeling Language (SysML), a dialect of the Unified Modeling Language 2 (UML 2), is an enabling technology for MBSE. It is a general-purpose graphical modeling language capable of supporting the specification, design, analysis, and verification phases of development due to its adequate semantic foundation for representing requirements, properties, structures, and behaviors of both systems and components.

SysML comprises seven diagrams that come from UML 2, and two exclusively developed diagrams: requirement and parametric. These diagrams, categorized in the so-called four pillars of SysML, are described in Table 3.5.

Another very important aspect of SysML is the possibility of cross-connecting different model elements that appear in different diagrams and thus be able to represent deeper relationships within the elements of the model. There are four main types of cross-connections between model elements described in Table 3.6.

With the aforementioned four pillars and cross-connecting relationships, the SysML modeling language is able to represent the system from different perspectives while maintaining at the same time consistency among those different views.

Although SysML is a relatively recent modeling language, it has already been used for supporting reliability and safety analysis by several authors such as Melani (2020), Mhenni, Nguyen, and Choley (2014, 2018), Hecht, Dimpfl, and Pinchack (2014), Suiphon, Simeu-Abazi, and Gascard (2014), Yakymets, Jaber, and Lanusse (2013), and Cressent, David, Idasiak, and Kratz (2011). The advantage of using a system model that conveys a considerable amount of information such as SysML in these analyses is that comparison and trade-offs between alternative designs can be done in an easier way.

TABLE 3.5 SysML diagrams.

Diagram	Shows	Pillar
Block definition	System structure: properties, operations, and relationships	Structure
Internal block	Internal structures of components: parts and connectors	
Package	Model organization in packages, views, and viewpoints	
Activity	System behavior as control flows and data flows	Behavior
Sequence	System behavior as interactions between system components	
State machine	System behavior in terms of components' states in response to some event	
Use case	System functions and their performers	
Requirement	System requirements and their relationships with other elements	Requirements
Parametric	Parametric constraints between structural elements	Parametric

TABLE 3.6 Cross-connections.

Relationship	Use
Allocate	Determine which block (or system component) is responsible for executing a given action or behavior
Satisfy	Determine which block is responsible for accomplishing a given requirement
Value binding	Link a value property from a block to systems equations
Verify	Represent how a test can be run to make sure a requisite is being satisfied

The main purpose of the SysML modeling language is to concentrate or unify all knowledge about a system in one place. Its nine diagrams make it possible to represent how a system works, what components it is made of, how it interacts with operators and maintainers, how it behaves under different operating conditions, etc. When properly used, several types of analyses can be carried out including fault diagnosis, reliability, availability, maintainability, efficiency, risk management, and asset management.

3.7 Additional relevant complex engineering systems design topics

3.7.1 Maintainability

Considering that the objective of design is to realize a system capable of being available and effectively operational, maintainability is a counterpart of reliability in achieving this goal. Maintainability can be defined as the ability of a system to be maintained, that is, to successfully undergo a series of actions aimed at keeping it at or restoring it to the intended operational state. It is also a design-dependent characteristic of a system. Therefore, it should be the purpose of systems engineering to design and develop systems that can be effectively, efficiently, economically, and safely maintained. In other words, it is convenient that systems be also designed for maintainability.

In much the same way as seen for reliability, after the requirements for the system have been established they should be deployed to the lower levels through maintainability allocation as part of the system requirements allocation process. In so doing, it should be remembered that maintainability parameters are related to reliability parameters and that sometimes maintainability and reliability allocations may be incompatible, requiring a strict feedback control between both activities.

Although perhaps mainly conceived as a means to accommodate future uncertainty, the increasing amount of modularity applied to engineering design in recent years has not only contributed to making complexity manageable, but also to furthering the cause of maintainability. Among other definitions, modularity can be thought of as a "*packaging of components such that they can be repaired via remove and replace action vs. on-board repair*" (U.S.A., 2010).

In order to take full advantage of modularity in design, however, it is necessary to adequately manage interfaces that exist internally and externally to systems. Thus, it is convenient to classify interfaces into key and nonkey interfaces. Besides, interface standards and interconnecting protocols should be well defined and used. The use of MBSE is now being considered for application in the development of modules (Albers et al., 2019; Scherer, Albers, & Bursac, 2017). It is also important to take into consideration that the need to design easy-to-produce parts should be balanced against the need to obtain parts that are easy to assemble. In addition, excessive modularization might not only result in increased interdependencies but also produce a fractured and difficult-to-optimize performance system.

Another approach devised for a different purpose, i.e., mainly with a view to environmental effectiveness, the design for disassembly that involves designing products for easier maintenance, repair, recovery, and reuse of components and materials with the intention of minimizing value loss at the end of their life cycle has also contributed in a certain measure for enhancing maintainability.

3.7.2 Interdependence

In recent years, it has been realized that in the design and operation of complex infrastructure systems it is important to take into consideration the implications of emerging interdependencies across the wider network, with a potential for cascading failures, particularly in the transport, water, energy, telecommunication, and financial sectors, all of which play roles of paramount importance in modern economies. Such is the case for example of electric power grids and information and communication technology (ICT) that have become increasingly coupled due to a combination of factors (Pacific Northwest and DOE, 2015).

The multidirectional relationships or interdependencies among infrastructure systems introduce a high degree of complexity in the system of systems of infrastructure networks. The adequate approach to problems of this kind seems to lie with resilience engineering, whose goal would be to enhance infrastructure systems' ability to reduce the magnitude and duration of disruptive events, that is, to improve performance with regard to robustness, resourcefulness, and rapid recovery. The resilience capacity of a system, that makes it capable of absorbing, adapting, and recovering from unexpected disturbances, can be considered as the combination of the three other capacities shown in Fig. 3.14.

So far, considerable work has been done to develop models and methods for analyzing interdependent infrastructure systems (Hickford et al., 2018). Empirical approaches investigate past events, while predictive approaches focus on modeling and simulation (Canzani, 2016; Jin-Zhu, 2019; Kong & Simonovic, 2018; Liu, Yin, Zhang, & Malaiya, 2016) in order to understand how disturbances cascade through complex infrastructure systems.

Since critical infrastructure networks are vital for modern societies and at the same time prone to disruptions that may be provoked by causes as diverse as component deterioration, natural disasters, and even intentional human actions, it is nowadays a matter of great concern the problem of restoration of systems of interconnected networks after disturbance events (Almoghathawi, Barker, & Albert, 2019; Karakoc, Almoghathawi, Barker, González, & Mohebbi, 2019; Ramachandran, 2015; Zhang, Kong, & Simonovic, 2018), as illustrated in Fig. 3.15.

Given the occurrence of a disruptive event "e_j," the network resilience can be defined as the time-dependent ratio of recovery over loss as expressed by Eq. (3.15).

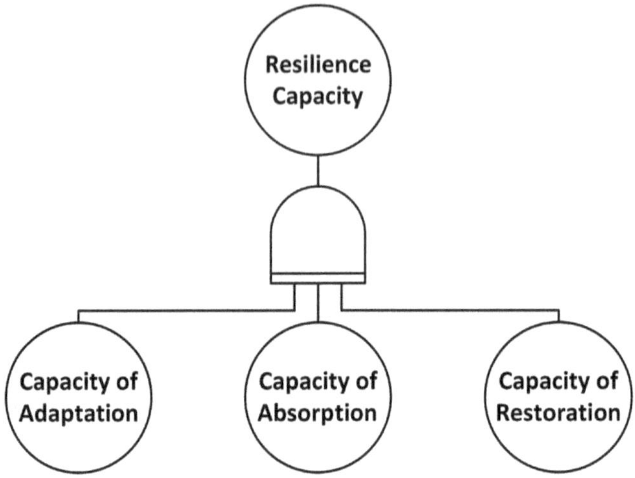

FIG. 3.14 Resilience capacity of complex infrastructure systems.

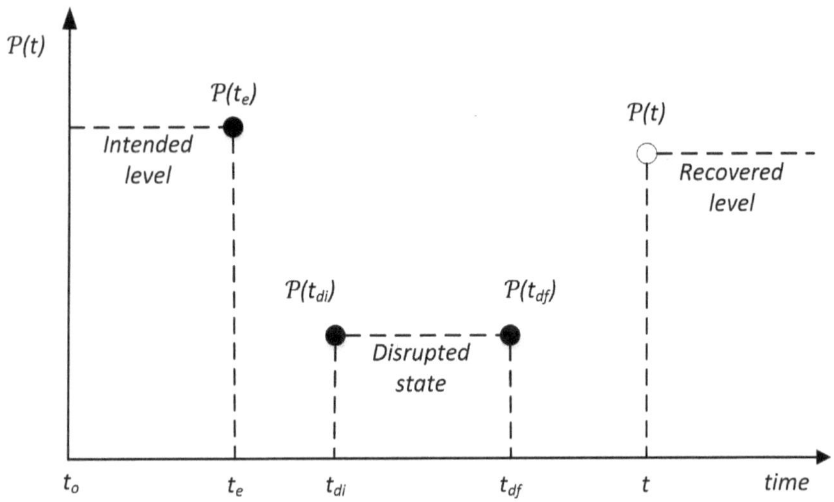

FIG. 3.15 Time-dependent network performance with disturbance.

$$\mathcal{R}(t/e_j) = \frac{P(t/e_j) - P(t_{di}/e_j)}{P(t_o/e_j) - P(t_{di}/e_j)} \tag{3.15}$$

where:

t_o is the initial time instant;
t_{di} is the time at the beginning of the disrupted state; and
t is the variable time until recovery.

It can be seen that network resilience values will vary between 0 and 1. When $\mathcal{R} = 1$, the network is said to be fully resilient.

It is probable, reasoning in accordance with the well-known Pareto principle that only a small number of failed components will be responsible for high-level failures in the interdependent "system of systems." However, pinpointing such components is generally not an easy and straightforward task. Therefore, as complexity increases it will be more and more important to develop models capable of describing the subtle mechanisms of infrastructure interactions in order to allow managers and decision makers to prevent such failures from occurring and at the same time to devise optimal restoration activities when they do occur.

References

Abdulkhaleq, A., Wagner, S., & Leveson, N. (2015). A comprehensive safety engineering approach for software-intensive systems based on STPA. *Procedia Engineering, 128*, 2–11. Elsevier Ltd.

Ackoff, R. L. (1990). *Ackoff's best: His classical writings on management*. John Wiley & Sons, Inc.

Albers, A., Bursac, N., Scherer, H., Birk, C., Powelske, J., & Muschik, S. (2019). Model-based systems engineering in modular design. *Design Science, 5*, 1–33. https://doi.org/10.1017/dsj.2019.15.

Almoghathawi, Y., Barker, K., & Albert, L. A. (2019). Resilience-driven restoration model for interdependent infrastructure networks. *Reliability Engineering and System Safety, 185*, 12–23. Elsevier.

Ardakan, M. A., & Hamadani, A. Z. (2014). Reliability–redundancy allocation problem with cold-standby redundancy strategy. *Simulation Modelling Practice and Theory, 42*, 107–118. Elsevier Ltd.

Ashby, W. R. (1956). *An introduction to cybernetics*. London: Chapman & Hall Ltd.

Ashraf, Z., et al. (2019). Type-2 fuzzy reliability-redundancy allocation problem and its solution using particle swarm optimization algorithm. *Granular Computing, 4*(2), 145–166.

Aziz-Alaoui, M. A., & Bertelle, C. (Eds.). (2009). *From system complexity to emergent properties* Springer.

Bivona, D. E. (2019). The emergence of emergence: G. H. Lewes, *Middlemarch*, and social order. *Dickens Studies Annual: Essays on Victorian Fiction, 50*, 66–80. https://doi.org/10.5325/DICKSTUDANNU.50.1.0066.

Blanchard, F. (2014). *Systems engineering and analysis*. Essex: Pearson Education Limited.

Canzani, E. (2016). Modeling dynamics of disruptive events for impact analysis in networked critical infrastructures. In *Proceedings of the ISCRAM 2016 conference, Rio de Janeiro, RJ, Brazil*.

Churchman, C. W. (1968). *The systems approach*. Dell Publishing Co., Inc.

Cressent, R., David, P., Idasiak, V., & Kratz, F. (2011). Dependability analysis activities merged with system engineering, a real case study feedback. In C. Berenguer, A. Grall, & C. G. Soares (Eds.), *Advances in Safety, Reliability and Risk Management ESREL 2011* (1st). CRC Press.

Cruz, J. A. (2016). *Applicability and limitations of reliability allocation methods*. Cleveland, OH: National Aeronautics and Space Administration—NASA, Glenn Research Center.

Damper, R. I. (2000). Emergence and levels of abstraction. Editorial for the special issue on 'emergent properties of complex systems'. *International Journal of Systems Science, 31*(7), 811–818.

Descartes, R. (1637). *Discourse on the method of rightly conducting one's reason and seeking truth in the sciences*. The Project Gutenberg EBook. Available at: https://www.gutenberg.org/files/59/59-h/59-h.htm (Accessed 6 May 2020).

Elegbede, A. O. C., Chu, C., Adjallah, K. H., & Yalaoui, F. (2003). Reliability allocation through cost minimization. *IEEE Transactions on Reliability, 52*(1), 106–111. https://doi.org/10.1109/TR.2002.807242.

Fleming, C. H., & Levenson, N. G. (2015). Integrating systems safety into systems engineering during concept development. In *25th annual INCOSE international symposium (IS2015) Seattle, WA, July 13–July 16, 2015.*

Gholinezhad, H., & Hamadani, A. Z. (2017). A new model for the redundancy allocation problem with component mixing and mixed redundancy strategy. *Reliability Engineering and System Safety, 164*, 66–73. Elsevier Ltd.

Goldstein, J. (2001). Emergence radical novelty, and the philosophy of mathematics. In W. Sulis, & I. Trofimova (Eds.), *NATO science series: Vol. 320. Nonlinear dynamics in the life and social sciences* (pp. 133–152). Amsterdam: IOS Press.

Hecht, M., Dimpfl, E., & Pinchack, J. (2014). Automated generation of failure modes and effect analysis from SysMl models. In *2014 IEEE international symposium on software reliability engineering workshops*.

Hickford, A. J., et al. (2018). Resilience engineering: Theory and practice in interdependent infrastructure systems. *Environment Systems and Decisions, 38*, 278–292. published online 5 September, 2018 https://doi.org/10.1007/s10669-018-9707-4.

Hollnagel, E. (2012). *FRAM: The functional resonance analysis method: Modelling complex socio-technical systems*. Farnham. Surrey: Ashgate Publishing Limited.

Jin-Zhu, Y. (2019). Modeling the resilience of interdependent infrastructure systems under uncertainty. In *13th international conference on applications of statistics and probability in civil engineering, ICASP13, Seoul, South Korea.*

Johnson, C. W. (2006). What are emergent properties and how do they affect the engineering of complex systems? *Reliability Engineering and System Safety, 91*(12), 1475–1481.

Karakoc, D. B., Almoghathawi, Y., Barker, K., González, A. D., & Mohebbi, S. (2019). Community resilience-driven restoration model for interdependent infrastructure networks. *International Journal of Disaster Risk Reduction, 38*, 101228. https://doi.org/10.1016/j.ijdrr.2019.101228. Elsevier Ltd.

Kim, H., & Kim, P. (2017). Reliability-redundancy allocation problem considering optimal redundancy strategy using parallel genetic algorithm. *Reliability Engineering and System Safety, 159*, 153–160. Elsevier Ltd.

Kim, K. O., Yang, Y., & Zuo, M. J. (2013). A new reliability allocation weight for reducing the occurrence of severe failure effects. *Reliability and Engineering System Safety, 117*, 81–88. Elsevier Ltd.

Kong, J., & Simonovic, S. P. (2018). A model of interdependence infrastructure system resilience. *International Journal of Safety and Security Engineering, 8*(3), 377–389. https://doi.org/10.2495/SAFE-V8-N3-377-389.

Levenson, N. G. (2011). *Engineering a safer world*. Cambridge, MA: The MIT Press.

Lewes, G. H. (1875). *Problems of life and mind. First series: The foundations of a creed. Vol. II*. Boston, MA: James R. Osgood and Company.

Li, J. (2016). Reliability calculation for dormant k-out-of-n systems with periodic maintenance. *International Journal of Mathematical, Engineering and Management Sciences, 1*(2), 68–76.

Li, R., Liu, X., & Huang, N. (2014). Availability allocation of networked systems using Markov model and heuristics algorithm. *Mathematical Problems in Engineering, 2014*, 315385. 9 pages https://doi.org/10.1155/2014/315385. Hindawi Publishing Corporation.

Li, G., et al. (2016). Reliability allocation method of production line based on fuzzy comprehensive evaluation. In *3rd international conference on materials engineering, manufacturing*

*technology and control (ICMEMTC 2016)*Atlantis Press. https://doi.org/10.2991/icmemtc-16.2016.263.
Liu, L., Yin, Y., Zhang, Z., & Malaiya, Y. K. (2016). Redundant design in interdependent networks. *PLoS ONE, 11*(10), e0164777. https://doi.org/10.1371/journal.pone.016477777.
Marais, K. (2005). *A new approach to risk analysis with a focus on organizational risk factors* (Thesis). Massachusetts Institute of Technology.
Massein, A., Daney, D., & Papegay, Y. (2016). Robust design of parameter identification. In *ARK 2016—Advances in robotics kinematics*. Grasse, France: HAL - Archive ouverte. Hal-01 405835.
Melani, A. H. A. (2020). *System fault diagnosis based on Bayesian networks and SysML*. São Paulo, SP: Escola Politécnica da Universidade de São Paulo.
Mhenni, F., Nguyen, N., & Choley, J. (2014). Automatic fault tree generation from SysML system models. In *2014 IEEE/ASME international conference on advanced intelligent mechatronics*.
Mhenni, F., Nguyen, N., & Choley, J. (2018). SafeSysE: A safety analysis integration in systems engineering approach. *IEEE Systems Journal, 12*(1), 161–172.
Mitchell, M. (2009). *Complexity a guided tour*. Oxford University Press, Inc.
Montgomery, D. C. (2012). *Design and analysis of experiments* (8th ed.). John Wiley & Sons, Inc.
Nakamura, Y. (2019). Violence as an emergent property. *Arts and Social Sciences Journal, 10*, 2. https://doi.org/10.4172/2151-6200.1000442.
National Defense Industrial Association (NDIA). (2011). *Final report of the model based engineering (MBE) subcommittee*. Arlington, VA: NDIA Systems Engineering Division, M&S Committee.
National Aeronautics and Space Administration—NASA. (2016). *Systems engineering handbook*. Washington, DC: Aeronautics Research Mission Directorate (ARMD).
Otto, K., Wang, J., & Uyan, T. (2019). Using open source code libraries for robust design analysis. In *Proceedings of the 22nd international conference on engineering design (ICED19), Delft, The Netherlands*. https://doi.org/10.1017/dsi.2019.179.
Pacific Northwest and U.S. Department of Energy. (2015). *The emerging interdependence of the electric power grid & information and communication technology*. Richland, WA: National Laboratory.
Pons, D. J., Pons, A. D., & Pons, A. J. (2013). Time: An emergent property of matter. *Applied Physics Research, 5*(6), 23–47. https://doi.org/10.5539/apr.v5n6p23.
Ramachandran, V. (2015). *Modeling supply chain interdependent critical infrastructure syst*. Missouri University of Science and Technology. Available at: https://scholarsmine.mst.edu/doctoral_dissertations/2392.
Rausand, M. (2014). *Reliability of safety-critical systems: Theory and applications*. Hoboken, NJ: John Wiley & Sons, Inc.
Rausand, M., & Høyland, A. (2004). *System reliability theory: Models, statistical methods, and applications*. Hoboken, NJ: John Wiley & Sons, Inc.
Ross, W. D. (1924). Aristotle's metaphysics. A revised text with introduction and commentary. *Mind, 34*(135), 351–361.
Salazar, D. E., & Rocco, C. (2007). Solving advanced multi-objective robust designs by means of objective evolutionary algorithms (MOEA): A reliability application. *Reliability Engineering and System Safety, 92*, 697–706. Elsevier Ltd.
Scherer, H., Albers, A., & Bursac, N. (2017). Model based requirements engineering for the development of modular kits. *Procedia CIRP, 60*, 145–150. https://doi.org/10.1016/j.procir.2017.01.032.
SEBoK Editorial Board. (2020). *The guide to the systems engineering body of knowledge (SEBoK)*. v. 2.2., Available at: www.sebokwiki.org (Accessed 10 June 2020).

Shahraki, A. F., & Noorossana, R. (2013). Combined algorithm for solving reliability-based robust design optimization problems. *Journal of Mathematics and Computer Science, 7*, 54–62.

Shahraki, A. F., & Noorossana, R. (2014). Reliability-based robust design optimization: A general methodology using genetic algorithm. *Computers & Industrial Engineering, 74*, 199–207. Elsevier Ltd.

Suiphon, B., Simeu-Abazi, Z., & Gascard, E. (2014). Premiers pas vers le diagnostic de défaillances par exploitation d'un modèle SysML. In *Congrès Lambda Mu 19 de Maîtrise des Risques et Sûreté de Fonctionnement, 21-13 Octobre 2014*.

Taguchi, G. (1986). *Introduction to quality engineering: Designing quality into products and processes*. Tokyo: Asian Productivity Organization.

Taguchi, G. (1987). *System of experimental design. Vol. 2*. White Plains, NY and Dearborn, MI: UNIPUB/Kraus International Publications and American Supplier Institute.

Tian, Z., & Zuo, M. J. (2008). Reliability-redundancy allocation for multi-series-parallel systems. *IEEE Transactions on Reliability, 57*(2), 303–310. https://doi.org/10.1109/TR.2008.920871.

United States of America. (1998). *DoD modeling and simulation (M&S) glossary*. Washington, DC: Department of Defense, Under Secretary of Defense for Acquisition Technology.

United States of America. (2001). *Systems engineering fundamentals*. Fort Belvoir, VA: Department of Defense, Systems Management College, Defence Acquisition University Press.

United States of America. (2010). *Defense acquisition guidebook*. Fort Belvoir, VA: Department of Defense, Defense Acquisition University.

von Bertalanffy, L. (1969). *General system theory: Foundations, development, applications*. New York: George Braziller.

von Bertalanffy, L. (1972). Zu einer allgemeinen Systemlehre. In K. Bleicher (Ed.), *Organisation als System. Organization und Führung* (pp. 31–45). Wiesbaden: Gabler Verlag. Aus: Biologia Generalis 1/1949, S. 114–129. https//doi.org/10.1007/978-3-322-86022-4_2.

Wang, P., et al. (2015). Reliability-based robust design of smart sensing systems for failure diagnostics using piezoelectric materials. *Computers and Structures, 156*, 110–121. Elsevier Ltd.

Wang, S., Li, Q., & Savage, G. J. (2015). Reliability-based robust design optimization of structures considering uncertainty in design variables. *Mathematical Problems in Engineering, 2015*. https://doi.org/10.1155/2015/280940, 280940.

Wang, Y., et al. (2020). Research on the reliability allocation method for a production system based on availability. *Mathematical Problems in Engineering, 2020*. https://doi.org/10.1155/2020/6159462, 6159462. 9 pages. Hindawi Publishing Corporation.

Wattanapongsakorn, N., & Levitan, S. P. (2004). Reliability optimization models for embedded systems with multiple applications. *IEEE Transactions on Reliability, 53*(3). https://doi.org/10.1109/TR.2004.833310.

Yadav, O. P., & Zhuang, X. (2014). A practical reliability allocation method considering modified criticality factors. *Reliability and Engineering System Safety, 129*, 57–65. Elsevier Ltd.

Yakymets, N., Jaber, H., & Lanusse, A. (2013). Model-based system engineering for fault tree generation and analysis. In *Proceedings of the 1st international conference on model-driven engineering and software development*.

Zhang, C., Kong, J., & Simonovic, S. (2018). Modeling joint restoration strategies for interdependent infrastructure systems. *PLoS One, 13*(4), e0195727. https://doi.org/10.1371/journal.pone.0195727.

CHAPTER

4

Analysis of systems' reliability and availability

Reliability is the precondition for trust.

Wolfgang Schäuble (1942–)

Products are developed to meet the needs of a market and must achieve the performance requirements imposed by that market. Products that perform better, including greater reliability, are recognized as the most trustful on the market.

Modern engineering systems are usually used in applications where performance and operational safety are very important requirements. Both requirements are associated with the definition of system reliability. The greater the design complexity, the more sophisticated the methods used to evaluate reliability in the system design phases. The accuracy of the reliability estimate is very important to guarantee customer satisfaction. The accurate reliability estimate shall be based on models more sophisticated than those presented in Chapter 2.

4.1 Introduction

The Reliability Block Diagram (RBD) presented in Chapter 2 can be used to model systems whose components are simply interconnected with independent failures. Based on that model, system reliability is calculated. Most modern engineering systems are composed of components interconnect in such a way that they cannot be broken down into groups of series and parallel components. In many cases, it is not easy to recognize which components are in series and which are in parallel in a complex system. As a consequence, they cannot be simplified to a single equivalent element or block through some combination of series and parallel reductions.

Reliability Analysis and Asset Management of Engineering Systems
https://doi.org/10.1016/B978-0-12-823521-8.00007-4

Systems that do not have simple interconnections are usually called complex for reliability analysis.

Another important characteristic of a modern engineering system is the possibility of functional reconfiguration aiming at keeping system performance in case of some components' failure or operational degradation. As an example of this type of system, one can mention the electronic control of the fuel injection system for a spark-ignition internal combustion engine. This system is capable of reconfiguring its functional architecture, mainly its control subsystem, in case of some sensor failure. The modeling of that system reliability, including reconfiguration, cannot be made using RBD.

The modern engineering systems may also present sequence-dependency between component failure events. For example, a failure of a component may cause an overload in other components influencing their reliability. In that case, the system reliability varies due to the occurrence of some component failures. The reliability is dependent on the timing and order of component failure events. The RBD model is not capable of modeling such dependence.

The aforementioned difficulties that may be faced during modern engineering system reliability analysis can be overcome through the use of more sophisticated techniques, such as the Markov model and Petri nets. Some of them are based on modeling the operational states that a system may present and on defining the probability that the system operates in each of those states.

The complex interconnections or functional dependencies between components not only affect reliability assessment but also the evaluation of the combination of failures that put the systems into critical operational conditions that may affect system safety. Usually used for that evaluation, the conventional Fault Tree Analysis (FTA) is not capable of modeling complex component interactions such as sequence-dependent failures. Dynamic gates are introduced to extend the conventional fault tree to model these complex interactions, and the analysis is named Dynamic Fault Tree (DFT). Since such analysis can no longer be done using minimal cut sets, some simulation methods, such as Markov chain or Petri nets, are required.

Besides the problem of reliability analysis, another important performance indicator that will be considered for complex engineering systems is availability as presented in Chapter 2. The availability analysis is influenced by the maintenance policy defined for the system, usually based on preventive or predictive practices. The availability evaluation models commonly consider the concept of "perfect maintenance actions" to estimate the index chosen for measuring availability. However, for modern engineering systems, the planned maintenance actions will rarely make the system as good as new since it is technically and/or financially almost impossible to define maintenance actions that replace/repair all degraded items. In addition, the human factor may also contribute to imperfections. So, the imperfections in maintenance actions affect repair rates during system operational life. The modeling of imperfect maintenance is also an important issue when modeling modern engineering systems' operational performance.

In this chapter, some techniques are presented that may overcome the complexities associated with modern engineering problems. The basic characteristics of these techniques will be presented aiming at guiding the selection of the most suitable technique to be used in a given system design. It is not the intention of this chapter to present a comprehensive review of the techniques used in the reliability and maintenance modeling of modern engineering systems.

4.2 Advanced techniques for reliability and availability analyses of complex systems

In addition to the techniques presented in Chapter 2, more advanced methods can be used to perform availability and reliability analyses. Accordingly, in this section, the Markov analysis, the Petri net, the Bayesian network, and the DFT will be presented. These techniques are capable of representing dynamic behaviors of a system that could not be considered through the techniques previously presented.

4.2.1 Markov analysis

Reliability problems are normally related to systems that are discrete in space, i.e., they can exist in one of a number of discrete and identifiable states, and continuous in time; that is, they exist continuously in one of the system states until a transition occurs which takes them discretely to another state in which they then exist continuously until another transition occurs.

For a given engineering system, with a discrete number of operational states, the following principle can be stated: the probability that the system will be in a certain state at time t_2 can be calculated from the knowledge of its state at an immediately preceding time t_1 and does not depend on the systems' states previous to t_1. A stochastic process that represents observations of an engineering system that satisfies this condition is a Markov process. This means that the behavior of the system must be characterized by a lack of memory; that is, the future states of a system are independent of all past states except the immediately preceding one (Billinton & Allan, 1992).

The Markov process may be discrete in time for which the state transition probabilities are constant. Reliability problems are normally concerned with systems that are continuous in time; that is, the system remains in one state until a transition is triggered which takes the system to another state in which it will remain until another transition is triggered. The Markov property is a form of memorylessness leading to the exponential distribution to model the transition probabilities between states. In a continuous-time Markov process, when a state is visited, the process stays in that state for an exponentially distributed length of time before moving to a new state (Billinton & Allan, 1992; Lewis, 1994).

To illustrate the application of Markov analysis, a two-component parallel system, as presented in Fig. 4.1, is considered in the analysis. To apply the Markov analysis, the system states must be initially defined. A two-component system has only four possible states as enumerated in Table 4.1. This analysis assumes that the failure rate of each component is independent of the state of the other component. The logic of the changes of states is best illustrated by a state transition diagram shown in Fig. 4.1.

The failure rates λ_a and λ_b for components A and B indicate the rates at which the transitions are made between states. Since $\lambda \Delta t$ is the probability that a component will fail between times t and $t + \Delta t$, given that it is

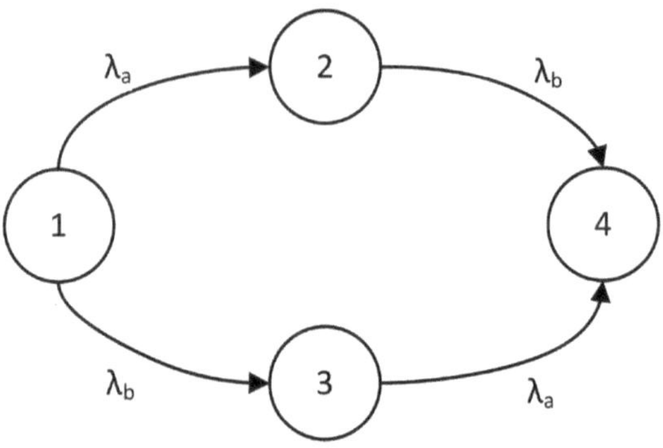

FIG. 4.1 Parallel active system-state transition diagram for reliability analysis.

TABLE 4.1 Parallel active system states.

	States			
Components	1	2	3	4
A	Operating	Failed	Operating	Failed
B	Operating	Operating	Failed	Failed

operating at t, the net change in the probability that the system will be in state 1, $P_1(t)$, is written as presented in Eq. (4.1):

$$P_1(t+\Delta t)-P_1(t)=-\lambda_a \Delta t P_1(t)-\lambda_b \Delta t P_1(t) \tag{4.1}$$

Eq. (4.1) is usually expressed in the differential form, as shown in Eq. (4.2) (Lewis, 1994):

$$\frac{dP_1(t)}{dt}=-\lambda_a P_1(t)-\lambda_b P_1(t) \tag{4.2}$$

Using the same modeling process, the differential equations that govern the probability of the system will be in other states, named $P_2(t)$, $P_3(t)$, and $P_4(t)$, may be written as shown in Eqs. (4.3)–(4.5):

$$\frac{dP_2(t)}{dt}=+\lambda_a P_1(t)-\lambda_b P_2(t) \tag{4.3}$$

$$\frac{dP_3(t)}{dt}=+\lambda_b P_1(t)-\lambda_a P_3(t) \tag{4.4}$$

$$\frac{dP_4(t)}{dt}=\lambda_a P_3(t)+\lambda_b P_2(t) \tag{4.5}$$

Eqs. (4.2)–(4.5) represent a system of four coupled differential equations in which the solution will define the probabilities over time that the system will be in a given state, presented in Eqs. (4.6)–(4.9).

$$P_1(t)=e^{-(\lambda_a+\lambda_b)t} \tag{4.6}$$

$$P_2(t)=e^{-\lambda_b t}-e^{-(\lambda_a+\lambda_b)t} \tag{4.7}$$

$$P_3(t)=e^{-\lambda_a t}-e^{-(\lambda_a+\lambda_b)t} \tag{4.8}$$

$$P_4(t)=1-e^{-\lambda_a t}-e^{-\lambda_b t}+e^{-(\lambda_a+\lambda_b)t} \tag{4.9}$$

The sum of these probabilities is equal to 1. Taking in view that the system is operating in states 1, 2, and 3, its reliability is expressed as presented in Eqs. (4.10) and (4.11):

$$R_S(t)=P_1(t)+P_2(t)+P_3(t) \tag{4.10}$$

$$R_S(t)=e^{-\lambda_a t}+e^{-\lambda_b t}-e^{-(\lambda_a+\lambda_b)t} \tag{4.11}$$

That is the same equation that can be derived from the analysis of the RBD, considering that the reliability of the components is exponentially distributed. It is important to note that if the system reaches state 4, there is no way to return to the other states. This state is called the absorbing state.

The Markov analysis can also be executed for the basic system reliability block configurations presented in Chapter 2. The great advantage of

Markov analysis appears in situations where complex interactions of components may occur causing difficulties in developing the RBD. In those cases, the definition of system states may be useful to support reliability analysis. The Markov analysis is also used in situations in which component failure rates can no longer be assumed to be independent of the system state. One of the common cases of dependence is in load-sharing components, where usually two components are used to execute a given function but, in case of failure of one component, the other component can perform the function but with an overload. The operation in an overload condition may increase the component failure rate.

The treatment of larger sets of components by Markov methods is simplified by expressing the coupled set of state transition equations in matrix form. For instance, for the transition diagram presented in Fig. 4.1, it is possible to rewrite the set of simultaneous differential equations (4.2–4.5) as presented in Eq. (4.12).

$$\begin{bmatrix} \frac{dP_1(t)}{dt} \\ \frac{dP_2(t)}{dt} \\ \frac{dP_3(t)}{dt} \\ \frac{dP_4(t)}{dt} \end{bmatrix} = \begin{bmatrix} -(\lambda_a+\lambda_b) & 0 & 0 & 0 \\ \lambda_a & -\lambda_b & 0 & 0 \\ \lambda_b & 0 & -\lambda_a & 0 \\ 0 & \lambda_b & \lambda_a & 0 \end{bmatrix} \cdot \begin{bmatrix} P_1(t) \\ P_2(t) \\ P_3(t) \\ P_4(t) \end{bmatrix} \tag{4.12}$$

Besides being used in reliability analysis, Markov analysis can also be used in system availability analysis. To illustrate that analysis, it is considered a parallel system composed of two components whose states were presented in Table 4.1. In the application, it is considered the existence of two repair crews, allowing repairs to be carried out on both units simultaneously, and the repair rate of component A is μ_a, and for component B, the repair rate is μ_b, which are constant. The components also have independent failures. The state transition matrix for availability analysis is presented in Fig. 4.2.

The set of simultaneous differential equations are presented in a matrix form in Eq. (4.13):

$$\begin{bmatrix} \frac{dP_1(t)}{dt} \\ \frac{dP_2(t)}{dt} \\ \frac{dP_3(t)}{dt} \\ \frac{dP_4(t)}{dt} \end{bmatrix} = \begin{bmatrix} -(\lambda_a+\lambda_b) & \mu_a & \mu_b & 0 \\ \lambda_a & -\mu_a-\lambda_b & 0 & \mu_b \\ \lambda_b & 0 & -\mu_b-\lambda_a & \mu_a \\ 0 & \lambda_b & \lambda_a & -(\mu_a+\mu_b) \end{bmatrix} \cdot \begin{bmatrix} P_1(t) \\ P_2(t) \\ P_3(t) \\ P_4(t) \end{bmatrix} \tag{4.13}$$

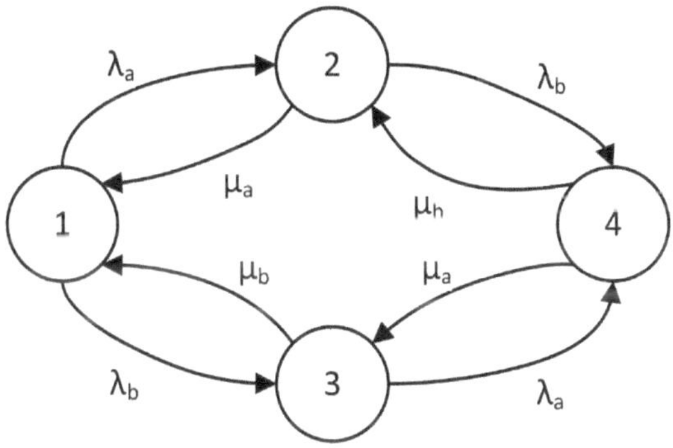

FIG. 4.2 Parallel active system-state transition diagram for availability analysis.

The analysis of the transition matrix indicated that in availability calculations, the system is always able to recover from any failed state. Similarly to the reliability analysis, the availability of the system is the sum of the state probabilities associated with the operational states, represented by Eq. (4.14):

$$A_S(t) = \sum_{i=1}^{3} P_i(t) \tag{4.14}$$

The probabilities $P_1(t)$, $P_2(t)$, $P_3(t)$, and $P_4(t)$ are being found in the operating state and failed state, respectively, as a function of time given that the system started at time $t = 0$ in the operating state. The solution of the differential equation is dependent on the system's initial state. For constant failure and repair rates, the solution of the system is usually obtained through the use of Laplace transformation. For systems with a great number of components, the analyst is more interested in steady-state (or asymptotic) availability than in the availability time dependence. The steady-state availability is calculated supposing $t \to \infty$ and the derivative on the right-hand side of Eq. (4.13) vanishes, deriving a time-independent relationship, presented in Eq. (4.15) (Billinton & Allan, 1992; Lewis, 1994):

$$\begin{bmatrix} 0 \\ 0 \\ 0 \\ 0 \end{bmatrix} = \begin{bmatrix} -(\lambda_a + \lambda_b) & \mu_a & \mu_b & 0 \\ \lambda_a & -\mu_a - \lambda_b & 0 & \mu_b \\ \lambda_b & 0 & -\mu_b - \lambda_a & \mu_a \\ 0 & \lambda_b & \lambda_a & -(\mu_a + \mu_b) \end{bmatrix} \cdot \begin{bmatrix} P_1(\infty) \\ P_2(\infty) \\ P_3(\infty) \\ P_4(\infty) \end{bmatrix} \tag{4.15}$$

The system is represented by four linearly dependent equations. To solve the system, a fifth relation may be used, as presented in Eq. (4.16):

$$\sum_{i=1}^{4} P_i(\infty) = 1 \tag{4.16}$$

The values of asymptotic state probability are presented in Eqs. (4.17)–(4.20):

$$P_1(\infty) = \frac{\mu_a}{(\lambda_a + \mu_a)} \cdot \frac{\mu_b}{(\lambda_b + \mu_b)} \tag{4.17}$$

$$P_2(\infty) = \frac{\lambda_a}{(\lambda_a + \mu_a)} \cdot \frac{\mu_b}{(\lambda_b + \mu_b)} \tag{4.18}$$

$$P_3(\infty) = \frac{\mu_a}{(\lambda_a + \mu_a)} \cdot \frac{\lambda_b}{(\lambda_b + \mu_b)} \tag{4.19}$$

$$P_4(\infty) = \frac{\lambda_a}{(\lambda_a + \mu_a)} \cdot \frac{\lambda_b}{(\lambda_b + \mu_b)} \tag{4.20}$$

The Markov model can be used to analyze systems of an arbitrary structure whose elements have constant failure and repair rates during the stay time (sojourn time) in every state. For a system with a great number of components, it may be difficult to define all possible states to support availability analysis (Kumamoto & Henley, 2000).

Modern engineering systems are composed of electromechanical, electronic, and control (including software) systems and the transition between states may not be modeled only with the use of constant failure and repair rates. The semi-Markov process is used to model systems in which the time interval between the transitions of states is not modeled by an exponential distribution. The Semi-Markov Process (SMP) is considered a generalization of the Markov process. One of the principal characteristics of SMP is that it can represent systems whose evolution of operational state is not only dependent on their last visited state but also on the elapsed time within that state. SMP can be defined by the transition probability matrix and the initial distribution. In a semi-Markov process, the time spent in any state after entering it is a random variable that can be described by an arbitrary distribution. In SMP, a process "holds" for a time t in a given state before it jumps to another state. This time *t* is called the sojourn time and has an arbitrary distribution. Future states of processes and their sojourn times do not depend on past states and their sojourn times if a present state is known.

As in the Markov model, the application of the semi-Markov model requires an estimate of the probability of the system assuming each of the previously defined states. Moura and Droguett (2008) present a comprehensive literature review regarding the application of the semi-Markov process in reliability and availability analyses. From that literature review, it is possible to highlight the study developed by Limnios and Oprişan (2001) that demonstrates some results and applications of SMPs in the context of reliability. Becker, Camarinopoulos, and Zioutas (2000) develop the mathematical formulation of SMPs described through transition rates.

In fact, the transition rates of an SMP may only depend on sojourn time in a state for the case of a homogeneous semi-Markov process, or both sojourn and process times for a nonhomogeneous semi-Markov process. Both references suggest the use of Laplace transforms to solve the set of differential equations.

This book does not intend to present the mathematical equations used by those references to calculate reliability and availability based on SMPs and invites the readers to consult the references mentioned earlier regarding applications of SMPs. When using SMPs, it is important to stress the need to previously define the systems' states and sojourn time in each of these states which can make that application time-consuming and based on expert knowledge.

4.2.2 Petri nets

Petri Nets (PNs) are a modeling technique that uses a formal graph to represent the dynamic behavior of a system and, because of that, have been used to perform reliability and availability analyses. Dynamic and complex systems, which change their mode of operation or their performance due to failures in their components, can be easily represented by PNs (Melani, Murad, Caminada Netto, Souza, & Nabeta, 2019).

Petri Nets were originally proposed by Carl Adam Petri in the 1960s. Although it was initially used for computer hardware and software analyses, this tool has been implemented in other application areas such as civil engineering, chemical industries, and system automation (Reddy, Murty, & Ghosh, 1993).

The PNs are composed of places, transitions, arcs, and tokens. Places are represented by circles, transitions are described by bars or boxes, arcs are represented by arrows connecting transitions to places and vice versa, and tokens are represented by a black dot. In a more formal way, a PN can be mathematically represented as a quintuple, presented in Eq. (4.21) (Murata, 1989):

$$\mathrm{PN} = (P, T, I, O, M) \tag{4.21}$$

where

- $P = \{p_1, p_2, \ldots, p_n\}$ is a finite set of places;
- $T = \{t_1, t_2, \ldots, t_m\}$ is a finite set of transitions;
- I is the input relation of the transitions, i.e., the arcs directed from places to transitions;
- O is the output relation of the transitions, i.e., the arcs directed from transitions to places;
- $M = \{m_1, m_2, \ldots, m_n\}$ is the marking of the PN, representing the number of tokens in each of the n places.

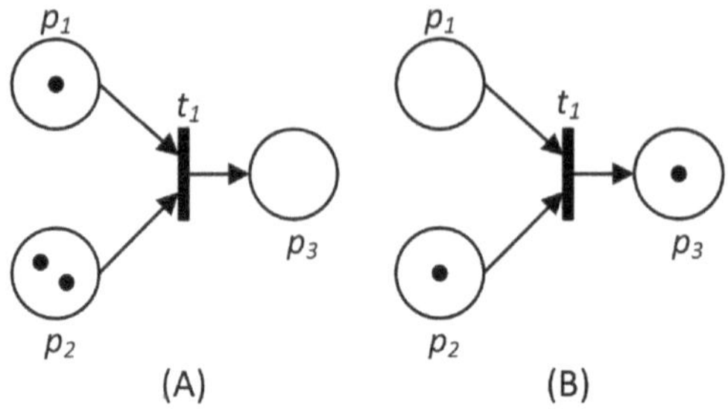

FIG. 4.3 Example of PN. (A) PN before firing the transition; (B) PN after firing the transition.

Fig. 4.3A presents an example of PN, comprised of three places (p_1, p_2, and p_3), a transition (t_1), and three tokens, one in p_1 and two in p_2. This shows that places can be occupied by one or more tokens (Schruben & Yucesan, 1994).

When modeling a given system through a PN, one must realize that the current state of the system is given by the current position of the PN tokens and that the dynamic behavior of the system is represented in the model by the movement of the tokens through transition firings. Transition firings cause tokens to move from one place to another connected by the given transition. Arcs determine the path that tokens take throughout the model. They can either enable or inhibit movement in the model, depending on their use (Melani et al., 2019). To exemplify, Fig. 4.3A and B shows the same PN before and after triggering transition t_1.

Although the PN proposed by Carl Petri did not consider the concept of time, that is, the transitions were triggered as soon as conditions were satisfied, it was not long before a timed PN was introduced. In the 1970s, the Stochastic Petri Nets (SPNs), an evolution of PN with timed transitions specified by random variables yielding probabilistic models, were presented (Nývlt & Rausand, 2012; O'Connor & Kleyner, 2012). Another evolution in the model, called Generalized Stochastic Petri Net (GSPN), allowed nondeterministic firing delays to coexist with traditional immediate transitions.

The ability to use stochastic processes by SPN and GSPN allowed such models to be used to compute the reliability and availability of systems. In fact, availability analysis has shown to be one of the most prominent areas of the application of Petri nets, given its capability of modeling and analyzing systems that include failure and repair processes, as well as a variability of the operating times (Melani et al., 2019).

Fig. 4.4A shows an example of an SPN that can be used to perform availability and reliability analyses. In this model, the system has only two operational states, "system operating" and "failed system," represented by places. The token shows the current state of the system and

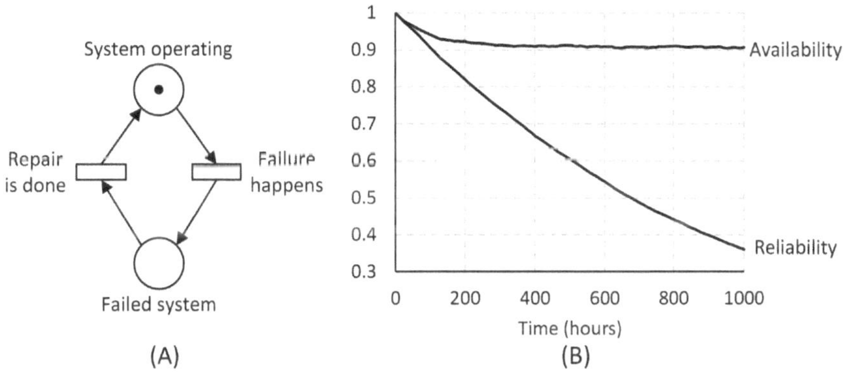

FIG. 4.4 An example of how PN can be used to computed reliability and availability.

the transitions are responsible for determining the change of state. Transitions are determined by the probabilistic distributions that represent the failure and repair of the system. When performing the model simulation, that is, when simulating the movement of the token in a given time interval, availability can be obtained by calculating the percentage of time in which the token is in the "system operating" place. Reliability, in turn, is estimated by calculating the probability that the token will make its first move. If, for example, both transitions are represented by exponential distributions determined by the parameters $\lambda = 0.001$ (failure rate) and $\mu = 0.01$ (repair rate), the obtained values for reliability and availability over time are shown in Fig. 4.4B.

The reliability and availability analyses by PNs are not limited to the example model in Fig. 4.4. Several authors have proposed different graphs to represent the particularities of the system under study. Melani, Souza, Murad, Caminada Netto, and Nabeta (2017), for example, proposed to model the various components of a cooling tower to see how their joint operation influences the reliability of the system as a whole. Melani et al. (2019) and Melani, Caminada Netto, Murad, Souza, and Nabeta (2018) then further extended the original model to consider the effect of predictive maintenance on system reliability, availability, and overall performance.

Several authors have also used GSPN to assess the availability of different systems taking into account particular aspects of each, such as the presence of wear-out and random failures (Beirong, Xiaowen, & Wei, 2012), common-cause failures (Thangamani, 2012), sequenced failures (Talebberrouane, Khan, & Lounis, 2016), different types of maintenance strategies (Leigh & Dunnett, 2016), and flexible production (Long, Zeiler, & Bertsche, 2017). This shows that PNs have been shown to be a promising tool that can be conveniently used to model and analyze any complex system.

4.2.3 Bayesian networks

The Bayesian Networks (BN) are a type of probabilistic graphical model that has become popular as a reliability analysis tool. Based on Bayes' theorem, it aims to model causal relationships through a direct acyclic graph composed of nodes and arcs. Each node represents a variable, discrete or continuous, of the universe under study. Each arc, or directed line, connects a pair of nodes and represents a conditional dependency between these variables.

Fig. 4.5 presents an example of a BN (Melani, 2020). It is possible to notice from the figure that a BN is composed of a qualitative and a quantitative component. The qualitative one is the graph itself, while the quantitative one is composed of the Conditional Probability Tables (CPTs), which quantifies the influence of parent nodes on child nodes.

Bayesian network allows the probability of an event, or series of events, to occur to be calculated. Suppose a BN that represents the dependency

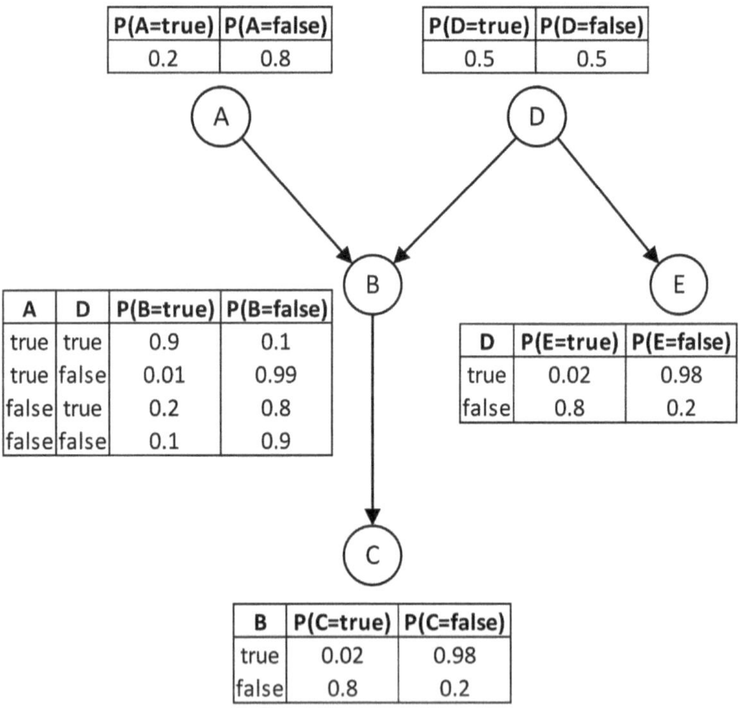

FIG. 4.5 Example of a Bayesian network.

relations of the n variables of a domain, $x_1, \ldots, x_n$. The joint probability can be computed by Eq. (4.22):

$$P(x_1, \ldots, x_n) = \prod_{i=1}^{n} P(x_i | x_{i-1}, \ldots, x_1) \tag{4.22}$$

Using Eq. (4.22), it is possible to use the BN to calculate the probability of a system to fail, since this probability is conditionally dependent on the failure of the subsystems and equipment that make up the system. The challenge, however, is to build the BN graph, as well as fill out its CPTs, so that the final model can properly represent the real system. That is why several authors propose to obtain a BN from another, better structured technique, such as Fault Trees (FTs). Several authors (Bobbio, Portinale, Minichino, & Ciancamerla, 2001; Hamada, Wilson, Reese, & Martz, 2008; Lampis & Andrews, 2009) described in detail how this conversion process should be done.

In general, for the conversion of an FT into a BN, the first step to be taken is the creation of a node for each FT event. The events that were connected through logical connectors in the fault tree must, in the Bayesian network, be connected directly through the directed arcs, as shown in Fig. 4.6. If eventually, any element (fault) appears more than once in the FT, it must be represented only once in the BN, a situation represented by node B in the figure cited. The last step in converting FT to BN is to give root nodes (which do not have parent nodes) their priori probabilities. As the basic events of FT correspond to the failure of components of the system under study, the a priori probabilities, in this case, are the

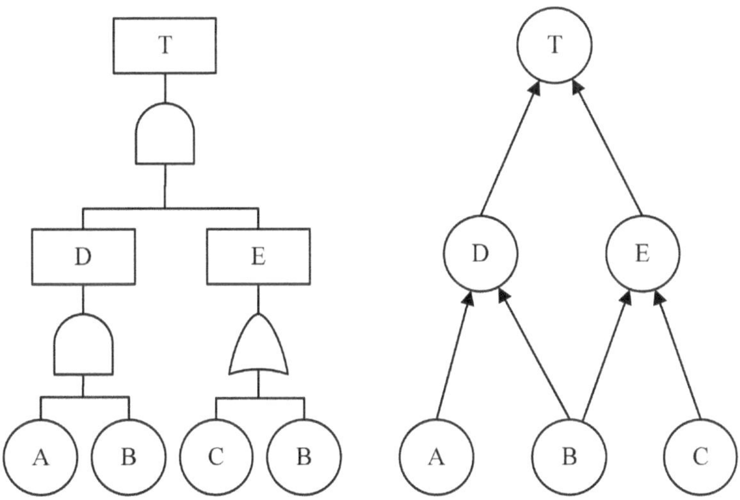

FIG. 4.6 An example of conversion of FT into BN.

probabilities of failure of these components. Such probabilities can be obtained through databases or from the opinion of experts. After the conversion, the BN allows the probability of the top event (the probability of system failure) to be calculated.

It is important to note that although traditional techniques normally support the development of Bayesian networks for reliability analysis, BNs have several characteristics that can optimize the results obtained. Three of these capabilities that stand out are uncertainty representation, performing inference, and multistate modeling. BN's ability to represent uncertainty overcomes the deterministic nature of more traditional techniques such as FT and block diagrams. The multistate modeling capability allows the BN to be able to represent other equipment states besides the functioning/failing dyad. Moreover, the ability to perform inference, which will be detailed in Chapter 6, allows the analysis to consider evidence or observations about the system to update the calculated probabilities.

Because of these capabilities from BN, it has often been used to perform reliability analysis. Several authors have already applied such techniques in a series of case studies, such as high-speed trains, LNG regasification systems, floating offshore wind turbines, and subsea blowout preventer control system (Chen, Qi, Chen, Guo, & Peng, 2017; Li, Guedes Soares, & Huang, 2020; Liu & Liu, 2019; Schleder, Martins, & Modarres, 2012). An extension of BN, called dynamic Bayesian networks, which allows adding temporality to the analysis, has also been frequently used (Jafari, Pouyakian, Khanteymoori, & Hanifi, 2020; Lee & Choi, 2020).

4.2.4 Dynamic fault trees

As already described in Chapter 2, fault trees (FTs) have become a widely used tool for studying and estimating the reliability of several types of systems. However, although FTA is an effective and highly utilized technique for reliability analysis, it is not able to capture dynamic behaviors of more advanced fault-tolerant systems such as sequence-dependent failures.

Consider, for example, a system composed of two engines, the main one and the spare one, and a switch responsible for activating the spare engine if the main one fails. A system failure would occur if the switch fails and, subsequently, the main engine fails, since it would not be possible to activate the spare one. However, if the main motor fails before the switch, the spare motor would be activated, avoiding a system failure. This type of behavior cannot be properly represented by a traditional FT.

In order to overcome this FT limitation, one of the most widely used alternatives is the DFT. DFT is able to more adequately represent temporal dependencies between the components' failures that may lead to a system failure. To do so, the DFT structure remains very similar to that of FT but adds several new types of gates to the list presented in Chapter 2. The main additional logic gates of the DFT are shown in Table 4.2. Note that the symbology of these gates may vary slightly between the various publications that use this technique. Although Table 4.2 shows the main logic gates used in the DFT, some authors have already proposed other specific gates that will not be covered in this book, such as "hot spare," "cold spare," "one-shot PDEP," and "persistent PDEP" (Dugan, Bavuso, & Boyd, 1990; Portinale, Raiteri, & Montani, 2010).

Just to illustrate how a DFT can be built and interpreted, consider the one shown in Fig. 4.7. Here, A is the trigger event for the FDEP gate. If it occurs, it forces the occurrence of events B and C, which lead to the occurrence of H. The basic events D and E, connected to the PAND gate, must occur in this order for the event G to occur. F, connected to the lower right rectangle of the SPARE gate, represents the spare part of E. If E occurs and,

TABLE 4.2 DFT additional gates.

Gate name	Gate symbol	Description
PAND (Priority AND)		In the PAND logic gate, the output event will occur only if all input events occur from left to right.
POR (Priority OR)		In the POR logic gate, the output event will occur only if the input event which is farthest to the left occurs before the other ones.
FDEP (Function Dependency)	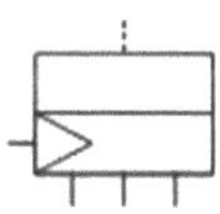	The FDEP gate represents an event where an element failure, called trigger, forces its dependent elements to fail.
PDEP (Probabilistic Dependency)	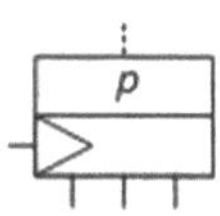	The PDEP gate is an extension of the FDEP gate. In it, when the trigger event happens, its dependent elements fail with a probability p. If $p = 1$, PDEP is equivalent to FDEP.
SPARE	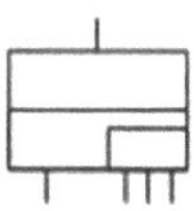	The SPARE gate represents a situation where a component can be replaced by a spare when it fails. The component can have multiple spares and the spares can be shared between several similar components in the system.
SEQ (Sequence Enforcing)		The SEQ gate guarantees that the input events only occur in sequence, from left to right.

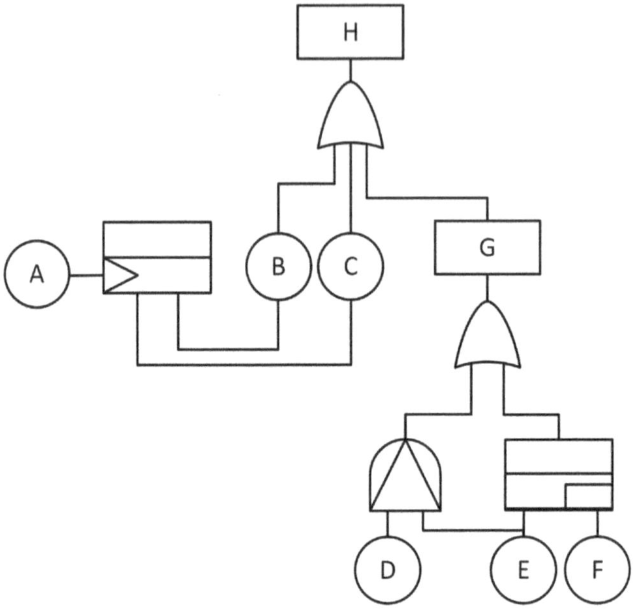

FIG. 4.7 An example of DFT.

after replacing it, F also occurs, the G event occurs. The top event H will occur if B, C, or G occurs. In fact, the relationship between events D, E, and F is the same between the main and spare motors and the switch mentioned already. E represents the main engine, F is the spare one, and D is the switch.

By now the reader may have already realized that DFT proposes to model features of fault-tolerant and safety-critical systems that a traditional FT is not capable of. By being able to model features such as order-dependent failures, sequential events, mutually exclusive events, redundancies, and spare management, DFT can provide more accurate results on the reliability analysis of engineering systems.

However, for a quantitative analysis to be carried out through the DFT, it is not enough to use the minimal cut sets technique used in traditional FT, since the history-dependent behavior of the DFT makes the generalization of such a technique insufficient. Therefore, DFT analysis is usually performed through stochastic processes. In fact, many authors have already proposed several methods for mapping DFTs in Markov chains, Petri nets, and Bayesian networks, for example (Boudali, Crouzen, & Stoelinga, 2007; Boudali & Dugan, 2005; Codetta-Raiteri, 2005; Dugan, Bavuso, & Boyd, 1992; Farsi, 2016; Herscheid & Troger, 2014; Kabir, Walker, & Papadopoulos, 2014; Montani, Portinale, Bobbio, & Codetta-Raiteri, 2006; Yevkin, 2016; Zhang, Miao, Fan, & Wang, 2009).

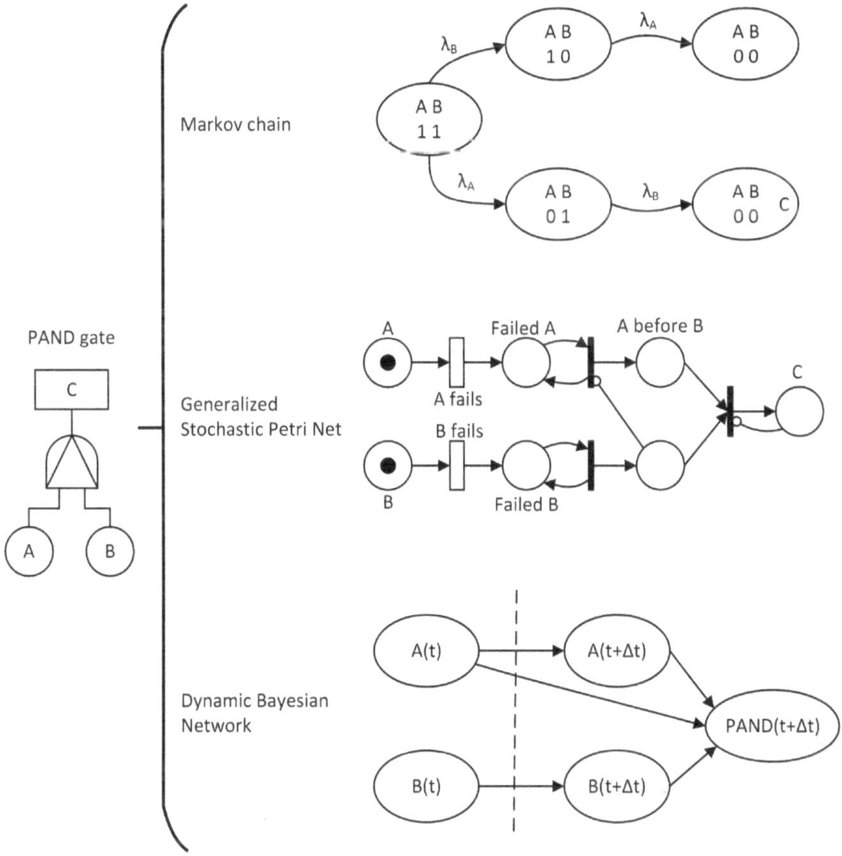

FIG. 4.8 Mapping a PAND gate into Markov chain, GSPN, and DBN.

Although the authors cited previously propose different ways of mapping DFT into other modeling techniques, they all follow a similar logic: each gate is transformed into a new structure that is expanded as new gates are added. Just as an example, Fig. 4.8 shows how a PAND logic gate can be mapped into a Markov chain, a GSPN, and a DBN. Once the DFT has been mapped, the quantitative reliability analysis can be performed.

It is important to mention that although DFT has several advantages associated with the ability to represent the system's dynamic behavior, the use of this technique ends up consuming a lot of development time, especially for large, complex systems. Similarly, quantitative analysis, which requires DFT mapping, can become computationally expensive. Even so, it is a technique that is becoming increasingly popular and has been used in several case studies (Ruijters & Stoelinga, 2015).

4.3 Comments on imperfect maintenance

The previously presented reliability and availability methods are capable of dealing with the analysis of complex engineering systems due to their ability to model failure dependencies and functional reconfiguration.

As with many engineering systems, modern engineering systems are not capable of performing their functions throughout the operational life without being submitted to maintenance actions. The maintenance actions intend to control critical devices' degradation or even restoring the system's operational performance in case of any device failure.

In Chapter 2, the main concepts regarding maintenance planning and the effects on system availability were presented. However, for complex engineering systems, there is the possibility of imperfections in the maintenance actions that can affect the system's operational performance, increasing the possibility of failure occurrence in a given operational campaign. Therefore, the subject of imperfect maintenance has been the object of research for the last decade in view of the development of models to support the evaluation of systems' reliability and availability.

Pham and Wang (1996) present some reflections regarding the preventive maintenance actions. Ideally, according to those authors, preventive maintenance should be an action that restores the system operating condition to as good as new. To achieve this goal, the maintenance action should be able to eliminate all components' degradation through repair or substitution. This goal is very difficult to be achieved in engineering systems not only due to the costs associated with eliminating all components degradation but also due to the lack of knowledge of all physical processes that may cause the degradation of components. Even the maintenance planning philosophies based on reliability and risk concepts intend to control the failure of critical components through the application of preventive and predictive maintenance and do not aim at eliminating all components degradation.

The imperfect preventive maintenance process involves the restoration of a system performance not to as good as new but younger than the condition just before the maintenance action. Usually, it is assumed that imperfect maintenance restores the system operating state to somewhere between as good as new and as bad as old, which represents a maintenance action that restores the system to the failure rate it had when it failed. Although possible, but not likely, the preventive maintenance action can increase the system failure rate in comparison with that value just before the action.

For the analysis presented in the sequence of the chapter, the corrective action is modeled as minimum repair, restoring the system to the same condition presented just before failure.

Suppose that the failures of repairable equipment, observed over time $t \geq 0$, occur at times $t_1 < t_2 < \cdots < t_k$. Time t_i ($i = 1, 2, \ldots, k$), which is measured from zero, is called the arrival time to the ith failure. If repair times are negligible, then the interarrival times $x_i = t_i - t_{i-1}$ (where $t_0 = 0$ and $i = 1, 2, \ldots, k$) represent the operating times between the $(i-1)$th and ith failures and constitute the failure pattern of the equipment (Pham & Wang, 1996).

The Poisson process is an event count, which can be defined as a collection of independent random variables $N(t)$, with $t \geq 0$, where $N(t)$ is the number of events in the interval $(0, T)$. $N(t)$ is characterized by independent increments and intensity function, which has the form presented in Eq. (4.23):

$$u(t) = \lim_{\Delta t \to 0} \frac{P[N(t + \Delta t) - N(t) = 1 | \mathcal{H}_t]}{\Delta t} \tag{4.23}$$

where $\mathcal{H}_t$ is the past of the failure process at time t, i.e., the set of all events occurred before t.

Due to the preventive maintenance actions, the systems may deteriorate or maintain the same pattern of failure occurrence throughout the operational life. If the preventive maintenance is perfect, returning the system to "as good as new" condition, the failure intensity function is constant and the process is named homogeneous Poisson process. For this case, the reliability analysis executed according to the concepts presented in Chapter 2 is still valid.

When failure intensity function is time-dependent, the most general counting processes class used is known as Nonhomogeneous Poisson Process (NHPP). More information about the Poisson process in the context of a repairable system can be obtained in Rigdon and Basu (2000).

The cumulative number of failures up to time t, $\{N(t), t \geq 0\}$, can be described by an NHPP. Often a counting process modeled by NHPP is specified by the average number of events in $(0, t]$, $m(t)$, the mean value function, calculated as presented in Eq. (4.24), that describes the behavior of the expected cumulative number of failures.

$$m(t) = E[N(t)] = \int_0^t u(x)dx \tag{4.24}$$

One of the most common and versatile parametric forms for failure intensity representation is the Power Law (PL) intensity function, proposed in Crow (1975). For this model, the intensity function is assumed to grow proportionally to the power of time, according to Eq. (4.25), where $u(t)$ is the failure intensity function and λ, β are model parameters ($\lambda > 0, \beta > 0$).

$$u(t) = \frac{1}{\lambda^\beta} \cdot \beta \cdot t^{(\beta - 1)} \tag{4.25}$$

It follows that the mean value function, $m(t)$, is given by Eq. (4.26):

$$m(t) = \frac{1}{\lambda^{\beta}} \cdot t^{\beta} \tag{4.26}$$

Malik (1979) introduces the concept of improvement factor in the maintenance scheduling problem that models the system's failure rate changes after each preventive intervention. The improvement factor includes a parameter in the intensity function to model the preventive maintenance (PM) effect. A commonly used parameterization improvement factor class is the one that considers system age reduction after maintenance action called virtual age of the system.

The parameterization of the improvement factor model considers three possibilities. The first considers a parameter to reduce the age of the system after the maintenance action, named proportional age reduction (PAR). The second parameterization uses a parameter to increase the failure intensity function of the equipment, named proportional rate increase (PRI). The third uses a hybrid formulation of the previous two, simultaneously considering the age reduction factor and increased failure rate. Several authors (Lin, Zuo, & Yam, 2000; Wu & Zuo, 2010; Zhou, Xi, & Lee, 2007) represented these parameterizations to model the fault behavior and optimization of maintenance time considering imperfect maintenance. Nakagawa (1988) and Zhao and Nakagawa (2012) described a model based on reducing age and increasing the failure rate.

In the PAR model, it is considered that the PM restores the system in such a way that the failure intensity function at time t is the same function, but with a failure intensity value referring to a shorter time. The idea is that PM actions reduce the age of the system by an a factor that is used to measure the effectiveness of an imperfect PM.

After the jth PM cycle, the age of the system is reduced by a factor $a \cdot T_j$ or the virtual age of the system is $(1 - a)T_j$.

Being the initial failure intensity function, in the jth PM cycle the failure intensity function is expressed as presented in Eq. (4.27):

$$u_j(t) = u_0\left(t - aT_{j-1}\right), t \geq Tj - 1 \tag{4.27}$$

In the model mentioned already, when $a = 1$, it corresponds to perfect maintenance, while $a = 0$ means totally imperfect maintenance that maintains the same age of the system. The case of $0 < a < 1$ corresponds to an imperfect PM. An unexpected but possible case would be a maintenance action that worsens the age of the equipment, with $a < 0$, that is, the equipment is worse than before the maintenance action, and the case in which $a > 1$ suggests a system restored to a better condition than new, that is, the equipment is modernized. In physical terms, factor a can be interpreted as an index of the effectiveness of the maintenance action.

In the proportional rate increase model, after a PM, the system returns to zero age, but the rate of growth of the failure intensity function is higher after each maintenance cycle. The action of PM rejuvenates the system but makes the rate of intensity increase faster (Nakagawa, 1988).

Using the b factor, it is possible to add an increase in function after an imperfect MP. Since $u_0(t)$ is the initial failure intensity function, in the jth PM cycle, the failure intensity function is expressed in Eq. (4.28):

$$u_j(t) = b^{j-1} \cdot u_0\left(t - T_{j-1}\right), t \geq Tj - 1 \tag{4.28}$$

If $b = 1$, the PM is perfect; if $b > 1$, the PM action is imperfect.

In order to predict future system's behavior, failure intensity functions based on improvement factor models can be used to compute reliability. Considering that the last observed failure time for a given system is t_n, it can be of interest to predict the time interval to the next failure, $t_{n+1} - t_n$, given the failure data up to time t_n. The following functions can describe the future behavior of the failure process and they might be useful for the system engineering team. The reliability function at time t is presented by Eq. (4.29):

$$R(t) = P(t_{n+1} - t_n > t | \mathcal{H}_t) = P(N(t_n, t_n + t) = 0 | \mathcal{H}_t) = exp\left[-\int_{t_n}^{t_n+t} u(w)\, dw\right] \tag{4.29}$$

This reliability function expresses the survival probability or the probability that the system will work without failure for a given period of time greater than t after t_n, given the past behavior of the failure process.

In addition to the study of imperfection in preventive maintenance actions, one can also model the problem of imperfection in corrective maintenance. According to Jack (1998), the same models used to define the effects of imperfect preventive maintenance can be used to define the effects of corrective maintenance. In this paper, the effects of both types of imperfect maintenance are modeled using the age reduction concept. The author defines the method used to split both effects in different age reduction concepts considering the failure intensity function. Pham and Wang (1996) also present a discussion regarding the coupled effect of imperfect preventive and corrective maintenance.

A common problem for the imperfect maintenance modeling presented in this session is the definition of the failure intensity function parameters. The maximum likelihood method is used to calculate those parameters.

The maintainability analysis shall be done using the same techniques presented in Chapter 2 aiming at modeling the repair rate associated with corrective maintenance.

The methods used to model imperfect maintenance may also be used to evaluate the reliability of devices or subsystems of engineering systems and generate failure rates that are variable with operational time and also dependent on the number of maintenance actions performed during an operational campaign. Thus, the reliability and availability estimates of the systems have to consider that variation and must be executed using the methods previously presented in this chapter.

4.4 Development of system reliability and availability analyses

When performing system reliability and availability analyses, the information obtained as a result can greatly contribute to the decision regarding the design, operation, or decommissioning of the system. Knowing, in advance, the value that these parameters can reach in the future not only can determine the economic viability and security of the system but also can promote positive changes in the system so that its operation is optimized.

But in order for the advantages of performing such an analysis to positively impact the functioning of the system, it is necessary that it be properly performed. It is known that any mathematical model, be it developed for any type of objective, has limitations and will never be able to perfectly mimic the behavior of the real system. Therefore, when conducting reliability and availability analyses, it is necessary to know and accept the origins of the limitations of the model used, as well as the impact they can cause on the final result.

To minimize the effects that the limitations of the model can have on the final results, Fig. 4.9 shows the four main steps for reliability and availability analyses to be properly developed.

The first step is to study the system. This study should focus on the information relevant to the analysis to be developed. The fundamental information to be acquired in this first development step is listed as follows:

- Operating modes/operating system settings;
- Components belonging to the system;
- The functions that the components must perform;
- The failure modes that these components can exhibit;
- The effect that failure modes have on equipment and the system;
- How does the system reconfigure its operation when it fails, if at all;
- Which failure modes or sequence of failure modes can make the system inoperable.

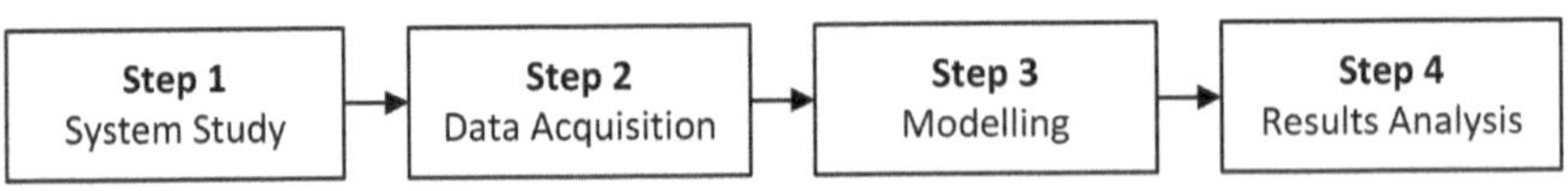

FIG. 4.9 Steps for the development of systems' reliability and availability analyses.

Other additional information can be collected in order to optimize the model's results, especially the availability results, which are influenced by the maintenance of the equipment. In this case, it may be relevant to study about:

- Spare parts management;
- Size and qualification of the maintenance crew; and
- Preventive maintenance actions are performed periodically.

As the amount of information to be collected is large, it is recommended to use techniques developed for this purpose. Some of the techniques that have been shown to be efficient for acquiring information about the system are:

- Functional tree: contributes to a hierarchical and functional survey of all subsystems and equipment in the system (Melani, Murad, Caminada Netto, Souza, & Nabeta, 2018);
- Failure Mode and Effect Analysis (FMEA): contributes to the survey of equipment failure modes, as well as their possible causes and effects (Melani, Murad, et al., 2018);
- Failure Mode and Symptom Analysis (FMSA): technique derived from FMEA focuses on the study of the symptoms that the occurrence of failure modes causes in the system (Murad et al., 2020); and
- System Modeling Language (SysML): modeling language from system engineering contributes to the specification, design, and analysis of a system (Melani, 2020).

Once the system study has been done, the next step consists of acquiring failure data and repairing the equipment. Such data must contain the operating times of each equipment until it presents a certain failure mode, as well as the time that the maintenance team took to repair or replace the equipment.

Once the system study has been done, the next step consists of acquiring failure data and repairing the equipment. Such data must contain the operating times of each equipment until it presents a certain failure mode, as well as the time that the maintenance team took to repair or replace the equipment. Such record contributes to evaluate parameters relevant to the equipment or the system such as Mean Time Between Failures (MTBF) and Mean Time To Repair (MTTR).

After performing the first two steps, it is already possible to develop a model of the system in order to assess its reliability and availability. The choice of the model type must take into account the characteristics of the available tools and the characteristics of the system to be modeled. Choosing between block diagrams, fault trees, DFTs, PNs, BNs, or any other modeling tool will depend on what types of system behavior you want to model, in addition to the availability of the appropriate software for analysis.

Finally, the model developed can be fed with data on failure, repair, spare parts, and all other information possibly relevant for calculating the availability and reliability of the system. From there, the results obtained contribute to decision-making regarding the design, operation, or decommissioning of the system. If the system is still being designed, for example, the results can show whether it will be economically viable or if some changes in its design can be made in order to improve its operation. If the system under study is already in operation, the availability and reliability estimate can show whether its performance in the following years justifies keeping it operating or not.

It is worth noting that availability and reliability are not the only parameters that can be estimated by the aforementioned techniques. It is also possible to obtain information such as the number of failures expected for each piece of equipment in a given period of time and the number of hours required for the maintenance team to carry out all the necessary repairs expected, which components most affect the system's operation. All of this information contributes to decision-making and is essential for asset management.

References

Becker, G., Camarinopoulos, L., & Zioutas, G. (2000). A semi-Markovian model allowing for inhomogenities with respect to process time. *Reliability Engineering & System Safety*, *70*(1), 41–48. https://doi.org/10.1016/S0951-8320(00)00044-2.

Beirong, Z., Xiaowen, X., & Wei, X. (2012). Availability modeling and analysis of equipment based on generalized stochastic petri nets. *Research Journal of Applied Sciences, Engineering and Technology*, *4*(21), 4362–4366.

Billinton, R., & Allan, R. N. (1992). *Reliability evaluation of engineering systems*. Boston, MA: Springer US. https://doi.org/10.1007/978-1-4899-0685-4.

Bobbio, A., Portinale, L., Minichino, M., & Ciancamerla, E. (2001). Improving the analysis of dependable systems by mapping fault trees into Bayesian networks. *Reliability Engineering & System Safety*, *71*(3), 249–260. https://doi.org/10.1016/S0951-8320(00)00077-6.

Boudali, H., Crouzen, P., & Stoelinga, M. (2007). Dynamic fault tree analysis using input/output interactive Markov chains. In *37th annual IEEE/IFIP international conference on dependable systems and networks (DSN'07)* (pp. 708–717). IEEE. https://doi.org/10.1109/DSN.2007.37.

Boudali, H., & Dugan, J. B. (2005). A new Bayesian network approach to solve dynamic fault trees. In *Annual reliability and maintainability symposium, 2005. Proceedings. IEEE* (pp. 451–456). https://doi.org/10.1109/RAMS.2005.1408404.

Chen, S., Qi, Z., Chen, D., Guo, L., & Peng, W. (2017). Investigation of Bayesian network for reliability analysis and fault diagnosis of complex systems with real case applications. *Advances in Mechanical Engineering*, *9*(10). https://doi.org/10.1177/1687814017728853. 168781401772885.

Codetta-Raiteri, D. (2005). The conversion of dynamic fault trees to stochastic petri nets, as a case of graph transformation. *Electronic Notes in Theoretical Computer Science*, *127*(2), 45–60. https://doi.org/10.1016/j.entcs.2005.02.005.

Crow, L. H. (1975). *Reliability analysis for complex, repairable systems*. Aberdeen, MD. Available at: https://apps.dtic.mil/sti/citations/ADA020296.

Dugan, J. B., Bavuso, S. J., & Boyd, M. A. (1990). Fault trees and sequence dependencies. In *Annual proceedings on reliability and maintainability symposium* (pp. 286–293). IEEE. https://doi.org/10.1109/ARMS.1990.67971.

Dugan, J. B., Bavuso, S. J., & Boyd, M. A. (1992). Dynamic fault-tree models for fault-tolerant computer systems. *IEEE Transactions on Reliability, 41*(3), 363–377. https://doi.org/10.1109/24.159800.

Farsi, M. A. (2016) Fault analysis of complex systems via dynamic Bayesian network. *AUT Journal of Mechanical Engineering, 2*(2), 10. https://doi.org/10.22060/ajme.2018.13711.5692.

Hamada, M. S., Wilson, A. G., Reese, C. S., & Martz, H. F. (2008). Bayesian reliability. *Springer series in statistics*. New York, NY: Springer New York. https://doi.org/10.1007/978-0-387-77950-8.

Herscheid, L., & Troger, P. (2014). Specification of dynamic fault tree concepts with stochastic petri nets. In *2014 eighth international conference on software security and reliability* (pp. 177–186). IEEE. https://doi.org/10.1109/SERE.2014.31.

Jack, N. (1998). Age-reduction models for imperfect maintenance. *IMA Journal of Management Mathematics, 9*(4), 347–354. https://doi.org/10.1093/imaman/9.4.347.

Jafari, M. J., Pouyakian, M., Khanteymoori, A., & Hanifi, S. M. (2020). Reliability evaluation of fire alarm systems using dynamic Bayesian networks and fuzzy fault tree analysis. *Journal of Loss Prevention in the Process Industries, 67*, 104229. https://doi.org/10.1016/j.jlp.2020.104229.

Kabir, S., Walker, M., & Papadopoulos, Y. (2014). Reliability analysis of dynamic systems by translating temporal fault trees into Bayesian networks. In *Model-based safety and assessment* (pp. 96–109). https://doi.org/10.1007/978-3-319-12214-4_8.

Kumamoto, H., & Henley, E. J. (2000). *Probabilistic risk assessment and management for engineers and scientists* (2nd ed.). New York, NY: Wiley-IEEE Press.

Lampis, M., & Andrews, J. D. (2009). Bayesian belief networks for system fault diagnostics. *Quality and Reliability Engineering International, 25*(4), 409–426. https://doi.org/10.1002/qre.978.

Lee, D., & Choi, D. (2020). Analysis of the reliability of a starter-generator using a dynamic Bayesian network. *Reliability Engineering & System Safety, 195*, 106628. https://doi.org/10.1016/j.ress.2019.106628.

Leigh, J. M., & Dunnett, S. J. (2016). Use of petri nets to model the maintenance of wind turbines. *Quality and Reliability Engineering International*. https://doi.org/10.1002/qre.1737.

Lewis, E. E. (1994). *Introduction to reliability engineering* (2nd ed.). New York: John Wiley & Sons.

Li, H., Guedes Soares, C., & Huang, H.-Z. (2020). Reliability analysis of a floating offshore wind turbine using Bayesian Networks. *Ocean Engineering, 217*, 107827. https://doi.org/10.1016/j.oceaneng.2020.107827.

Limnios, N., & Oprişan, G. (2001). *Semi-Markov processes and reliability*. Boston, MA: Birkhäuser Boston. https://doi.org/10.1007/978-1-4612-0161-8.

Lin, D., Zuo, M. J., & Yam, R. C. M. (2000). General sequential imperfect preventive maintenance models. *International Journal of Reliability, Quality and Safety Engineering, 07*(03), 253–266. https://doi.org/10.1142/S0218539300000213.

Liu, Z., & Liu, Y. (2019). A Bayesian network based method for reliability analysis of subsea blowout preventer control system. *Journal of Loss Prevention in the Process Industries, 59*, 44–53. https://doi.org/10.1016/j.jlp.2019.03.004.

Long, F., Zeiler, P., & Bertsche, B. (2017). Modelling the flexibility of production systems in Industry 4.0 for analysing their productivity and availability with high-level Petri nets. *IFAC-PapersOnLine, 50*(1), 5680–5687. https://doi.org/10.1016/j.ifacol.2017.08.1118. Elsevier B.V.

Malik, M. A. K. (1979). Reliable preventive maintenance scheduling. *A I I E Transactions, 11*(3), 221–228. https://doi.org/10.1080/05695557908974463.

Melani, A. H. A. (2020). *System fault diagnosis based on Bayesian networks and SysML*. University of São Paulo.

Melani, A. H. A., Caminada Netto, A., Murad, C. A., Souza, G. F. M., & Nabeta, S. I. (2018). Petri net based reliability analysis of thermoelectric plant cooling tower system: Effects of operational strategies on system reliability and availability. In *Proceedings of the joint ICV-RAM ISUMA UNCERTAINTIES conference. Florianópolis, SC, Brazil.*

Melani, A. H. A., Murad, C. A., Caminada Netto, A., Souza, G. F. M., & Nabeta, S. I. (2018). Criticality-based maintenance of a coal-fired power plant. *Energy, 147*, 767–781. https://doi.org/10.1016/j.energy.2018.01.048.

Melani, A. H. A., Murad, C. A., Caminada Netto, A., Souza, G. F. M., & Nabeta, S. I. (2019). Maintenance strategy optimization of a coal-fired power plant cooling tower through generalized stochastic petri nets. *Energies, 12*(10), 1951. https://doi.org/10.3390/en12101951.

Melani, A. H. A., Souza, G. F. M., Murad, C. A., Caminada Netto, A., & Nabeta, S. I. (2017). Petri Net based reliability analysis of thermoelectric plant cooling tower. ABCM. https://doi.org/10.26678/ABCM.COBEM2017.COB17-2203.

Montani, S., Portinale, L., Bobbio, A., & Codetta-Raiteri, D. (2006). Automatically translating dynamic fault trees into dynamic Bayesian networks by means of a software tool. In *First International conference on availability, reliability and security (ARES'06)* (pp. 6–809). IEEE. https://doi.org/10.1109/ARES.2006.34.

Moura, M. C., & Droguett, E. L. (2008). A continuous-time semi-Markov Bayesian belief network model for availability measure estimation of fault tolerant systems. *Pesquisa Operacional, 28*(2), 355–375. https://doi.org/10.1590/S0101-74382008000200011.

Murad, C., Melani, A. H. A., Michalski, M. A. C., Caminada Netto, A., de Souza, G., & Nabeta, S. I. (2020). Fuzzy-FMSA: Evaluating fault monitoring and detection strategies based on failure mode and symptom analysis and fuzzy logic. *ASCE-ASME Journal of Risk and Uncertainty in Engineering Systems, Part B: Mechanical Engineering*. https://doi.org/10.1115/1.4045974.

Murata, T. (1989). Petri nets: Properties, analysis and applications. *Proceedings of the IEEE, 77*(4), 541–580. https://doi.org/10.1109/5.24143.

Nakagawa, T. (1988). Sequential imperfect preventive maintenance policies. *IEEE Transactions on Reliability, 37*(3), 295–298. https://doi.org/10.1109/24.3758.

Nývlt, O., & Rausand, M. (2012). Dependencies in event trees analyzed by Petri nets. *Reliability Engineering & System Safety, 104*, 45–57. https://doi.org/10.1016/j.ress.2012.03.013.

O'Connor, P., & Kleyner, A. (2012). *Practical reliability engineering* (5th ed.). Wiley.

Pham, H., & Wang, H. (1996). Imperfect maintenance. *European Journal of Operational Research, 94*(3), 425–438. https://doi.org/10.1016/S0377-2217(96)00099-9.

Portinale, L., Raiteri, D. C., & Montani, S. (2010). Supporting reliability engineers in exploiting the power of Dynamic Bayesian Networks. *International Journal of Approximate Reasoning, 51*(2), 179–195. https://doi.org/10.1016/j.ijar.2009.05.009. Elsevier.

Reddy, G. B., Murty, S. S. N., & Ghosh, K. (1993). Timed Petri net: An expeditious tool for modelling and analysis of manufacturing systems. *Mathematical and Computer Modelling, 18*(9), 17–30. https://doi.org/10.1016/0895-7177(93)90140-T.

Rigdon, S. E., & Basu, A. P. (2000). *Statistical methods for the reliability of repairable systems*. New York, NY: Wiley-Interscience.

Ruijters, E., & Stoelinga, M. (2015). Fault tree analysis: A survey of the state-of-the-art in modeling, analysis and tools. *Computer Science Review, 15–16*, 29–62. https://doi.org/10.1016/j.cosrev.2015.03.001.

Schleder, A. M., Martins, M. R., & Modarres, M. (2012). The use of Bayesian networks in reliability analysis of the LNG regasification system on a FSRU under different scenarios. In *Proceedings of the twenty-second (2012) international offshore and polar engineering conference. Rhodes, Greece* (p. 8).

Schruben, L., & Yucesan, E. (1994). Transforming petri nets into event graph models. In *Proceedings of the 1994 winter simulation conference. Orlando, FL, USA* (pp. 560–565).

Talebberrouane, M., Khan, F., & Lounis, Z. (2016). Availability analysis of safety critical systems using advanced fault tree and stochastic Petri net formalisms. *Journal of Loss Prevention in the Process Industries*. https://doi.org/10.1016/j.jlp.2016.09.007.

Thangamani, G. (2012). Generalized stochastic petri nets for reliability analysis of lube oil system with common-cause failures. *American Journal of Computational and Applied Mathematics*, *2*(4), 152–158. https://doi.org/10.5923/j.ajcam.20120204.03.

Wu, S., & Zuo, M. J. (2010). Linear and nonlinear preventive maintenance models. *IEEE Transactions on Reliability*, *59*(1), 242–249. https://doi.org/10.1109/TR.2010.2041972.

Yevkin, O. (2016). An efficient approximate Markov chain method in dynamic fault tree analysis. *Quality and Reliability Engineering International*, *32*(4), 1509–1520. https://doi.org/10.1002/qre.1861.

Zhang, X., Miao, Q., Fan, X., & Wang, D. (2009). Dynamic fault tree analysis based on Petri nets. In *2009 8th international conference on reliability, maintainability and safety* (pp. 138–142). IEEE. https://doi.org/10.1109/ICRMS.2009.5270223.

Zhao, X., & Nakagawa, T. (2012). Optimization problems of replacement first or last in reliability theory. *European Journal of Operational Research*, *223*(1), 141–149. https://doi.org/10.1016/j.ejor.2012.05.035.

Zhou, X., Xi, L., & Lee, J. (2007). Reliability-centered predictive maintenance scheduling for a continuously monitored system subject to degradation. *Reliability Engineering & System Safety*, *92*(4), 530–534. https://doi.org/10.1016/j.ress.2006.01.006.

CHAPTER

5

Engineering systems fault detection methods

If you can see, look. If you can look, observe.

José Saramago (1922–2010) In: Blindness

The world-renowned Nobel Prize winner Portuguese writer, José Saramago, playing with the meaning of the words "see," "look," and "observe" in his best-seller "Blindness," gives a glimpse to the reader about what seems to be one of the greatest difficulties of mankind: being able to really see beyond the superficial. While "looking" comes with the ability to see, "observing" comes with the ability to examine critically, seeking to analyze a specific situation. Looking is extended to the perception field showing how supposedly open and developed societies appear to be so obtuse and short-sighted. Saramago shows with his metaphor that it is not enough to have the correct and functional instrument, our eyes, to analyze the world; it is necessary to go further and understand the mechanisms that raise and drive the observed events, using the human being's logical reasoning capacity and illuminating the darkness enshrined by functional blindness.

Regarding Reliability Analysis and Asset Management of Engineering Systems, an efficient process for fault detection has a similar paradigm to Saramago's metaphor: fault detection in engineering systems does not depend only on the correct installation of a large number of appropriate sensors and probes used to monitor such a system; the presence of emergent properties and behaviors in modern engineering systems, related to their nonlinearity, requires more sophisticated techniques that go beyond human predictive capacity. Besides, time is a key factor in the fault analysis process and must be a concern. Fault detection is actually the first activity of a more complex process that includes diagnosis and prognosis. Therefore, the more premature the fault detection is, the more effective its diagnosis and prognosis can be.

https://doi.org/10.1016/B978-0-12-823521-8.00003-7

The objective of this chapter is to present a brief overview of detection methods used for the health monitoring of engineering systems. The fundamental concepts of fault detection are introduced, followed by the classification of fault detection methods, such as analytical model-based and data-driven methods. The main features, their relation to Artificial Intelligence (AI) requirements, and the limitations of those methods are discussed, as well as the main data sources for fault detection addressed in this chapter. The Principal Components Analysis (PCA), a statistical multivariate method, is analyzed in detail in view of its advantages for modern systems analysis aiming at the fusion of a great amount of monitoring data. Finally, the application of sampling-based methods aiming to define variability and uncertainty in a fault detection process associated with variation in system operating conditions and manufacturing characteristics is also presented.

5.1 Fault detection fundamental concepts

In order to introduce the fundamental concepts of fault detection, the main purpose of this section, firstly it is interesting to present to the reader, or remind those already initiated in this knowledge field of an even more fundamental concept: what is a fault? Formally, a "fault" can be defined as a condition within a system in which an unexpected or unpermitted deviation of at least one characteristic property or parameter of this system occurs, taking as reference its acceptable, usual, or standard behavior.

Extrapolating the concept of fault, one can understand fault detection as the observation of such anomalous condition in a system, that is, to recognize that a fault is happening (Isermann, 2011). In fact, canonically, fault detection is defined as the determination of faults present in a system. The time in which the detection is performed is also a fundamental information expected from this activity.

Fault detection is essential to preserve the functions of a system, as well as to protect its integrity. If the detection of an ongoing fault does not occur, protection measures will eventually need to be taken, or otherwise, the system may, in addition to fail, be damaged, generating not only financial, but also environmental and personal losses, often irreparable. Therefore, the early detection of system faults becomes essential for Engineering Systems Asset Management, preventing the progression of abnormal events, the occurrence of failures, and the unscheduled shutdown of equipment, playing an important role in high-cost and safety-critical processes. The ability to detect a fault in its early stages is a critical part of the development and implementation of effective systems diagnosis and prognosis technologies, allowing counteractions and decisions to be taken from the information obtained, such as operations changes, system reconfiguration, maintenance planning, or immediate repair.

The activities that follow fault detection in a Health Monitoring and Fault Detection (HMFD) process are fault diagnosis and fault evaluation. Fault diagnosis may be understood as finding the cause and location of the fault and can be subdivided into two subactivities: fault isolation and fault identification. The former determines the type and location of the fault, while the latter determines its magnitude and time-variant behavior. Some authors also include the detection activity as a subactivity in the fault diagnosis process, considering the detection event as the bottom line of the fault classification process. In its turn, fault evaluation includes the system's condition and failure prognosis, the fault/failure severity analysis, and the Remaining Useful Life (RUL) forecasting, assisting in the decision-making process (Isermann, 2006; Miljković, 2011; Pagiatakis, Dritsas, Chatzarakis, Todorov, & Stoev, 2017; Vachtsevanos, Lewis, Roemer, Hess, & Wu, 2006).

Fig. 5.1 shows a diagram based on the P-F curve (Moubray, 1997), in which "P" refers to the Potential failure and "F" refers to the asset's Functional failure, with a conceptual overview of the Fault Detection and Diagnosis (FDD) process.

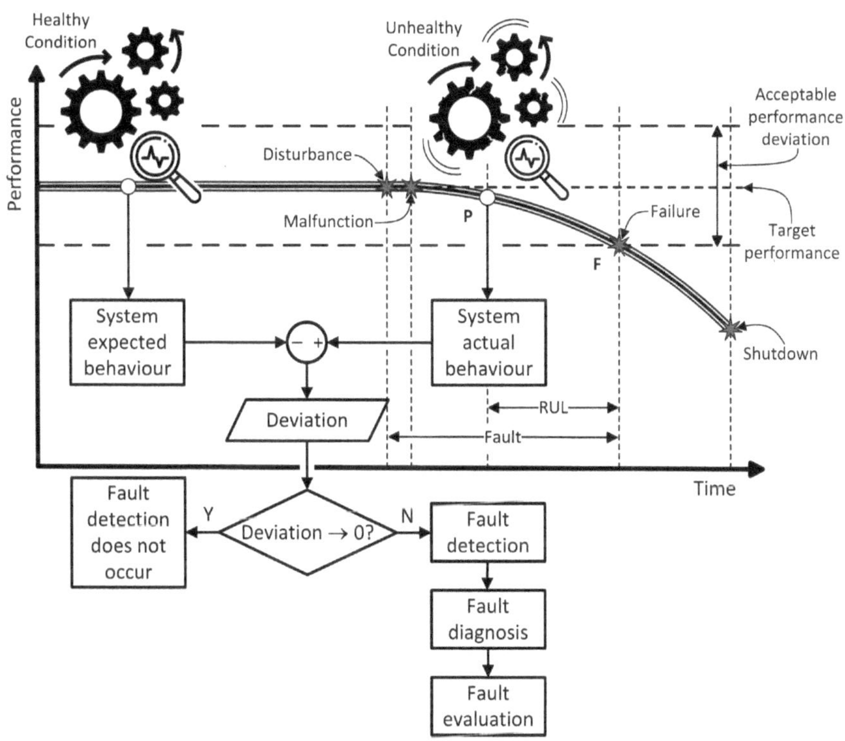

FIG. 5.1 Fault detection conceptual process.

As shown in Fig. 5.1, the origin of the fault is generally a "disturbance," an unknown and uncontrolled input acting on the observed system. Human error, both during the system operation and in previous maintenance activities, as well as manufacturing issues, assembling errors, and even the natural wear-and-tear process, can be included among these disturbances.

However, the fault detection process begins even before any type of disturbance to the system during an operational campaign; the starting point for successful detection is the recognition of the system behavior in a stable and healthy condition, that is, without any fault ongoing in the system. Defining the system expected behavior in these normal conditions is essential, as will be seen later.

Although the evolution of a fault from a disturbance does not necessarily impact the correct operation of a system immediately, it can cause a reduction or loss in its ability to perform a specific function. This partial deficiency or irregularity, intermittent or not, in fulfilling at least one of the desired functions of a system defines a "malfunction."

On the other hand, a permanent interruption of the system's ability to perform a required function within the specified limits under specified operating conditions can be understood as a "failure." Malfunctions and failures may also be referred as "defects" by some authors, encompassing both types of incapacity to perform a function (total and partial), and may not be noticed in all operating conditions of a system (Isermann, 2011; Miljković, 2011; Sartori, Amaro, De Souza, & Embiruçu, 2012).

Fig. 5.2 shows the possible developments of the fault "condition" into the events of "failure" or "malfunction." Although the evolution of the fault from a disturbance to a malfunction and a failure is represented gradually in Fig. 5.1, it is possible that under certain circumstances the evolution of a system fault will cause it to fail instantly, as, for example, when a screw fractures due to a peak overload.

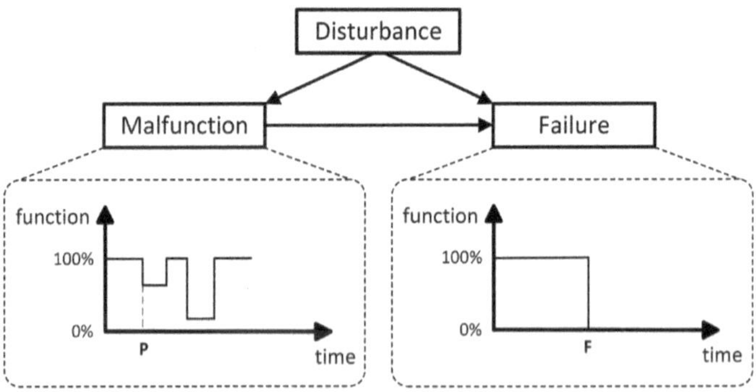

FIG. 5.2 Development of the fault condition into the events of "malfunction" and "failure."

Note that points "P" and "F" in Fig. 5.2 are the same presented on the P-F curve, in Fig. 5.1. Point P is the point at which potential failure is recognized, while point F is the point when functional failure is reached if the system continues to degrade. Only from point P, it is possible to detect, diagnose, and evaluate the system faults. Before this point, variations in the system behavior are not detectable. The operating time interval between the fault detection and the F point defines the system RUL.

In a generic and conceptual way, the detection of a fault occurs when a "deviation" is observed in the current behavior of the monitored system in relation to its normal or expected behavior for the same operational and parametric conditions, as presented in Fig. 5.1. Thus, prior collection of this information is essential.

Usually, this deviation can be called "error," when a difference is computed between the value of an output variable of the system and the real, specified, and theoretically correct value of the same variable, under the same conditions; or "residue," when the difference is computed between measurements and values based on calculations from models and/or equations (Isermann, 2006, 2011; Sartori et al., 2012).

It is important to consider that a robust fault detection process is able to distinguish a real fault from a "perturbation," that is, an input acting on the system resulting in a temporary departure from its steady-state condition. In this case, it can be said that the system in a steady-state condition has at least two attractors[a]: one, which returns the system to a steady-state condition after a perturbation, and another that eventually takes the system to a fixed point (fixed point attractor) due to the occurrence of a failure from the evolution of a fault.

In the next subsections, some aspects that influence the failure detection process in a system will be addressed.

5.1.1 Fault classification

The concept of fault was presented in the introduction to this section, but it must be considered that faults can present themselves in different ways in a system and are therefore classified accordingly.

Considering the different pieces of equipment that compose an engineering system, such as controllers, actuators, process components, and sensors, a fault can be classified according to the equipment in which it develops, as shown in Fig. 5.3.

[a]In the field of mathematics and dynamic systems, an attractor is defined as a set of numerical values toward which a system tends to evolve, from a wide variety of initial conditions of the system. In this case, the system values that are close enough to the attractor values remain close, even if slightly disturbed (Milnor, 1985).

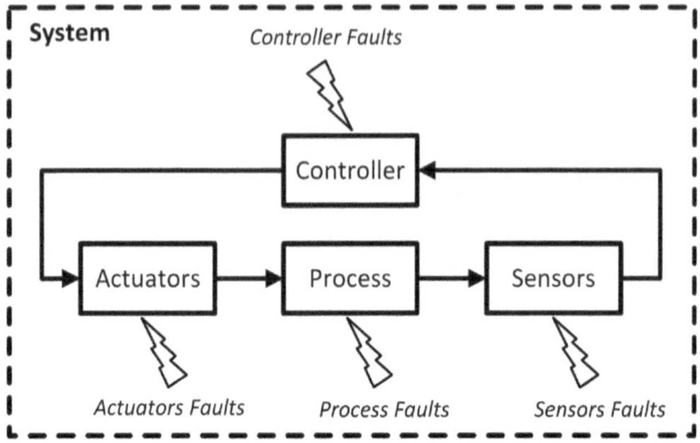

FIG. 5.3 Fault classification based on faulty component.

Although the identification of the faulty component is an assignment of the diagnosis process, which will be properly addressed in the next chapter, it is important to keep in mind that different components may require different detection methods due to the characteristics of its failure modes. Besides, the same failure mode can occur in different components of a system, affecting them, however, in a specific way. For example, faulty wiring in a sensor will have entirely different effects if the same fault develops in an actuator. Thus, classifying a system's faults according to the type of component in which the fault may develop can be extremely beneficial for the entire FDD process, as will be presented in Chapter 7 of this book.

Another way to classify faults is considering its time dependence, that is, the speed and way at which the fault progresses to a malfunction or a failure. This feature is extremely important for the FDD process: if the progression from a fault to a failure, for example, happens in a short period of time, the distance between the points P and F of the P-F curve shown in Fig. 5.1 will be very small, hampering maintenance actions that could be taken to prevent, mitigate or even avoid the consequences of the functional failure (Moubray, 1997). Thereby, faults can be classified into three types according to their time dependence, as shown in Fig. 5.4: abrupt fault (stepwise), incipient fault (drift type), or intermittent fault (Isermann, 2011; Miljković, 2011).

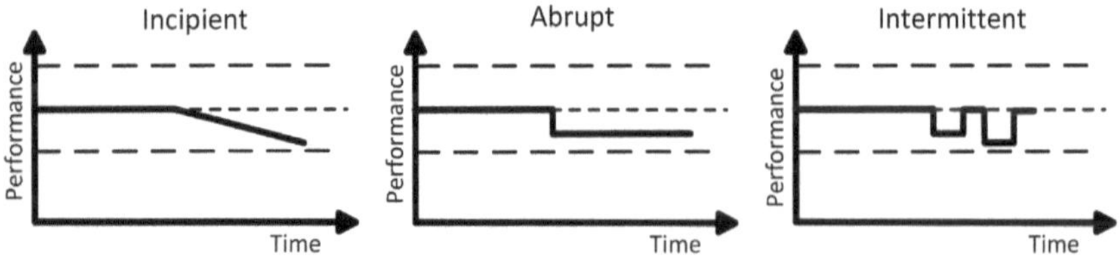

FIG. 5.4 Time dependency of faults.

Once again, recognizing the temporal characteristics of a fault is of great importance for the detection process. If, on the one hand, abrupt failures require detection methods with a speed of response directly proportional to the fault's evolution speed, on the other hand, incipient faults that evolve slowly and gradually require detection methods that are sensitive enough to detect them as soon as possible while being robust enough to distinguish them from temporary disturbances. The same logic applies to intermittent failures, easily mistaken for disturbances, and which, if not properly identified, can progress by leading the equipment to a failure.

Regarding the way faults impact the system behavior, they can be usually classified as additive or multiplicative faults. In general, additive faults affect a system as unknown inputs to the processes, changing the system output magnitude without any changes being observed in the system dynamics. Sensor faults (such as sensor bias or offsets), actuator malfunctions, or leakages in pipelines can be classified as additive faults. Multiplicative faults, on the other hand, usually have important effects on systems dynamics. These faults reveal themselves as changes in parameters or structural properties of the system (such as damping, stiffness, impedance, and capacitance). Shaft cracks, pipe clogging, and surface oxidation, for example, are classified as multiplicative faults. Fig. 5.5 presents these fault models in an illustrative way (Huang, Li, & Jeng, 2007; Isermann, 2011; Miljković, 2011).

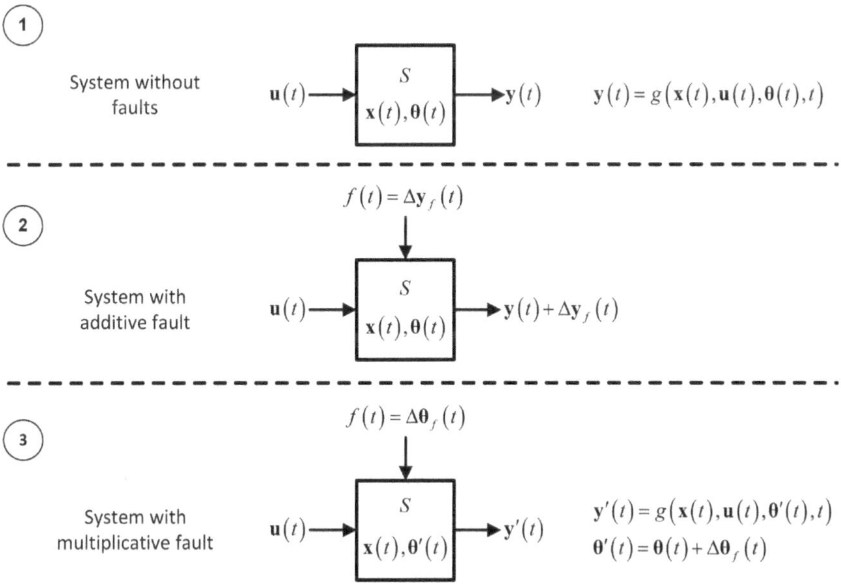

FIG. 5.5 Illustrative additive and multiplicative fault models.

The three scenarios presented in Fig. 5.5 considered different operational conditions for the system S, for which a set of parameters and states[b] are defined, represented respectively by the vectors $\boldsymbol{\theta}(t)$ and $\mathbf{x}(t)$. In the first scenario (1), the system inputs are represented by the vector $\mathbf{u}$(t) and its measured outputs by the vector $\mathbf{y}(t)$. Furthermore, $\mathbf{y}(t)$ is defined by the function $g(.)$, having as input variables the vectors $\mathbf{x}(t)$, $\mathbf{u}(t)$, and $\boldsymbol{\theta}(t)$, besides the independent variable, t. In the second scenario (2), presenting the system S with an additive fault, a set of elements of the output vector is directed affected by the fault, represented by the vector $\Delta\mathbf{y}_f(t)$ and added to the system original output, $\mathbf{y}(t)$. In the third scenario (3), which presents the system S with a multiplicative fault, the fault affects a set of system's parameters, represented by the vector $\Delta\boldsymbol{\theta}_f(t)$, changing the system output. In this case, the actual system's output is represented by the vector $\mathbf{y}'(t)$, a function of the vectors $\mathbf{x}(t)$, $\mathbf{u}(t)$, and the vector of modified parameters, $\boldsymbol{\theta}'(t)$. The latter, in turn, is the result of the vector of original parameters plus the effects of the fault.

It is important to distinguish between additive and multiplicative faults since, as previously shown, each type of fault impacts the system in a specific way. And, consequently, the fault detection process must be adapted to each scenario. In both cases, only a group of elements of the system's output vector, or parameters, is impacted by the fault. However, in the case of additive faults, it can be said that the entire FDD process begins by observing the $\Delta\mathbf{y}_f(t)$ portion of the system's output vector. In the multiplicative case, in its turn, the FDD process must observe and isolate the $\Delta\boldsymbol{\theta}_f(t)$ portion of the system's parameter vector, $\boldsymbol{\theta}'(t)$.

The reader should note at this point that in this conceptual representation of additive and multiplicative faults, multiple failure modes are not being considered under development in the system. In this way, both $\Delta\mathbf{y}_f(t)$ and $\Delta\boldsymbol{\theta}_f(t)$ are related to a single failure mode, for each case. Thus, it would not be a mistake to say that for each different failure mode that develops in the system S, there will be a single vector $\Delta\mathbf{y}_f(t)$, or $\Delta\boldsymbol{\theta}_f(t)$, which will represent the variations caused by the fault in the system. Identifying this vector means identifying the fault. However, in order to develop and adequately base this premise, some concepts must be considered, as they will be presented in the next subsection of this chapter.

[b]The state of a system is the smallest set of variables—called state variables—that allows a complete description of the system; that is, once its dynamic equations and respective inputs are known, its future states can be predicted (Maybeck, 1979).

5.1.2 Excitability, detectability, and the fault detection process

As previously presented, faults classified differently affect the same system differently. Understanding the particularities of each type of fault is of great importance to define the best fault detection strategy, since each failure mode correlates in a particular way with two major properties of a system: the system excitability and the system detectability (under a faulty condition).

The notion of excitability has its origins in biology and physiology in an attempt to understand the behavior of neuron electrical pulses and their electronic simulators, having become relevant to other fields of science such as chemical reactions, liquid crystals, optical systems (including lasers and photonic crystals), and economic processes. One of the major characteristics related to the excitability of a system is a discontinuity in its response with respect to the intensity of the disturbance inserted into the system. An excitable system remains in a steady-state condition if it is not disturbed, producing a proportional response to a "small" disturbance, but a large transient response when the disturbance exceeds a certain threshold. The system has an asymptotic state, but the return from its excited condition to its steady state will depend on the magnitude of the disturbances (Bashkirtseva, Pisarchik, Ryashko, & Ryazanova, 2016; Hjelmfelt & Ross, 1994; Wieczorek, Ashwin, Luke, & Cox, 2011).

Typically, a disturbance can be a rapid and large enough change in at least one of the system's parameters (such as an abrupt fault); a stochastic fluctuation in system variables (such as an intermittent fault); or a ramp in the form of a constant, slow, and monotonic change in at least one of the system's parameters (such as an incipient fault). That is, the different types of disturbances are totally correlated with the different types of faults in relation to their time dependence.

From a purely phenomenological point of view, it can be concluded that an excitable system must have the following properties: a steady-state condition (the normal and "healthy" operational condition), an excitability threshold above which the system leaves the steady-state condition and migrates to an excitable response (the fault condition), and a feedback mechanism that specifies the type of excitable response (the health monitoring and diagnosis process). Once the disturbance is mitigated or eliminated (by means of the approach adopted by the maintenance management team, including maintenance actions), the system returns to its stable (Wieczorek et al., 2011).

When understanding the concept of excitability, the following conclusion is reached: it is only possible to detect that a system is in a fault condition if it is excited by this fault. If a system with a given failure mode in progress has low excitability, it will be less likely to detect the associated fault in time to take some action to prevent the failure. That is, in the case of a fault that affects the behavior of a system in a minor way, it would be necessary to use a more sensitive detection method, which leads us to another concept, i.e., the detectability of a system.

Detectability is a concept that comes from the control theory and can be defined as a measure related to how well the unstable state variables of a system can be estimated from the observation of the system's outputs. On the other hand, the variables that cannot be observed are considered stable. In this way, the detectability can be understood as a slightly weaker concept (less strict) if compared to the observability concept. As the reader will notice, observability implies detectability, but not necessarily the other way around (Przyłuski, 1997).

In control theory, observability is a measure of the quality of the inference of a system's internal states from the knowledge of its external outputs, as introduced by the engineer and mathematician Rudolf E. Kalman for linear dynamic systems (Kalman, 1960, 1963). Considering a physical system modeled as a state-space representation, it is said that the system will be observable if, for any possible evolution of the components of the state vector and control over time, the current state can be estimated using only the information from the outputs (which, physically, generally correspond to the information obtained by the sensors). In other words, the behavior of the entire system can be determined from its outputs; that is, the vectors $\mathbf{x}(t)$ and $\mathbf{y}(t)$, shown in Fig. 5.5, have the same dimension. On the other hand, if the system is not completely observable, there are state trajectories that are not distinguishable just by measuring their outputs.

However, the observability of all system's state variables is not always necessary. In some applications, knowing only a few components of the state vector is sufficient. When the differences between additive and multiplicative failures were addressed in the previous subsection, an aspect of great relevance emerged as follows: the detection of a fault would theoretically be a simple observation of a set of the measured variables or system parameters affected by this fault. In other words, the idea of a "partial observation" of the output vector translates perfectly into the concept of detectability, provided that the set of observed system's output variables are related to the state variables affected by the fault in progress.

At this point, it is possible to present to the reader the correlation between excitability, detectability and the fault detection process. A faulty system may be considered detectable if, in some way, the fault changes the system response; that is, the system must be excited by the fault. Also, the system's state components that become unstable must be observable in such a way that the difference between the values of this set of variables in the "healthy" and "unhealthy" conditions can be measured (i.e., there is a measurable and nonzero deviation, as presented in Fig. 5.1).

In other words, for a given fault, the greater the excitability of a system, the faster and more intense the effects of the failure mode development on that system will be. At the same time, the greater the system's detectability, the greater the likelihood of observing the effects of the fault on the state variables sensitive to it.

That said, it can be concluded that greater excitability implies greater detectability which implies the need for a less sensitive detection method. In turn, a less sensitive method allows a reduction in the False Alarm Rate (FAR)[c] and an increase in the Fault Detection Rate (FDR),[c] making the method more robust and accurate. The opposite conclusion is also valid: less excitability implies less detectability which implies the need for a more sensitive detection method, increasing the likelihood of a higher FAR and a lower FDR, which makes the method less robust and less accurate.

As excitability and detectability concepts are defined, the next foundation for fault analysis can be addressed: the monitoring of system output variables.

5.1.3 Data monitoring

The ability to detect and assess faults in a system is not only based on the system excitability and detectability but is also grounded on the ability to monitor the system. In simple words, monitoring can be defined as a continuous and real-time task to determine the conditions of a physical system, recording information, and being aware of possible signs of abnormalities.

An "abnormal" condition can indicate future unavailability of the system when associated with a fault, or a single and temporary change in a parameter of one of its components, when related to a disturbance. Observations of changes in the system's measurements over time must be recorded within a data collection system, a database, which will allow the construction of a history of the behavior of a given variable or parameter of the system. This process is carried out by means of a network of sensors and other hardware (such as signal conditioners), installed at

[c]False Alarm Rate (FAR) and Fault Detection Rate (FDR) are two parameters conventionally used to evaluate the performance of fault detection methods. FAR is characterized by the rate at which false alarms occur; i.e., a detection event is characterized, without a fault in the system. On the other hand, FDR can be considered as the detection method success rate, i.e., the rate at which detections are characterized in the presence of a failure mode. In both cases, such parameters can be defined in the form of probabilities, being FAR the probability of a detection occurring without a failure and FDR the probability of a fault that is actually present in the system be detected (Ammiche, Kouadri, Halabi, Guichi, & Mekhilef, 2018; Chen et al., 2017; Zhang, Hao, Chen, Ding, & Peng, 2015).

strategic points of the monitored system equipment, allowing its continuous observation. From digital recorders, all sensor information can be stored, usually in a binary system, allowing the construction of the history of the behavior of a parameter or variable over time (Alvarez, 2020).

At this point, the reader will have noticed that if observability is the ability to infer the internal states of a system, then monitoring is the measurement actions involved in providing observability (and, consequently, detectability), that is, observing the quality of the system's performance over a period of time. In this way, monitoring with its tools and processes can describe the performance, health, and relevant characteristics of the internal states of a system. Monitoring refers, therefore, to the process of translating the dynamics observed in the state variables and parameters of a system into meaningful and actionable insights (Vachtsevanos et al., 2006).

In this way, sensors and measurements—the information collected with sensors, including not just system's outputs but also its inputs—form the foundation of the fault detection, diagnosis, and evaluation processes. Strategic issues that arise with sets of sensors used to collect data that will eventually lead to the online realization of diagnostic and prognostic algorithms are associated with the type, number, and location of the sensors, as well as their size, weight, cost, dynamic range, and other characteristic properties.

However, even when the system is correctly monitored (with the appropriate quantity, position, and type of sensors), if the fault does not bring significant changes to the operating status of the system it may not be detected, staying in a hidden (or invisible) fault state that can still influence the functioning state of a system (Zhou et al., 2019). Also, analyzing many data points continuously will generate large volumes of unnecessary alerts, data, and false alarms. In this case, it will be difficult to analyze the behavior of the system, unless the correct metrics are assessed and unnecessary noise is filtered.

In fact, the observability of a system (and, therefore, its detectability) depends on the complexity of the system's structure, the insightful representation of the performance metrics, and the ability of the monitoring tools to identify the correct metrics. This combination produces the insights needed to illustrate an accurate representation of internal states, despite the inherent complexity of a system. Thus, it is clear that there is a commitment between the data monitoring, detectability, and excitability that must be taken into account. Fig. 5.6 represents the fault analysis foundations in a symbolic relationship between analysis, monitoring, detectability, and excitability.

This commitment starts with the choice of sensors to be used in the system. Thus, in the next subsection, some topics on sensor sets that may be useful to the reader will be presented.

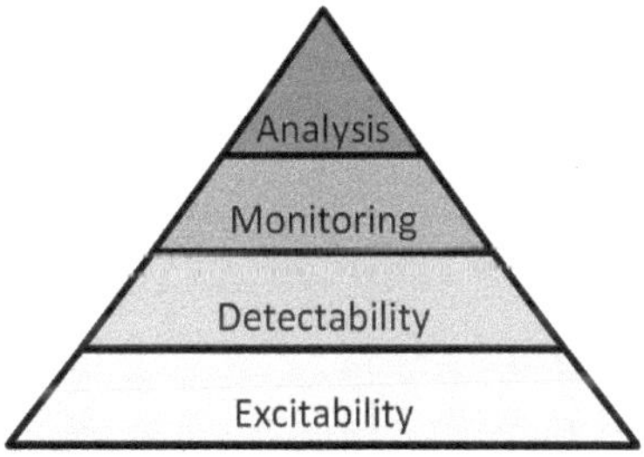

FIG. 5.6 Fault detection and analysis foundations.

5.1.3.1 Sensor sets

Generally, the sensor sets used for system monitoring are specific to the application domain for which they are intended. Based on physical phenomena or physical properties to be measured, several sensors are designed and used to collect different types of data. Acoustic emission sensors, for example, can detect transient elastic waves produced by sudden stress redistribution in a material due to physical damage or the beginning of crack propagation, making them very efficient tools for monitoring damage expansion. Accelerometers are some of the most used sensors in dynamic structures and are used to measure overall information, such as frequency and vibration modes. Strain gauges not only provide localized measurements but are also capable of capturing static or dynamic measurands at a relatively low rate of change. Optical fiber sensors can measure local deformation, are immune to electromagnetic interference, and are suitable for long-distance data transport.

Some sensors are specifically intended to measure variables that are directly related to the failure modes that can be identified in the system, such as tension meters, ultrasonic sensors, proximity devices, electrochemical fatigue sensors, and interferometers, for example. On the other hand, there are sensors for multipurpose applications, such as temperature, speed, and flow sensors, designed to monitor process variables for control and performance evaluation.

Table 5.1 contains a summary comparison of the different types of sensors used in machinery faults' detection and diagnosis (Chen, Wang, Qiao, & Chen, 2018).

Nowadays, cables are still the most used means of communicating sensor data. Transporting data via cables is a very stable technique in which users do not have to worry about bandwidth and packet loss. However, many difficulties and potential issues can be found in the displacement of cables, switches and replacements, which in turn restrict the use of this technology, especially in an industrial environment. Large extensions of cabling may be necessary for the placement of transmission networks for traditional monitoring equipment, which increases installation and maintenance costs. In addition, if the cables are damaged, the replacement process is often very complicated and may even be impossible in some cases.

TABLE 5.1 Comparison of different sensor technologies used in HMFD.

Sensor technology	Major application	Pros	Cons
Ultrasonic	Flaw detection; distance and thickness measurements	Exact and efficient	Limited by the shapes of surfaces and the density or consistency of the material
Acoustic emission	Crack growth, friction, delamination, and matrix cracking detection	Detect, locate, and characterize damage	Changes in physical properties only
Magnetic	Speed, motion, and position measurements; crack or large deformation detection with magnetic leakage	Efficient	Magnetic field required; expensive
Eddy current	Displacement, distance and position, oscillation, and vibration measurements	Useful for demanding industrial environments; high resolution and temperature stability; high-frequency response	Only for electrically conductive materials; expensive
Accelerometer	Shock, vibration, and acceleration measurements	High-frequency response; simple and reliable	Expensive for some applications
Strain gauge	Deformation and strain measurements	Cheap	Low-frequency response
Shape memory alloy	Deformation detection; active control	Fast response to temperature changes	Low-frequency response; structural fatigue and functional fatigue
Laser interferometer	Derivation or displacement measurements	High precision	Very expensive
Optical fiber	Strain, displacement, pressure, and temperature measurements	Compact; high precision; immune to electromagnetic interference	Expensive
Electromagnetic acoustic	Flaw detection; thickness measurements	Useful for automated inspection, and hot, cold, clean, or dry environments	Limited to metallic or magnetic products; low transduction efficiency

TABLE 5.1 Comparison of different sensor technologies used in HMFD—cont'd

Sensor technology	Major application	Pros	Cons
Piezoelectric lead zirconate titanate	Active sensor; vibration and crack detection	High-frequency response; cheap	Cannot be used for truly static measurements; internal resistance drop and sensitivity at elevated temperatures
Piezoelectric polymer-ceramic composites' film	Vibration and crack detection	High-frequency response; cheap	Internal resistance drop and sensitivity at elevated temperatures
Laser Doppler velocimetry	Velocity measurement	Absolute, linear with velocity and requires no precalibration; noncontact measurement	Expensive
Digital image correlation	Deformation, displacement, strain, and optical flow measurements	Ease of implementation and use; noncontact measurement	Cannot measure preexisting damage

However, with the rapid technological progress and declining costs of monitoring equipment that have been taking place in the last few decades, a new technique for collecting and transmitting signals may provide an alternative solution for cost-effective data communication in fault detection and diagnosis of engineering systems: wireless sensor networks. The advantages of this type of technology are well known: support for fixed networks is not necessary, numerous wireless sensors can be flexibly arranged, and their allocation in remote locations is much simpler for both installation and maintenance (Lynch & Loh, 2006).

Regardless of the technology applied or the type of measurement performed, basically a good sensor set must have the following capabilities and features (Chen et al., 2018):

- Good sensitivity to measured variables,
- Insensitivity to any other variables that may be encountered during the measurement,
- Physical resistance to the influence of measured variables.

On the other hand, the main issues that remain in this field are the position of the sensors and the detection of their own faults. Some techniques are available to assist in optimizing positioning and determining the type and number of sensors for a given system configuration (Yang, Liang, & Zhang, 2020). However, the techniques of detection and validation of sensor failures are delayed when compared to those used for other components of a system, such as actuators, structures, and process equipment (Vachtsevanos et al., 2006).

It is of great importance for the fault detection activity that the set of sensors used is validated and reliable in such a way that the collected data are trustworthy; that is, the sensors themselves must not be subject to a fault condition (Vachtsevanos et al., 2006). Operators, maintainers, and engineers must be able to distinguish when there is a fault or failure in the measurement system and when there is a "de facto" fault under development in the monitored system. Thus, it is interesting to note that sensors, hardware, and software for data acquisition, conditioning, processing, and storage once installed in a system become an integral part of this system. They should be considered as assets and managed as such. And it is important to keep this aspect in mind because, in an attempt to optimize the process of monitoring and diagnosing faults in a system, one may choose to install new sensors and monitoring devices, which can create negative feedback increasing the fault likelihood of the system as a whole (Isermann, 2006).

However, with the increase in automated processes in Industry 4.0 and the use of new technologies currently available, such as wireless sensor networks and smart sensors, the gap between sensors' fault detection and process fault detection has been reduced (Goelles, Schlager, & Muckenhuber, 2020; Sadough Vanini, Meskin, & Khorasani, 2014).

In fact, smart sensors are currently a hot subject. They can be very advantageous in certain applications, and today, they must be considered as a strong option when specifying a system for monitoring and processing data and signals (Alvarez, 2020; Vachtsevanos et al., 2006).

5.1.4 From measurements to symptoms

As mentioned earlier, sensors and measurements are the basis of the FDD process, regardless of the methods used. However, from the acquisition of raw data there is a flow of information that, with regard to fault detection, culminates in the symptoms. This information flow between the system's data acquisition process and a fault diagnosis is shown in Fig. 5.7.

From Fig. 5.7, it is noted that four groups of variables can be used in the detection process: system output variables, input variables, parameters, and state variables. The first two are obtained from the data acquisition process and can be called measurements in general. The other two are not obtained from data monitoring, but from model estimates. As these are specific to each system, their acquisition process is more objective. However, in

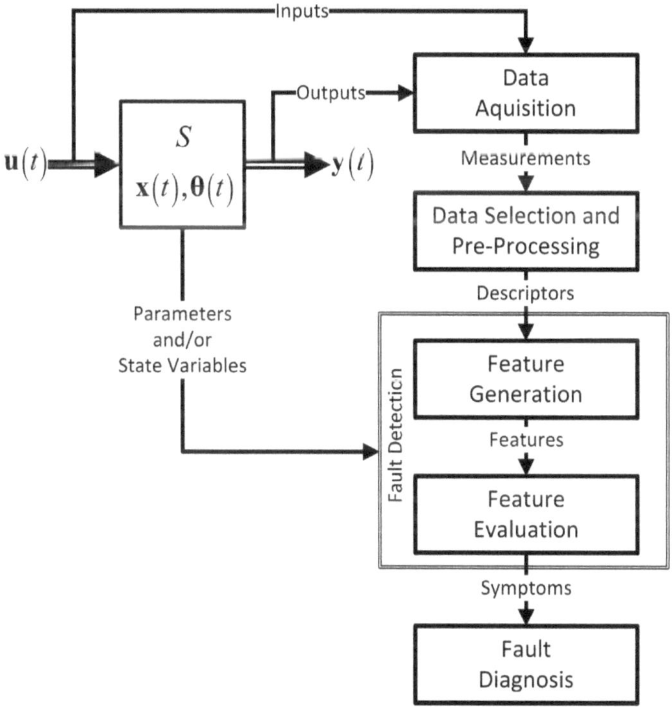

FIG. 5.7 Information flow between the system's data acquisition and fault diagnosis.

the case of measurements, as shown in the previous Data Monitoring subsection, several types of sensors can be used, generating several possibilities for collecting measurement data. Besides, according to the ISO 13379-1:2012 standard (BSI Standards Publication, 2012) all measurements used in system's condition monitory are generally suitable for the FDD process.

Table 5.2 gives, as an example, a set of several measures and parameters used for machine condition monitoring and diagnosis.

However, measurements are rarely useful in their raw form. It is important to have the collected data "noise-free" and appropriately formatted so that features (or signatures) of "healthy" or "unhealthy" conditions can be properly generated. This is one of the most significant steps in the Condition-Based Maintenance (CBM) and Prognostics Health Management (PHM) architecture. The vector of these extracted characteristics will serve as an essential input for fault diagnosis algorithms, regardless of the applied FDD method (Vachtsevanos et al., 2006).

In this way, measurements must be properly preprocessed, that is, selected, filtered, compressed, and correlated, so that the noise levels can be reduced or even removed, the volume of data can be compressed, and the extraction of useful information can be performed. The result of

TABLE 5.2 Example of measurements and parameters used for fault detection.

Performance	Mechanical	Electrical	Oil analysis, product quality, and others
Power consumption	Thermal expansion	Current	Oil analysis
Efficiency	Position	Voltage	Ferrography and wear debris analysis
Temperature	Fluid level	Resistance	Product dimensions
Pressure	Temperature	Inductance	Product physical properties
Infrared thermography	Vibration displacement	Capacitance	Product chemical properties – color – visual aspect – smell – nondestructive testing
Flow	Vibration velocity	Magnetic field	
	Vibration acceleration	Insulation resistance	
	Audible noise	Partial discharge	
	Ultrasonic waves	Infrared thermography	
	Infrared thermography		

this preprocessing may be a reduced version of the original data, preserving, however, as much as possible the characteristic properties or indicators of faults present in the system. Such data sets maintain the necessary attributes for further analysis of system faults and can be called descriptors.

Descriptors are preferred to raw measurements since they offer greater selectivity concerning faults. The more selective the descriptors are, the more selective the fault symptoms will be and, therefore, the more efficient the FDD process will become. That is, the descriptor selectivity reduces the number of failure modes' hypotheses when inferring from the symptoms to the fault. Some examples of descriptors are the first harmonic amplitude of a shaft vibration displacement, crest factor of the vibration acceleration, oil total acid number, rotational speed, rolling element bearing damage factor, and temperature gradient on an infrared thermography trace (BSI Standards Publication, 2012).

Measurements can be also selected as operational parameters in the FDD process. Operational parameters can be used both for establishing some descriptors and the operating conditions (circumstance) in which the symptoms appear. Some care is needed when considering measurements as operational parameters in order to avoid using an operational condition as a descriptor. For example, the power generated by a turbogenerator is a descriptor when monitoring and diagnosing generator faults. However, it becomes an operational condition when monitoring and diagnosing bearing faults, as it has an influence on their work, but is not descriptive in relation to their failure modes.

Descriptors and operational parameters are inputs for feature generation. Examples of feature generation can be special signal processing, state estimation, parameters' identification and estimation, or parity relations (Isermann, 2006). The generated features are the input for the feature evaluation task, which basically consists of verifying and analyzing the deviations found in the generated features, resulting in the fault symptoms. Feature generation and feature evaluation compose the fault detection process.

Symptoms are the result of the fault detection process. A symptom can be expressed in the following terms: the considered feature, time characteristic (the time constant, expressed in a quantitative or qualitative way, of the feature evolution), the type of evolution and magnitude change (also expressed in a quantitative or qualitative way), its location (where the symptom is observable on the system, considering the origin of the features), and the observation circumstance (the operation condition in which the detection occurs).

It is important to note that the generation of symptoms that may be too dependent (highly correlated) must be avoided since the assessment of dependent symptoms does not add new information to the diagnosis process and, therefore, does not allow the diagnosis to advance (BSI Standards Publication, 2012).

5.2 Fault detection methods and applications

Fault detection and diagnosis is an active research area that has been encouraging the development of a wide range of methods and heuristics (Michalski et al., 2020). Many authors have approached FDD methods, presenting review articles and comparative studies in several areas of application, including modern engineering systems.

Habibi, Howard, and Simani (2019), for example, presented a review about model-based fault detection and fault-tolerant control techniques applied to wind turbines. Wind farms can be modeled as complex engineering systems, especially when it comes to offshore units. Improving its reliability in power generation is the key issue that can turn wind power into one of the main sources of energy to respond to current world demands (Habibi et al., 2019; Tchertchian & Millet, 2014).

On the other hand, Yin, Wang, and Karimi (2014) present a robust data-based fault detection approach applied to a wind turbine benchmark. The authors consider that the main challenges of fault detection in these systems are their nonlinearity, the unknown disturbances to which wind turbines are subjected, as well as a significant noise intrinsic to the measurements.

Another industrial segment to which special attention has been given with the development and application of modern methods of FDD is the chemical industry. Chemical processes are generally risky, high-volume operations that can result in a colossal loss of life and property in the event of an accident. Thus, ensuring efficient monitoring and early FDD is essential to prevent such losses. Alauddin, Khan, Imtiaz, and Ahmed (2018) present a bibliometric review and analysis of data-driven FDD methods for process systems in which they point out that more than 73% of the articles with applications of these methods published between the years 2015 and 2017 are related to chemical processes.

Chemical processes, when analyzed in their total scope, usually take place in engineering systems, with a large amount of measurement data available and an inherent difficulty in modeling physical phenomena that can occur in such a system (Tidriri, Chatti, Verron, & Tiplica, 2018). Many of these studies use Tennessee Eastman process simulations as case studies for plant-wide control, statistical process monitoring, detecting sensor faults, and identifying network models based on data. There are many implementations of this process with different levels of detail, always being an unstable nonlinear open-loop process. It typically consists of five main process units: a two-phase reactor where an exothermic reaction occurs, a separator, a stripper, a compressor, and a mixer (Martin-Villalba, Urquia, & Shao, 2018).

An industrial benchmark of the Tennessee Eastman process is also used as a background by Yin, Ding, Haghani, Hao, and Zhang (2012) to compare the performance and illustrate the efficiency of several basic data-based methods for process monitoring and fault diagnosis. Recent developments, original ideas, implementation conditions, offline design, online computing algorithms, and the computing complexity of these methods are discussed in detail in this paper.

In recent decades, the detection and diagnosis of faults in synchronous permanent magnet machines, widely used in electric vehicle traction motors, has been also a topic of great relevance for research. Krichen, Naourez, Chaieb, and Neji (2017), for example, discuss some fault detection and diagnosis methods that have been applied to permanent magnet synchronous motors, having selected a signal-based method as the simplest and most efficient approach to be used in future works in this area.

Alzghoul, Backe, Löfstrand, Byström, and Liljedahl (2014) presented a comparison of a knowledge-based and data-driven fault detection and diagnosis methods to query the capacity and performance of low-speed, high-torque hydraulic drive systems. These components are usually applied in sectors such as mining, recycling, pulp and paper, marine, and construction, often being Maintenance Significant Items (MSI) of modern engineering systems. Since they are generally subject to great stress conditions in hostile environments, there is interest from users and manufacturers in improving the availability of these units from fault monitoring and diagnosis.

These works are just a sample of what can be found in a literature review. It is interesting to note, however, that although the purpose of the applications is basically the same, the path that each author chooses to achieve this goal can vary significantly. Thus, the objective of this section is to present in a comprehensive way the main aspects of the detection methods used in monitoring the health of engineering systems.

As presented in Chapter 3, the study of engineering systems is an emerging scientific field, distinguished by being a discipline with applications within many different areas of scientific study. A modern engineering system can be defined as a system with several interacting components, whose collective behavior cannot be simply inferred from the behavior of the individual components. The interest of these systems is how their parts give rise to the collective behaviors of the whole and how the system interacts with the environment in which it is involved (Zanoli, Astolfi, & Marczyk, 2012).

However, the complexity of such systems and their failure mechanisms make the process of detecting and locating faults very difficult. Because they have faults with dynamic characteristics, diversity of location distribution, and related epistemic uncertainties, the challenges in the detection and diagnosis of faults increase significantly. Such challenges require a significant paradigm shift that is now clearly happening in the world of maintenance and support for modern engineering systems.

Vachtsevanos et al. (2006) present an interesting analogy between the change being made possible by the intelligent diagnosis and prognosis of system faults in today's world of complex equipment maintenance and how the former coal miners used canaries to monitor the health of their mines.

Coal miners used to take canaries with them into the mines. Since canaries are more sensitive to dangerous gases than humans, if the canary died, miners knew they were in the presence of dangerous gases and thus left the mine. This "old" approach even gave rise to an idiomatic expression, "canary in the coal mine," used when something works as a warning sign for some kind of problem. The modern approach, however, would be to use available technologies and resources to continuously monitor the canary's health and obtain an early indication that the mine's air quality is starting to deteriorate. This approach provides early indication and prognosis of air quality in the mine, allowing significantly better management of workers' health, as well as keeping the canary alive.

With the development of science and technology, the functional requirements and the level of modernization of current equipment are increasingly greater, which makes systems increasingly complex, almost comparable to biological systems. This complexity becomes evident when contributing to cascade failure events in the real world, becoming a challenge for fault detection.

Early detection of anomalies in engineering systems is essential not only for preventing catastrophic cascading failures, but also for improving system performance and availability. However, the high dimensionality of the data obtained from a large number of sensors and the large amount of data obtained during the operation of the machine create real difficulties in solving the problem of fault detection and diagnosis. In this way, an appropriate method of "information compression" can be very useful for efficiently saving and processing large data. Thus, it will be a necessary respective method for fault detection and diagnosis that is able to work with this "compressed information."

The most used approaches for this task are the model-based approach and the data-based approach with many methods and algorithms developed (Vachkov, 2006). These algorithms extend from simple level and trend checking to complex algorithms based on process and signal-based models, capable of capturing the different nonlinear effects in a system caused by the presence of a fault. Some of these models take into account the deterministic behavior caused by the fault. A more general class consists of models capable of dealing with random effects associated with a fault. In addition, models can be grouped as linear or nonlinear based on their ability to handle a mathematical relationship between physical quantities. Depending on the problem in question, the most suitable models must be chosen (Jovanovic, 2014).

The model-based fault detection and diagnosis approach deals with parameter estimation and system state estimation. In this case, the parameters of the current system model are compared with those of the "healthy system" to make a decision on the system's fault state. However, based only on the fundamental principles of physics, accurate and computationally treatable modeling of the dynamics of modern engineering systems is often not feasible. One may have to rely on semiempirical models whose parameters can be identified using time series data generated from sensors and other sources of information. Thus, decision and control systems may have to be formulated based on such semiempirical models and real-time data obtained through system monitoring.

The data-based approach, on the other hand, uses certain system responses from different operating conditions and attempts to classify them into groups (clusters or rules) that clearly represent some defective conditions. Although this approach is easier to apply, as it does not need a rigorous model of the system, it has the disadvantage of being highly dependent on data. This means that it tends to have a memory related to the specific data set used for training and, therefore, lacks good generalizability (Duan, Hu, & Lin, 2017; Surana, Ray, & Chin, 2003; Vachkov, 2006).

For both approaches, considering complex industrial processes subject to different operating conditions, two aspects must be identified at first:

- Process variables can have typical nonstationary characteristics with the correlations of linear or nonlinear variables.
- Although some of the underlying characteristics of the process are nonstationary, the others can remain stationary.

Based on these acknowledgements, the different characteristics of the underlying process must be distinguished from each other. It includes two important elements, that is, how to separate the characteristics of the stationary and nonstationary process in a more comprehensive way and how to capture changes in its different distribution structures so that incipient changes can be detected in time (Zhao & Huang, 2018).

Considering the aspects discussed until now, several efficient methods of fault detection and condition monitoring have been proposed, playing an important role in the safe and faultless operation of modern engineering systems. Some of these methods will be covered next. However, first, a brief presentation will be made of how fault detection methods can be classified.

5.2.1 Fault detection methods' classification

In general, fault detection methods can be classified according to three aspects: the input data source, the feature generation method, and the algorithm data label. Depending on the bibliography considered, some details in relation to this classification may vary, but in a broad way these three covered aspects are, in fact, the most relevant to be addressed.

5.2.1.1 Fault detection and the input data source

A fundamental aspect of any approach to fault detection and diagnosis is the selection of the method input data. These input data are usually measurements of the system collected by sensors that are part of the analyzed system and provide information about the system's operations, as discussed in the previous sections of this chapter. However, the input data can also include data generated by performing experimental tests in the physical process, or from simulation data generated by computational models—first-principles simulations and modeling (Alauddin et al., 2018)—with sufficient fidelity for that purpose. In other words, one can categorize the input data for the fault detection methods into two main groups: measured data and simulated data (Khorasgani, Farahat, Ristovski, Gupta, & Biswas, 2018).

Measured data can be subdivided into two groups: field data and experimental data. The so-called field data are sensor data generated during the operation of the analyzed system. The major objective of any health monitoring system is to detect and isolate faults in real operational situations, often in real time during the failure mode development. Therefore, it is important to use field data for the design and validation of FDD solutions, regardless of the selected method. On the other hand, sensor data generated during a specific test or experimental study can be called experimental data. Such tests and experiments can be done to understand the faults' effect and evolution over time. These data can be used to design fault detection and diagnosis methods, as well as to test preexisting or already known approaches.

Simulation data are generally generated from a computational approach considering a physical model of the system, such as state-space equations, to simulate situations that may occur in real industrial conditions. Data from actual system operations under "unhealthy" conditions may not always be available, and, in addition, in many cases, system data in real "healthy" conditions are also not available for all operational conditions considered. In these cases, system models can be used to generate data for both conditions considered ("healthy" and "unhealthy"), which can then be used to design and validate fault detection and diagnosis methods, as in the case of experimental data.

Hybrid approaches can also be used to combine different data sources specially to address the lack of data in unhealthy conditions in real-world fault detection and diagnosis applications. Although field and experimental data are generally more reliable than simulation data, as models are not always completely accurate, for situations where field or experimental data are not available, simulated data can be used to enrich the data set. It is important to consider, however, that simulated data are generally generated at different levels of detail if compared to field and experimental data. If, on the one hand, in industrial systems, the number, position, sampling rate, and type of probes and sensors are not a free choice, being dependent on the design of the equipment, geometry, and operational issues, on the other hand in simulations virtually any variable in any desired sample rate can be obtained. Thus, to combine simulation data with real-world data, one can use the variables associated with the sensors in the real system and apply a resampling approach to create consistent sample rates between data from different sources.

5.2.1.2 Fault detection and algorithm data labeling

Fault detection algorithms use generated features to detect and isolate faults. According to the need for a priori information about the data, the methods can be classified into two main groups: supervised and unsupervised approaches. Supervised approaches require training data to "learn" how to classify the analyzed conditions as healthy and unhealthy. These approaches are generally applied to classification methods, such as Artificial Neural Networks (ANN) or Bayesian Networks (BN) to map operating modes' features. Unsupervised approaches, on the other hand, do not require a priori labeled data. These methods usually make the implicit assumption that normal instances are much more frequent than anomalies in the test data set and use hypothesis tests or clustering approaches to classify data groups according to normal or abnormal conditions.

Supervised and unsupervised techniques are not limited to data-driven methods. Model-based techniques can also use classifiers or clustering algorithms to map residual outputs to healthy and unhealthy conditions. When training data for normal and abnormal conditions are available, a classifier can be used to distinguish failure modes' development from steady-state operation conditions. The classifier input can be the feature set while the output is the system's operating modes. The use of classifiers improves diagnostic performance; however, fault condition data are not always available. When only healthy condition data are available, or it is possible from a model to generate data under these same conditions, the hypothesis test to compare the features' nominal values can be applied against the values actually observed, identifying whether the system is in a fault condition if tests indicate that one or more features are out of their nominal operating ranges (Khorasgani et al., 2018).

5.2.1.3 Fault detection and feature generation methods

Feature generation methods can broadly be classified into two categories: model-based and data-based. Both solutions employ system measurements for FDD. However, model-based approaches imply the rigorous development of a robust and reliable system model based on physical laws or device implementation knowledge, rather than primarily on data. Model-based techniques use, therefore, inconsistency among the measurements and the model simulations' predictions (the so-called residuals) for the feature generation.

Although these methods have been widely studied, developed, and applied in several processes, they are not easily implemented for the early detection of faults in modern engineering systems due to the difficulty of representing complexities and nonlinearities of the system in these models, which also makes the task of keeping them up to date throughout the entire life cycle of the system very laborious. Additionally, they require sound quantitative or qualitative knowledge a priori about the analyzed system. Accordingly, reliable models of these systems are not always available. And, even when these models are available, they are often incomplete and troubled by uncertainties regarding the system's behavior. This condition can lead to high FARs or low FDRs which is, in both cases, an issue (Khorasgani et al., 2018).

On the other hand, data-based methods are based on available measurements that require minimal knowledge of the process. A set of relevant attributes from the measurement data or a subset of measurements that are sensitive to the faults (generally called clusters) is necessary for the feature generation. When dealing with very noisy or irrelevant measurements in the analyzed data set, detecting and isolating faults by monitoring only the raw data become a very difficult, if not impossible, task. As part of the development of a detection and diagnosis approach, in these cases, it is necessary to devise methods to define new resources or to select a new subset of measurements that are sensitive to system failures as a resource (Alauddin et al., 2018; Khorasgani et al., 2018; Tidriri, Chatti, Verron, & Tiplica, 2016).

One major issue regarding this approach, however, is the recurring lack of historical data. In real industrial applications, sufficient training data are generally not available to determine the nominal system behavior in all operating conditions. Having access to labeled data for unhealth conditions is even more rare, expensive, and difficult. This can lead to poor performance for data-based FDD methods. In addition, the quality and scope of the data can play a critical role in the performance of the machine learning methods employed to derive models from these data (Khorasgani et al., 2018).

Both model-based and data-based methods can be classified as quantitative or qualitative methods (also known as knowledge-based or domain knowledge methods). Knowledge-based methods can also play an important role in feature selection. These methods can identify important attributes for detecting and isolating each fault. Moreover, they can help to define new features by providing critical information about the nominal behavior of each measurement regarding the data set. Signal processing techniques, a subclass of qualitative data-based methods, for example, are well developed and widely used in practice, especially considering rotating machinery faults (Adams Jr., 2001; Muszynska, 2005). Fig. 5.8 shows some subclasses of these families of methods according to the feature generation.

Associative approaches created by combining features extract from knowledge-based, model-based, and data-driven methods can also be used for FDD (Khorasgani et al., 2018). Each of these families has its relative strengths and weaknesses, being a difficult task to define which method is the most suitable for a given application. Objectively, one can define some requirements to support such a decision, such as quick detection and diagnosis ability, isolability, robustness, adaptability, and multiple fault identifiability (Venkatasubramanian, Rengaswamy, Kavuri, & Yin, 2003; Venkatasubramanian, Rengaswamy, Yin, & Kavuri, 2003; Venkatasubramanian, Rengaswamy, & Kavuri, 2003). Even so, it can be concluded that no single method has all the desirable features that a complete method must hold, boosting future developments that point to hybrid solutions for FDD, in which two or more methods are integrated, complementing each other, and overcoming the limitations of single-method approaches.

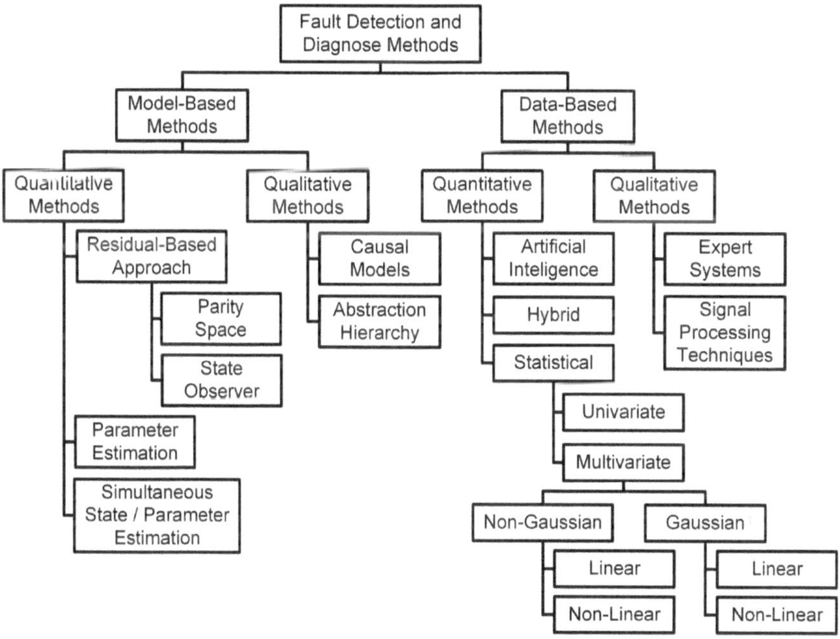

FIG. 5.8 Fault detection methods.

5.2.2 Model-based fault detection methods

The model-based approach requires the development of an explicit graphical or mathematical model, which becomes more difficult as the analyzed system complexity increases. A model validation step is needed since the model usually relies on assumptions. However, when an accurate model that integrates the physical understanding of the process can be developed, this approach outperforms data-based methods (Tidriri et al., 2016, 2018).

There are several possible strategies within this family of methods, and, therefore, the purpose of this section is to present some of the possible approaches. It should be noted, however, that none of these methods are intended to be presented in detail, for two main reasons: firstly, because the idea of this chapter is to bring a conceptual presentation in which the reader becomes aware of the main issues involving the fault detection in engineering systems and, from here, can go deeper in this subject; secondly, because there are several works in the literature presenting in a very detailed way the methods discussed here.

Model-based methods can be classified basically into quantitative and qualitative methods, as presented in Fig. 5.8. The first are methods in which the models are built from physical properties, constitutive relationships, and differential equations. Quantitative models are also more accurate and specific about a system, but require a lot of effort in building the model. The latter, on the other hand, are methods that require little or even no mathematical formulations, being usually developed based on some fundamental understanding of the process. These methods are useful in situations where quantitative models are impossible to develop due to lack of knowledge about the system or are prohibitively expensive. They can also correspond to common sense knowledge better than abstract mathematical models (Venkatasubramanian, Rengaswamy, & Kavuri, 2003).

Currently, the use of digital models has become more evident with the development of Digital Twins, that is, a digital model of a specific physical element or a process with data connections that allow convergence between the physical and virtual states at an appropriate synchronization rate. Different technologies related to Industry 4.0 such as the Internet of Things (IoT), Cloud Systems, and Big Data Analytics contribute to the creation of what is the Digital Twin of a physical process, that is, a mathematical model capable of describing the process, the product, or the service so as to carry out analyses and apply the desired approaches. These solutions can integrate artificial intelligence, machine learning, and software analysis with data collected from production plants to create digital simulation models that are updated when production process parameters or working conditions change, and developed to create a consistent improvement in efficiency, minimize failure rates, shorten development cycles, and open up new business opportunities. It is assumed that the development of a reference model becomes necessary and strategic to increase the safety levels of the staff involved in the maintenance phases. They also, through a virtual representation of physical assets, allow the application of predictive policies in the management and maintenance of the plant. The principal advantages of using digital models in the industry are as follows (Bevilacqua et al., 2020):

- Virtual models driven by real-time data are able to provide a very accurate representation of the real system and, as such, can be built to support immersive interactions between operators, maintainers, and machines;
- Digital Twin data can integrate real and simulated data, allowing the user to obtain a comprehensive description of the entire system and a deeper understanding of physical entities. Finally, digital models provide concise, intuitive, and easy-to-use platforms, such as software and mobile applications, which can be adopted by users in different areas.

In the next subsections, some of the quantitative and qualitative model-based fault detection methods most found in the literature will be covered shortly.

5.2.2.1 *Residual-based approach*

The residual-based approach represents one of the most usual quantitative techniques for detecting different faults in a system. This approach is based on a residual signal obtained by the comparison of the output measurements of a system and its corresponding duplicate, that is, a model that represents the system behavior and can carry any probable fault information. In a model-based residual generation, despite the techniques of physical redundancy, redundant measurements can be obtained through mathematical models or through the design of an appropriate observer. By adopting an adequate residual evaluation approach, the fault can be detected. The most usually found fault detection techniques based on this type of approach are the parity relation method and observer design.

The widely exploited residual assessments in the fault detection residual-based approach are simple geometric logic (e.g., limit checking), statistical scheme (e.g., statistical feature extraction), and Bayesian approaches. In the first case, an adaptive or fixed threshold value is stipulated, with which the absolute value of the residue is compared. If the value of the residue is greater than the stipulated threshold, the fault is detected; otherwise, the system is considered fault-free. It should be noted that the presence of measurement errors, disturbances, and modeling uncertainties can lead to false alarms, increasing the value of the FAR. In fact, these effects can cause the obtained residue to be greater than the threshold value established even when there is no fault in the system. In addition, when selecting a very large threshold in order to avoid false detection, some faults that do not bring significant changes to the system's behavior may not be detected, reducing the method FDR and bringing consequences for the system's availability. Therefore, residual statistical analysis can be considered to provide a more accurate assessment. The generalized likelihood ratio test, the cumulative variance index, and the use of counters are some methods of statistical analysis that can be used for this purpose (Habibi et al., 2019).

5.2.2.1.1 Parity space approach

The parity residual generation approach can be adopted for a system or some of its subsystems and components. This approach uses a mathematical model fed with the same inputs as the system. In a healthy condition, both the system and its model generate the same outputs. Therefore, by comparing these two outputs, the residue can be calculated and if it deviates considerably from zero, the fault can be detected. However, the nonlinear dynamics of engineering systems, as well as variables operating conditions, make it difficult to build a high-fidelity model that is capable of simulating the entire behavior of the system. In addition, the design of

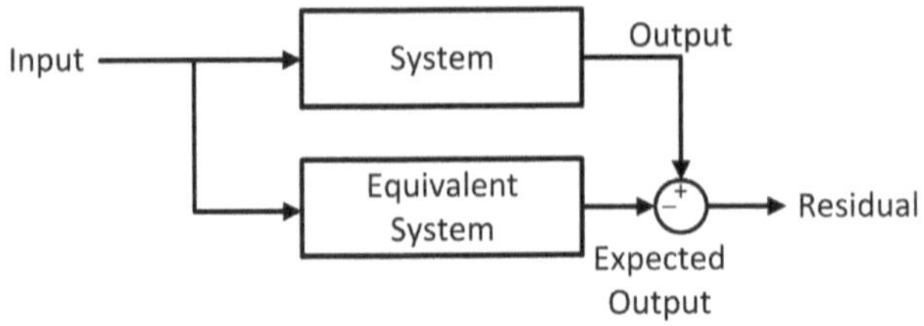

FIG. 5.9 Model-based fault detection—Parity space approach.

different models for each subsystem or component can increase the complexity of the final scheme. Also, some input variables may not be accurately measured. In the case of systems subjected to very adverse environmental conditions, which bring great influence to their behavior, as in the case of offshore wind turbines, the sensors' noise can be tremendous. In these cases, the use of filters can be considered, making the residue more robust, but still sensitive to the considered faults. The adoption of adaptive limits can also be a solution in these cases (Gao, Cecati, & Ding, 2015; Habibi et al., 2019). Fig. 5.9 presents a schematic representation of this approach.

5.2.2.1.2 State observer approach

One of the most accurate ways for residual estimation is the use of appropriate state observers. This approach depends on residual signals that are decoupled from unknown inputs, noise, or other sources of uncertainty. In addition, in the observer's design, the observation error can be feedback into the system model to reduce the model output error, adopting an appropriate observation gain. In this way, using the observed states, the residue can be obtained. The state observer is designed considering a nonfault model, and its only entrenched clause is that the model of the system under consideration (or a part of it) is observable.

The observer design of a modern engineering system can include several types of Kalman filters and Unknown Input Observer (UIO) schemes. In the case of Kalman filters as the optimal observer, the noise reduction in the system's output measurements in the observed states can be optimally guaranteed. The main disadvantage of fault detection methods based on state observers is that a linear model of the analyzed system must be used to directly apply the design theories of linear observers already developed. Sliding mode observers (Drakunov & Utkin, 1995) can bring interesting results when implemented in nonlinear models. Linear Parameter-Varying (LPV) models (Lovera, Bergamasco, & Casella, 2013) can be considered in the observer's design, bringing acceptable performance to different fault scenarios. In addition, through $H\infty$ optimization (Kwakernaak, 1993), the observer's robustness is improved against

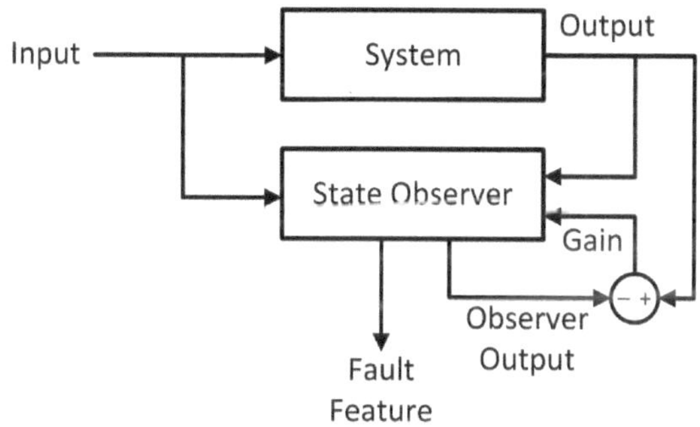

FIG. 5.10 Model-based fault detection—State estimation approach.

additive noise. Other techniques, such as the extended Kalman filter and banks of several observers, are also alternatives found in the literature (Gao et al., 2015; Habibi et al., 2019). Fig. 5.10 presents a schematic representation of this approach.

5.2.2.2 Parameter estimation approach

The parameter estimation approach is a quantitative model-based technique that depends on a suitable estimator that is used for the purpose of detecting a fault. In addition, a bank of estimators can be designed to isolate the faults that affect the different components of the analyzed system. In this way, each bank's estimator is designed for a single, specific failure. Despite the observer's approach, in the estimation techniques, the parameters related to a fault can be directly obtained. The main step in the estimator design is the selection of the parameters to be estimated. In fact, in the case of proper selection of these parameters, the need for a threshold is eliminated since it is sufficient to compare the estimated value of the parameter with its nominal value. The estimator's structure is designed as a dynamic system, as in the case of Kalman filters, or a static estimator, such as the Least Squares Filter (LSF).

It should be noted that, considering the nonlinear dynamics of some engineering systems and their different sources of disturbance, adaptive filters can be applied to obtain better results. The advantage of the adaptive estimator relates to its robustness, which can be guaranteed theoretically. This feature leads to a reduction in false and missed detection rates. Parameter estimator designs can also exploit diffuse sliding mode estimators, as found in some more recent works (Gao et al., 2015, Habibi et al., 2019). Fig. 5.11 presents a schematic representation of this approach.

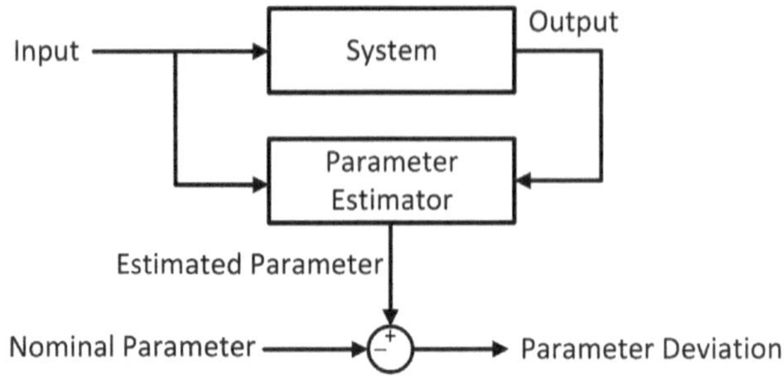

FIG. 5.11 Model-based fault detection—Parameter estimation approach.

5.2.2.3 *Causal models*

Causal models are classified as qualitative mode-based methods. Qualitative causal models are very important and used extensively to identify functional changes that have resulted in the malfunction of a process from a description of the system in conjunction with the observations made of the process during its malfunction.

Cause-effect relationships or models can be represented in the form of Signed Digraphs (SDG). A digraph is a graph with arcs directed between the nodes, the SDG in particular, a graph in which the directed arcs have a positive or negative sign attached to them. The directed arcs lead from the "cause" nodes to the "effect" nodes. Each node in the SDG corresponds to the deviation from the steady state of a variable. SDGs have nodes that represent events or variables and edges that represent the relationship between nodes. They are much more compact than truth tables, decision tables, or finite-state models.

The SDGs provide a very efficient way to represent qualitative models graphically. These digraphs have been the most widely used form of causal knowledge for the detection and diagnosis of process faults (Venkatasubramanian, Rengaswamy, & Kavuri, 2003).

5.2.2.4 *Abstraction hierarchy*

Abstraction hierarchy is another qualitative mode-based method. In abstraction hierarchies, a process system is decomposed into its process units. This decomposition seeks to draw inferences about the behavior of the entire system only from the laws that govern the behavior of its subsystems, which means, in a way, to observe the whole from the parts. However, the reverse statement is not foreseen. In a hierarchical description, a generic description of a class of process units can be represented. Governing equations that describe an entire class of process units can make assumptions about the class as a whole, but they cannot make assumptions about the behavior of particular units. Another important principle for the decomposition of systems is the principle of locality: the laws of one party cannot specifically refer to any other party. This principle allows behavior to be predicted based on local information only.

The decomposition of a process system into subsystems can be performed at several levels of abstraction and allows a general representation of the functionality of a system in terms of the input-output relationships of its units. If the level of abstraction refers to control systems, for example, then these subsystems represent several individual control loops. If the level of abstraction is the units of the system, the subsystems represent individual units. This leads to the concept of abstraction hierarchies as the structural or functional description of a system.

The structural decomposition specifies the connectivity information for a unit, representing the connectivity information for the system and its subsystems. Structural decomposition is an efficient decomposition in systems where there is a general equivalence between structure and function. On the other hand, functional decomposition specifies the output of a unit as a function of its inputs (and possibly state information), representing the means-end relationships between a system and its subsystems (Venkatasubramanian, Rengaswamy, & Kavuri, 2003).

5.2.3 Data-based fault detection methods

Data-based techniques have been developed over the past two decades and are widely used in numerous industrial sectors these days. The center idea of data-based techniques is to take full advantage of huge amounts of available data, looking for information applications in a useful way. Data-driven models built from historical data can be used for monitoring, fault diagnosis, optimization, and control, for example. Compared with model-based approaches, data-based techniques provide an efficient and alternative solution for different industrial systems that operate under several conditions, especially when an accurate model development of the analyzed system is difficult or even impossible to be built. In addition, the analysis of multiple operational conditions from models generally requires significant computational cost and requires a response time that can hamper the use of this approach in real-time applications. Another significant advantage of data-driven methods is that they are based on the available system's measurements, requiring minimal knowledge of the process (Alauddin et al., 2018; Yin, Li, Gao, & Kaynak, 2015).

In a generalized and simplified way, fault detection using data-based methods is carried out by applying a method of this class to two different groups of data, with the same dimensions and variables. Considering that one of the data groups is related to a healthy condition and the other to an unhealthy condition of a system, the comparison of the results obtained will give rise to a residue that can be used as a feature to detect the system fault. Fig. 5.12 presents a schematic representation of this generalization.

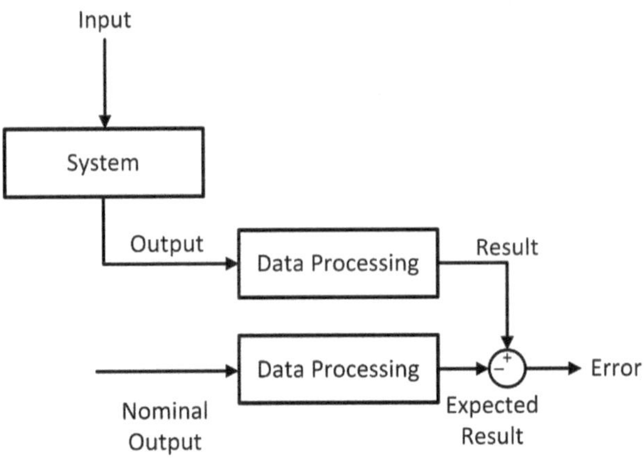

FIG. 5.12 Data-based fault detection generalized representation.

As previously presented for model-based methods, data-based methods for fault detecting can also be classified as qualitative or quantitative. For example, Expert Systems (ES) and Qualitative Trend Analysis (QTA) are two well-known qualitative techniques. In the former, data are represented as trend-setting primitives, a unique combination of signs for the signal and its derivatives (Villez, 2014). This method provides relatively accurate results but becomes computationally unsuitable for complex processes.

Quantitative methods can be classified into two categories: statistical or nonstatistical. Support Vector Machine (SVM) and ANN, for example, are the two most commonly used nonstatistical methods. These methods are also known as supervised learning methods. They are able to perform the FDD process by building learning models of labeled historical data and comparing the models with current process data. The main disadvantage of these supervised techniques, as already mentioned, is the need for a substantial number of samples of data labeled for training. On the other hand, statistical methods, such as Principal Component Analysis (PCA) and Partial Least Squares (PLS), use data projection into a smaller dimension space than the original data space to perform the fault detection process.

Statistical data-based methods can also be classified as univariate or multivariate methods. In the first case, each variable is monitored individually by limit or trend checking using conventional Statistical Process Control (SPC) charts, such as Shewhart control charts, Cumulative Sum (CUSUM) control charts and the mean Exponentially Weighted Moving Average (EWMA) control charts. Although the univariate control strategy is still dominant in the process industries, its performance deteriorates when the variables are highly correlated. In addition, they cause so-called alarm floods, which are becoming a major challenge for the safety of many process systems. In contrast, multivariate methods are based on the cumulative effects of a group of variables, taking into account the correlation between variables and being able to discern operational changes from process failures in a better way. In addition, they result in fewer features than univariate methods, avoiding alarm floods. According to Alauddin et al. (2018), multivariate methods have received more attention than univariate methods in recent years.

Techniques such as PCA and PLS have been well established, and, besides, many new techniques are emerging such as Independent Component Analysis (ICA) and BN. In the past 30 years, PCA, ANN, Fuzzy Logic (FL), BN, and PLS were the most usual methods found in the FDD techniques' literature. On the other hand, Hidden Markov's Methods (HMM), Gaussian Mixture Model (GMM), ICA, and Support Vector Machine (SVM) have gained more attention in the last 5 years. Some of these techniques will be presented in more detail next.

5.2.3.1 Principal component analysis

In the data-based fault detection methods' family, PCA is classified as a linear Gaussian multivariate statistical method and is perhaps one of the earliest, most popular, and better-established data-based approach for the past two or three decades. Besides being very versatile, PCA presents some important aspects: its efficiency in dimension reduction and latent feature extraction. Also, PCA considers correlations between the variables, rather than just the average value, which is essential in pattern analysis (Abdi & Williams, 2010; Jiang, Yan, & Huang, 2016; Yin et al., 2015).

PCA is applied in many fields, especially those where it is necessary to find patterns in multivariate data of high-dimensional problems, such as big data engineering, machine learning, medical diagnostic techniques, face recognition, and image compression. In relation to modern engineering systems, PCA (and its variations) can be very useful for detecting abnormal conditions. Considering the analysis of a data set composed of several variables, the fundamental idea is to use the method ability to synthesize while preserving as much as possible of the variation of the original data (Bosisio, Berizzi, Dinh-Duong, Fabio, & Giorgio, 2019; Jolliffe, 2002; Ma & Jiang, 2012; Sarkar, Saha, & Agrawal, 2014; Pinzón, Audivet, Torres, Alexander, & Sanjuán, 2017).

In general, PCA presents good results when applied to linear systems that run under stationary operating conditions. When evaluating FDR and FAR, PCA presents very similar results to other techniques. Also, the method generally presents some advantages: it is stable, robust, and very likely amenable to a simple computational implementation. These features can be a great advantage in the case of automated implementations (Yin et al., 2012).

Considering a data set of correlated variables in a matrix form, where lines are the observations and columns the variables, PCA uses an orthogonal transformation to convert this data set into a group of linearly uncorrelated values, the principal components. In other words, PCA transforms discrete variables into uncorrelated coefficients. These new uncorrelated coefficients can be interpreted geometrically as the projections of the observations toward the principal components (Abdi & Williams, 2010).

PCA is defined in such a way that the principal components are presented in order, as a function of its variance, from the component with the highest variance to the component with the lowest variance. The first component, with the greatest variation, contains most of the data set information. The second component, orthogonal to the first, holds most of the remaining information. The other components are estimated in the same way, successively. Since PCA focuses on variances, not disregarding covariations and correlations though, generally only the first two or three components with the greatest variance are considered in the subsequent analyses (Jolliffe, 2002).

Basically, six steps are required to apply the PCA to a data set:

- Step 1: get the data set, $\mathbf{X}_{nxm}$, that means n samples of m variables.
- Step 2: calculate the mean of each variable in the data set, as presented in Eq. (5.1).

$$\overline{x}_j = \frac{1}{n}\sum_{i=1}^{n} X_{ij} \tag{5.1}$$

- Step 3: subtract the average from each of the data set dimensions. This leads to a new matrix, **B**, composed of re-centered data with mean equals to zero, whose element related to the i-th sample and j-th variable is presented in Eq. (5.2).

$$B_{ij} = X_{ij} - \overline{x}_j \tag{5.2}$$

- Step 4: calculate the covariance matrix, $\mathbf{S_B}$, presented in Eq. (5.3), of the new matrix **B**. Since the data are m dimensional, the covariance matrix will have order m.

$$\mathbf{S_B} = \frac{1}{n-1}\mathbf{B}^T\mathbf{B} \tag{5.3}$$

- Step 5: calculate the eigenvectors, $\mathbf{v_B}$, and eigenvalues, $\lambda_\mathbf{B}$, of the covariance matrix, $\mathbf{S_B}$, as presented in Eq. (5.4).

$$\mathbf{S_B}\mathbf{v_B} = \lambda_\mathbf{B}\mathbf{v_B} \tag{5.4}$$

- Step 6: order the eigenvectors obtained from the covariance matrix by eigenvalues, highest to lowest, creating a new data set matrix arranged so that the first vector corresponds to the largest eigenvalue and the last vector corresponds to the lowest eigenvalue. This way, the components are given in order of significance.

For each principal component, all original variables have a set of corresponding eigenvectors. The higher the eigenvector component value, the more important is the variable for the component. As each eigenvector is related to an eigenvalue, there is an eigenvalue for each component. Its value is a direct indication of the component importance in the global analysis of the data set variation. The eigenvectors matrix can be called as the load matrix. The product of the original data set by the load matrix originates the score matrix, which is the projection of the original data set into the principal component subspace (Yang, 2004).

A PCA model can be established based on the historical data collected when the analyzed system is at a steady-state condition. Then, two statistical measures are generally used to monitor the process: T^2 score and the Squared Prediction Error (SPE, also known as the Q statistic). The control limits of these two statistical measures are calculated based on the reference historical data, and new observations can be compared to these control limits to detect any abnormal events. The detection by either one of these two methods (T^2 score or Q statistic) is enough to declare that a sample is possibly out-of-control. That is, in basic PCA applications, just monitoring these two values, rather than several process parameters, enables faults' detection.

Due to its simplicity of implementation, robustness, and proven positive results as found in several works in the literature, the PCA will be the method used in Chapter 9 of this book for the detection of failures in the example cases.

5.2.3.2 *Partial least squares*

Like PCA, PLS is also classified as a linear Gaussian multivariate statistical method, basic multivariate statistical method, widely used for model building and fault detection. Using the trained offline correlation model and online measurements, the prediction of latent faults in a system can be performed. For the purposes of monitoring the process, PLS is able to detect faults that occurred at the input variables of the process by using the information contained in the input-output correlation. PLS generates a set of orthogonal components that clarifies the input and output variables and provides predictive dependence between them. With appropriate test statistics to monitor the mentioned subspaces, successful process monitoring can be achieved accordingly. Like PCA, T^2 score and SPE statistical tests are usually applied for online monitoring. However, for the successful application of PLS-based approaches, the input and output variables must also follow the Gauss multivariate distribution. In effect, PLS analysis appends the operation of PCA with Canonical Correlation Analysis (CCA) and Multivariate Linear Regression (MLR).

Although considered an effective technique, PLS is not suitable for applications in nonlinear, non-Gaussian, or dynamic systems. Thus, dynamic PLS can be used as an alternative to deal with the time-varying nature of these systems. In addition, Recursive Partial Least-Squares (RPLS) algorithms for online process modeling to adapt process changes and offline modeling to handle a large number of data samples can be used as well. PLS does not perform well when the input space has a significant orthogonal variation to the output space. In this case, Orthogonal Projections to Latent Structures (OPLS) that removes variation descriptor variables that are not correlated to property variables can be used. Other alternative approaches are Total Projection to Latent Structure (TPLS), Multiway Partial Least-Squares (MPLS), Partial Least-Squares Structural Equation Modeling (PLS-SEM), and prescribed Multiway Interval Partial Least-Squares (MIPLS) (Alauddin et al., 2018; Yin, Ding, Xie, & Luo, 2014).

5.2.3.3 Independent component analysis

ICA is a linear non-Gaussian multivariate statistical method, therefore being considered an optimal method for non-Gaussian data which are frequently encountered in process systems. It is based on Higher-Order Statistics (HOS) such as kurtosis and negative entropy. Unlike PCA, ICA vectors are not orthogonal, and each component is equally important. ICA can be achieved by three approaches: perform PCA and subsequently fix the remaining degrees of freedom in HOS, work directly with the HOS, or exploit a combination of second-order and higher-order statistics.

Several algorithms have been developed on the basis of the above approaches. For example, the Fast-ICA algorithm is based on and projection to the nonorthogonal weight vectors. In this case, the optimal weight vectors are iteratively determined by maximizing kurtosis. Fast-ICA algorithms are parallel, distributed, computationally simple, and cubic convergent. However, being based on the Newton-Raphson method, Fast-ICA can get trapped at local minima. The other well-known ICA algorithms include Simultaneous Third-Order Tensor Diagonalization (STOTD), Joint Approximate Diagonalization of Eigenmatrices (JADE), Higher-Order Eigenvalue Decomposition (HOEVD), Maximal Diagonality (MD), and Maximum Likelihood (ML).

A major drawback of the conventional ICA approach, however, is that the independent components cannot be ranked with respect to the amount of variance explained by them, as in the PCA approach. In addition, ICA is limited to the linear systems and uses kernel-based mapping for dealing with nonlinear systems. However, it becomes computationally expensive with the increasing number of samples. Many researchers came up with modifications in the existing algorithm to deal with these limitations (Alauddin et al., 2018).

5.2.3.4 *Gaussian mixture model*

Gaussian Mixture Model (GMM) is one of the more recent algorithms to deal with non-Gaussian data, being classified as a linear non-Gaussian multivariate statistical method. It is a statistical method based on the weighted sum of probability density functions of multiple Gaussian distributions. The number of Gaussian components (along with the statistical parameters) is approximated by the Expectation-Maximization (EM) algorithm which is composed of two steps: the Expectation step (E-step) and a Maximization step (M-step). The E-step determines the posterior probability, given all the parameters and the prior probabilities, while M-step finds all the parameters. The maximum likelihood solution is obtained by repeating the E-step and the M-step in an iterative procedure.

GMM can also be based on Bayesian Information Criteria (BIC). In this case, a GMM is built from the normal operating data of a system with multiple steady states. The optimal number of Gaussian components is determined by iterative estimation. Then, a contribution index is calculated for each Gaussian component and, finally, the major faulty variables are identified by integrating the multiple local contribution indices into a single global contribution index by using Bayesian inference (Alauddin et al., 2018).

5.2.4 Neural networks and fuzzy inference system approaches

Neural networks and fuzzy inference systems can be classified as artificial intelligence quantitative data-based methods and can be applied in engineering systems for FDD being generally divided into two different approaches: the input-output representation and the fault feature generation, as will be presented next (Habibi et al., 2019).

5.2.4.1 *Input-output representation*

Neural networks can provide one of the best tools to represent the nonlinear and partially known behavior of dynamic engineering systems. In this approach, the designed neural network is fed with the actual and the estimated inputs to generate the redundant outputs. Fig. 5.13 presents a schematic representation of this approach.

The multilayer perceptron networks and radial basis function neural networks are the most usually adopted structures for FDD purposes considering this family of methods. The main step in this approach is the offline training to tune the neuron weights to the optimal ones.

The adaptive neuro-fuzzy interface system is an online fast learning adaptive training approach, which takes advantage of the neural network robustness, the learning and training capabilities, and the fuzzy inference system interpretability. In both offline and online schemes, a properly

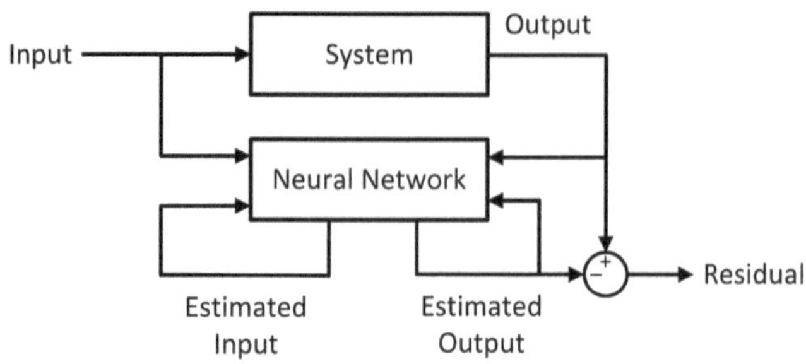

FIG. 5.13 Neural networks input-output representation approach.

large data set should be available as a priori knowledge to train the network in the fault-free case. Accordingly, a data-driven learning scheme can also be proposed (Habibi et al., 2019).

5.2.4.2 *Fault feature generation*

In contrast to the input-output approach, in the fault feature generation method the fault information can be directly extracted or inferred, relying on the design of an accurate a priori knowledge-based network, such as an Adaptive Neuro-Fuzzy Inference System (ANFIS) or a Fuzzy Inference System (FIS), as presented in Fig. 5.14. Consequently, the expert knowledge must be included in the design, whether as the numerical rules or fuzzy "if-then" linguistic rules (also known as classifiers). One of the advantages of fuzzy logic and fuzzy membership representation is that measurements' uncertainties can be directly applied (Habibi et al., 2019).

5.3 Dealing with uncertainties

Process perturbations, measurement noise, model nonlinearities, and other sources of uncertainty, such as parametric and epistemic uncertainty, make fault detection a challenging activity that is further puzzled by the increasing complexity of industrial systems (Paulson, Raimondo, Findeisen, Braatz, & Streif, 2014).

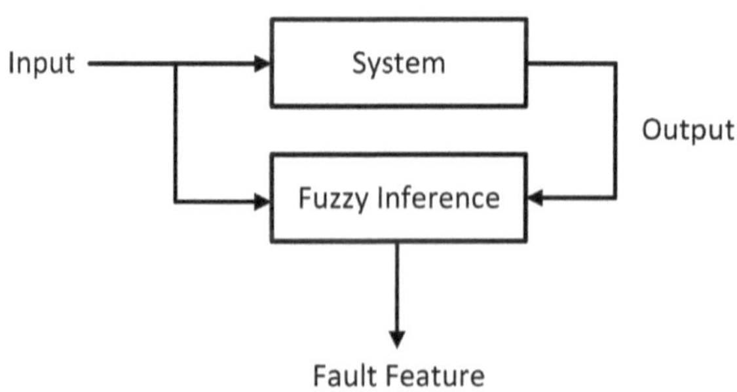

FIG. 5.14 Fuzzy fault feature generation approach.

Some of these uncertainty sources will always be present, regardless of the fault detection method applied or the system analyzed. Disturbances, transient conditions, and noise in the measured signals may vary in magnitude depending on the system, but must always be taken into account. Emergent properties and nonlinearities will also be constant in engineering systems. In these cases, the designed detection method must be robust enough not to be influenced by such uncertainties, while remaining sensitive enough to detect faults in advance. If this precaution is not properly taken, the effect of such uncertainties will spread to the fault symptoms, impairing not only their detection but also its diagnosis and evaluation. If the detection method is too sensitive, there will be a high rate of false alarms. On the other hand, if the sensitivity of the method is very low, the FDD capacity will be impaired, and if the failure mode in progress is detected, it will probably be in a very advanced condition, not allowing remediation actions to be taken in time, leading the system to fail.

In the specific case of detection methods that require training data, there will be an opposite effect to that described in the case of excessive noise levels in the measurement data. In these cases, such methods will have their sensitivity reduced since small variations will not be noticed during training. On the other hand, model-based quantitative methods, such as state observers and Kalman filters, may not be so impaired since the inclusion of measurement noise is part of its design. In such cases, however, the results will be adversely affected due to modeling (parametric and epistemic) uncertainties. For the best efficiency of these methods, it is important that the system is well known so the computational model may be as accurate as possible. Such a condition, however, becomes more difficult in the case of modern engineering systems.

In all cases, however, it is interesting to assess the sensitivity of the designed method to variations in system inputs as well as to the uncertainties inherent to it. In fact, uncertainty and sensitivity analysis are essential parts of analyses for modern engineering systems. Some methods are found in the literature capable of accomplishing this task. For example, sampling-based methods, including differential analysis, response surface methodology, Monte Carlo analysis, and variance decomposition procedures, have become very popular for uncertainty and sensitivity analysis. These methods involve the generation and exploration of a mapping from uncertain analysis inputs to uncertain analysis results (Helton & Davis, 2002).

Uncertainty analysis refers to the determination of the uncertainty in analysis results that derives from uncertainty in analysis inputs. In contrast, sensitivity analysis refers to the determination of the contributions of individual uncertain analysis inputs to the uncertainty in analysis results.

The uncertainties found in a system can be basically classified into two groups: epistemic uncertainty, also known as the state of knowledge, subjective, reducible, and type B uncertainty; and aleatory uncertainty, also designed as variability, stochastic, irreducible, and type A uncertainty. If, on the one hand, epistemic uncertainties result from a lack of knowledge about the appropriate value to be used for a quantity that is presumed to have a fixed value in the context of a particular analysis, on the other hand, random uncertainty arises from inherent randomness in the behavior of the system under study. In the conceptual and computational analysis organization, these two uncertainties are generally considered to be distinct (Helton, Johnson, Sallaberry, & Storlie, 2006).

References

Abdi, H., & Williams, L. J. (2010). Principal component analysis. *Wiley Interdisciplinary Reviews: Computational Statistics*, *2*(4), 433–459. https://doi.org/10.1002/wics.101.

Adams, M. L., Jr. (2001). *Rotating machinery vibration: From analysis to troubleshooting*. Marcel Dekker, Inc.

Alauddin, M., Khan, F., Imtiaz, S., & Ahmed, S. (2018). A bibliometric review and analysis of data-driven fault detection and diagnosis methods for process systems. *Industrial and Engineering Chemistry Research*, *57*(32), 10719–10735. https://doi.org/10.1021/acs.iecr.8b00936.

Alvarez, G. P. (2020). Real-time fault detection and diagnosis using intelligent monitoring and supervision systems. In F. P. García Márquez (Ed.), *Fault detection, diagnosis and prognosis*. IntechOpen. https://doi.org/10.5772/intechopen.90158.

Alzghoul, A., Backe, B., Löfstrand, M., Byström, A., & Liljedahl, B. (2014). Comparing a knowledge-based and a data-driven method in querying data streams for system fault detection: A hydraulic drive system application. *Computers in Industry*, *65*(8), 1126–1135. https://doi.org/10.1016/j.compind.2014.06.003.

Ammiche, M., Kouadri, A., Halabi, L. M., Guichi, A., & Mekhilef, S. (2018). Fault detection in a grid-connected photovoltaic system using adaptive thresholding method. *Solar Energy*, *174*, 762–769. https://doi.org/10.1016/j.solener.2018.09.024.

Bashkirtseva, I., Pisarchik, A., Ryashko, L., & Ryazanova, T. (2016). Excitability and complex mixed-mode oscillations in stochastic business cycle model. *Advances in Complex Systems*, *19*(1–2), 1–16. https://doi.org/10.1142/S0219525915500277.

Bevilacqua, M., Bottani, E., Ciarapica, F. E., Costantino, F., Donato, L. D., Ferraro, A., et al. (2020). Digital twin reference model development to prevent operators' risk in process plants. *Sustainability (Switzerland)*, *12*(3), 1–17. https://doi.org/10.3390/su12031088.

Bosisio, A., Berizzi, A., Dinh-Duong, L., Fabio, B., & Giorgio, G. (2019). Improving DTR assessment by means of PCA applied to wind data. *Electric Power Systems Research*, *172*, 193–200. https://doi.org/10.1016/j.epsr.2019.02.028.

BSI Standards Publication (2012). *Condition monitoring and diagnostics of machines—Data interpretation and diagnostics techniques—Part 1: General guidelines*. BSI Standards Publication BS ISO.

Chen, X., Wang, S., Qiao, B., & Chen, Q. (2018). Basic research on machinery fault diagnostics: Past, present, and future trends. *Frontiers of Mechanical Engineering*, *13*(2), 264–291. https://doi.org/10.1007/s11465-018-0472-3.

Chen, Z., Zhang, K., Shardt, Y. A. W., Ding, S. X., Yang, X., Yang, C., et al. (2017). Comparison of two basic statistics for fault detection and process monitoring. *IFAC-PapersOnLine*, *50*(1), 14776–14781. https://doi.org/10.1016/j.ifacol.2017.08.2586.

Drakunov, S., & Utkin, V. (1995). Sliding mode observers. Tutorial. In *Proceedings of 1995 34th IEEE conference on decision and control, New Orleans, LA* (pp. 3376–3378). https://doi.org/10.1016/0005-1098(93)90122-A.

Duan, R., Hu, L., & Lin, Y. (2017). Fault diagnosis for complex systems based on dynamic evidential network and multi-attribute decision making with interval numbers. *Eksploatacja i Niezawodnosc—Maintenance and Reliability, 19*(4), 580 589. https://doi.org/10.17531/ein.2017.4.12.

Gao, Z., Cecati, C., & Ding, S. X. (2015). A survey of fault diagnosis and fault-tolerant techniques—Part I: Fault diagnosis with model-based and signal-based approaches. *IEEE Transactions on Industrial Electronics, 62*(6), 3757–3767. https://doi.org/10.1109/TIE.2015.2417501.

Goelles, T., Schlager, B., & Muckenhuber, S. (2020). Fault detection, isolation, identification and recovery (FDIIR) methods for automotive perception sensors including a detailed literature survey for lidar. *Sensors (Switzerland), 20*(13), 1–21. https://doi.org/10.3390/s20133662.

Habibi, H., Howard, I., & Simani, S. (2019). Reliability improvement of wind turbine power generation using model-based fault detection and fault tolerant control: A review. *Renewable Energy, 135*, 877–896. https://doi.org/10.1016/j.renene.2018.12.066.

Helton, J. C., & Davis, F. J. (2002). Illustration of sampling-based methods for uncertainty and sensitivity analysis. *Risk Analysis, 22*(3), 591–622. https://doi.org/10.1111/0272-4332.00041.

Helton, J. C., Johnson, J. D., Sallaberry, C. J., & Storlie, C. B. (2006). Survey of sampling-based methods for uncertainty and sensitivity analysis. *Reliability Engineering and System Safety, 91*(10–11), 1175–1209. https://doi.org/10.1016/j.ress.2005.11.017.

Hjelmfelt, A., & Ross, J. (1994). Pattern recognition, chaos, and multiplicity in neural networks of excitable systems. *Proceedings of the National Academy of Sciences of the United States of America, 91*(1), 63–67. https://doi.org/10.1073/pnas.91.1.63.

Huang, H.-P., Li, C.-C., & Jeng, J.-C. (2007). Multiple multiplicative fault diagnosis for dynamic processes via parameter similarity measures. *Industrial and Engineering Chemistry Research, 46*(13), 4517–4530. https://doi.org/10.1021/ie061118c.

Isermann, R. (2006). *Fault-diagnosis systems: An introduction from fault detection to fault tolerance* (pp. 1–475). Berlin Heidelberg: Springer. https://doi.org/10.1007/3-540-30368-5.

Isermann, R. (2011). *Fault-diagnosis applications—Model-based condition monitoring: Actuators, drives, machinery, plants, sensors, and fault-tolerant systems*. Springer Nature. https://doi.org/10.1007/978-3-642-12767-0.

Jiang, Q., Yan, X., & Huang, B. (2016). Performance-driven distributed PCA process monitoring based on fault-relevant variable selection and Bayesian inference. *IEEE Transactions on Industrial Electronics, 63*(1), 377–386. https://doi.org/10.1109/TIE.2015.2466557.

Jolliffe, I. T. (2002). *Principal component analysis*. Springer. https://doi.org/10.1007/b98835.

Jovanovic, D. P. (2014). *Fault detection in complex and distributed systems*. University of Queensland.

Kalman, R. E. (1960). On the general theory of control systems. *IFAC Proceedings Volumes, 1*(1), 491–502. https://doi.org/10.1016/S1474-6670(17)70094-8.

Kalman, R. E. (1963). Mathematical description of linear dynamical systems. *Journal of the Society for Industrial and Applied Mathematics Series A Control, 1*(2), 152–192. https://doi.org/10.1137/0301010.

Khorasgani, H., Farahat, A., Ristovski, K., Gupta, C., & Biswas, G. (2018). A framework for unifying model-based and data-driven fault diagnosis. *Proceedings of the Annual Conference of the PHM Society, 10*(1), 1–10. https://www.phmpapers.org/index.php/phmconf/article/view/530.

Krichen, M., Naourez, B., Chaieb, M., & Neji, R. (2017). Fault detection and diagnosis methods in permanent magnet synchronous machines: A review. In T. Bouktir, & R. Neji (Eds.), *Proceedings of the international conference on recent advances in electrical systems—ICRAES 2017* (pp. 229–237). Hammamet, Tunisia: IEEE. https://www.researchgate.net/profile/Manel_Krichen/publication/328610743_Fault_Detection_and_Diagnosis_Methods_in_

Permanent_Magnet_Synchronous_Machin_Detection-and-Diagnosis-Methods-in-Permanent-Magnet-Synch.

Kwakernaak, H. (1993). Robust control and H∞-optimization—Tutorial paper. *Automatica, 29*(2), 255–273. https://doi.org/10.1016/0005-1098(93)90122-A.

Lovera, M., Bergamasco, M., & Casella, F. (2013). LPV modelling and identification: An overview. *Lecture Notes in Control and Information Sciences, 437*, 3–24. https://doi.org/10.1007/978-3-642-36110-4_1.

Lynch, J. P., & Loh, K. J. (2006). A summary review of wireless sensors and sensor networks for structural health monitoring. *The Shock and Vibration Digest, 38*(2), 91–128. https://doi.org/10.1177/0583102406061499.

Ma, J., & Jiang, J. (2012). Detection and identification of faults in NPP instruments using kernel principal component analysis. *Journal of Engineering for Gas Turbines and Power, 134*(3), 1–6. https://doi.org/10.1115/1.4004596.

Martin-Villalba, C., Urquia, A., & Shao, G. (2018). Implementations of the Tennessee Eastman process in Modelica. *IFAC-PapersOnLine, 51*(2), 619–624. https://doi.org/10.1016/j.ifacol.2018.03.105.

Maybeck, P. S. (1979). *Stochastic models, estimation and control* (Vol. 1). Academic Press, Inc.

Michalski, M. A. C., Melani, A. H. A., Silva, R. F., Souza, G. F. M., Nabeta, S. I., & Hamaji, F. H. (2020). Applying moving window principal component analysis (MWPCA) for fault detection in hydrogenerator. In P. Baraldi, F. Di Maio, & E. Zio (Eds.), *Proceedings of the 30th European Safety and Reliability conference and the 15th Probabilistic Safety Assessment and Management conference—ESREL2020-PSAM15* (pp. 1–8). Research Publishing.

Miljković, D. (2011). Fault detection methods: A literature survey. In *MIPRO 2011—Proceedings of 34th international convention on information and communication technology, electronics and microelectronics—Proceedings* (pp. 750–755).

Milnor, J. (1985). On the concept of attractor. *Communications in Mathematical Physics, 99*(2), 177–195. https://doi.org/10.1007/BF01212280.

Moubray, J. (1997). *Reliability centered maintenance* (2nd ed.). Industrial Press.

Muszynska, A. (2005). Rotordynamics. In L. L. Faulkner (Ed.), *Mechanical engineering* (Issue 38). Taylor & Francis.

Pagiatakis, G., Dritsas, L., Chatzarakis, G., Todorov, G., & Stoev, B. (2017). Introducing concepts and methodologies of fault detection into electrical engineering education: The induction machine example. In *Proceedings of IEEE global engineering education conference, EDUCON* (pp. 381–388). https://doi.org/10.1109/EDUCON.2017.7942876.

Paulson, J. A., Raimondo, D. M., Findeisen, R., Braatz, R. D., & Streif, S. (2014). Guaranteed active fault diagnosis for uncertain nonlinear systems. In *Proceedings of 2014 European control conference (ECC), Strasbourg, France* (pp. 926–931). https://doi.org/10.1109/ECC.2014.6862510.

Pinzón, H., Audivet, C., Torres, M., Alexander, J., & Sanjuán, M. (2017). A comparative assessment on static and dynamic PCA for fault detection in natural gas transmission systems. In *Proceedings of ASME 2017 11th international conference on energy sustainability—ES2017, Charlotte* (pp. 1–8).

Przyłuski, K. M. (1997). A note on detectability, observability and stability of implicit linear discrete–time systems. *IFAC Proceedings Volumes, 30*(6), 399–402. https://doi.org/10.1016/S1474-6670(17)43397-0.

Sadough Vanini, Z. N., Meskin, N., & Khorasani, K. (2014). Multiple-model sensor and components fault diagnosis in gas turbine engines using autoassociative neural networks. *Journal of Engineering for Gas Turbines and Power, 136*(9), 1–16. https://doi.org/10.1115/1.4027215.

Sarkar, J., Saha, S., & Agrawal, S. (2014). An efficient use of principal component analysis in workload characterization—A study. *AASRI Procedia, 8*, 68–74. https://doi.org/10.1016/j.aasri.2014.08.012. Secs.

Sartori, I., Amaro, C. A., De Souza, M. B., & Embiruçu, M. (2012). Detecção, diagnóstico e correção de falhas: Uma proposição consistente de definições e terminologias. *Ciencia y Engenharia/Science and Engineering Journal, 21*(2), 41–53. https://doi.org/10.14393/19834071.2012.13183.

Surana, A., Ray, A., & Chin, S. (2003). Anomaly detection in complex systems. *IFAC Proceedings Volumes (IFAC—Papers Online), 36*(5), 1119–1124. https://doi.org/10.1016/S1474-6670(17)36643-0.

Tchertchian, N., & Millet, D. (2014). Eco-maintenance for complex systems: Application on system of renewable energy production. In *Proceedings of 3rd international symposium on environmental friendly energies and applications (EFEA)* (pp. 1–7). https://doi.org/10.1109/EFEA.2014.7059988.

Tidriri, K., Chatti, N., Verron, S., & Tiplica, T. (2016). Bridging data-driven and model-based approaches for process fault diagnosis and health monitoring: A review of researches and future challenges. *Annual Reviews in Control, 42*, 63–81. https://doi.org/10.1016/j.arcontrol.2016.09.008.

Tidriri, K., Chatti, N., Verron, S., & Tiplica, T. (2018). Model-based fault detection and diagnosis of complex chemical processes: A case study of the Tennessee Eastman process. *Proceedings of the Institution of Mechanical Engineers. Part I: Journal of Systems and Control Engineering, 232*(6), 742–760. https://doi.org/10.1177/0959651818764510.

Vachkov, G. (2006). Fault diagnosis of complex systems based on modular knowledge base and information compression. In *ECMS 2006—Proceedings 20th European conference on modelling and simulation, Bonn, Germany* (p. 6).

Vachtsevanos, G., Lewis, F., Roemer, M., Hess, A., & Wu, B. (2006). *Intelligent fault diagnosis and prognosis for engineering systems* (pp. 1–434). John Wiley and Sons. https://doi.org/10.1002/9780470117842.

Venkatasubramanian, V., Rengaswamy, R., & Kavuri, S. N. (2003). A review of process fault detection and diagnosis. Part II: Qualitative models and search strategies. *Computers and Chemical Engineering, 27*(3), 313–326. https://doi.org/10.1016/S0098-1354(02)00161-8.

Venkatasubramanian, V., Rengaswamy, R., Kavuri, S.N., & Yin, K. (2003). A review of process fault detection and diagnosis. Part III: Process history based methods. *Computers and Chemical Engineering, 27*(3), 327–346. https://doi.org/10.1016/S0098-1354(02)00162-X.

Venkatasubramanian, V., Rengaswamy, R., Yin, K., & Kavuri, S. N. (2003). A review of process fault detection and diagnosis. Part I: Quantitative model-based methods. *Computers and Chemical Engineering, 27*(3), 293–311. https://doi.org/10.1016/S0098-1354(02)00160-6.

Villez, K. (2014). Qualitative trend analysis for process monitoring and supervision based on likelihood optimization: State-of-the-art and current limitations. *IFAC Proceedings Volumes*, 7140–7145. https://doi.org/10.3182/20140824-6-za-1003.01596.

Wieczorek, S., Ashwin, P., Luke, C. M., & Cox, P. M. (2011). Excitability in ramped systems: The compost-bomb instability. *Proceedings of the Royal Society A: Mathematical, Physical and Engineering Sciences, 467*(2129), 1243–1269. https://doi.org/10.1098/rspa.2010.0485.

Yang, C., Liang, K., & Zhang, X. (2020). Strategy for sensor number determination and placement optimization with incomplete information based on interval possibility model and clustering avoidance distribution index. *Computer Methods in Applied Mechanics and Engineering, 366*(113042), 1–21. https://doi.org/10.1016/j.cma.2020.113042.

Yang, Q. (2004). *Model-based and data driven fault diagnosis methods with applications to process monitoring*. Case Western Reserve University.

Yin, S., Ding, S. X., Haghani, A., Hao, H., & Zhang, P. (2012). A comparison study of basic data-driven fault diagnosis and process monitoring methods on the benchmark Tennessee Eastman process. *Journal of Process Control, 22*(9), 1567–1581. https://doi.org/10.1016/j.jprocont.2012.06.009.

Yin, S., Ding, S. X., Xie, X., & Luo, H. (2014). A review on basic data-driven approaches for industrial process monitoring. *IEEE Transactions on Industrial Electronics, 61*(11), 6418–6428. https://doi.org/10.1109/TIE.2014.2301773.

Yin, S., Li, X., Gao, H., & Kaynak, O. (2015). Data-based techniques focused on modern industry: An overview. *IEEE Transactions on Industrial Electronics*, *62*(1), 657–667. https://doi.org/10.1109/TIE.2014.2308133.

Yin, S., Wang, G., & Karimi, H. R. (2014). Data-driven design of robust fault detection system for wind turbines. *Mechatronics*, *24*(4), 298–306. https://doi.org/10.1016/j.mechatronics.2013.11.009.

Zanoli, S. M., Astolfi, G., & Marczyk, J. (2012). Complexity-based methodology for fault diagnosis: Application on a centrifugal machine. In *Proceedings of IFAC conference on analysis and control of chaotic systems—The international federation of automatic control June, Cancún, México* (pp. 51–56). https://doi.org/10.3182/20120620-3-MX-3012.00060.

Zhang, K., Hao, H., Chen, Z., Ding, S. X., & Peng, K. (2015). A comparison and evaluation of key performance indicator-based multivariate statistics process monitoring approaches. *Journal of Process Control*, *33*, 112–126. https://doi.org/10.1016/j.jprocont.2015.06.007.

Zhao, C., & Huang, B. (2018). Incipient fault detection for complex industrial processes with stationary and nonstationary hybrid characteristics. *Industrial and Engineering Chemistry Research*, *57*(14), 5045–5057. https://doi.org/10.1021/acs.iecr.8b00233.

Zhou, Z., Feng, Z., Hu, C., Han, X., Zhou, Z., & Li, G. (2019). A hidden fault prediction model based on the belief rule base with power set and considering attribute reliability. *Science China. Information Sciences*, *62*(10), 1–16. https://doi.org/10.1007/s11432-018-9620-7.

CHAPTER

6

Engineering systems' fault diagnosis methods

Reasoning draws a conclusion, but does not make the conclusion certain, unless the mind discovers it by the path of experience.

Roger Bacon (1214–1294) In: Opus Majus

In arguing for the study of nature through empiricism, the medieval English philosopher Roger Bacon, one of the earliest advocates of the modern scientific method, refers to one of the most important capabilities of the human mind—reasoning. And although humans now share this ability with computers, the inherent uncertainty in the conclusions drawn by reasoning, highlighted by the aka Doctor Mirabilis, still needs to be dealt with.

It is inspired by the words of Roger Bacon that this chapter intends to discuss fault diagnosis, a reasoning process that has attracted the attention of operators and maintainers of engineering systems because of its ability to assist in maintenance management decision-making. The first section of this chapter proposes to present the basic concepts related to fault diagnosis and to contextualize it in an increasingly complex industrial sector. In the second section, the main methods currently used to solve diagnostic problems will be discussed. The third section will present the main limitations addressed in the area, as well as the main desirable characteristics in a fault detection and diagnosis (FDD) system. Finally, the fourth section will present the main steps to be followed when developing a diagnostic system, contributing to the choice of the best method to be implemented.

https://doi.org/10.1016/B978-0-12-823521-8.00006-2

6.1 Fault diagnosis in the context

There always has been, especially since the second industrial revolution, a large push in the industrial sectors to seek greater efficiency in its manufacturing and production processes. At the same time, there has been pressure, from a society that is increasingly aware of the environment and the well-being of workers, for industries to comply with stringent environmental and safety regulations. To meet such high standards, several technological and managerial solutions have been developed and implemented in industrial processes, always aiming at obtaining higher quality products, reducing waste generation, and increasing its safety, reliability, and availability.

The main solutions to these challenges began to become evident in the 1980s, with an exponential increase in automation and monitoring of industrial processes. While automation allowed industrial production to maximize efficiency and decrease waste generation, monitoring ensured that the entire process could be supervised both by its operators and by an integrated supervisory control system. Additionally, monitoring also allowed a new maintenance strategy to be adopted by the industry, condition-based maintenance (CBM).

CBM is a strategy that aims to monitor the physical assets' current condition, allowing maintenance actions to be performed only when a loss of performance or an impending failure is detected (Melani, Murad, Caminada Netto, Souza, & Nabeta, 2018). Extensive data availability, given in large part by extensive monitoring of physical assets, became an opportunity for maintenance improvement through the development of condition-based strategies. CBM, then, became popular in the industrial sectors by contributing for maintenance actions to be carried out at the right time, reducing its total cost, and increasing the system's reliability. Furthermore, for the CBM strategy to be used by the maintenance team, an FDD system needs to be implemented in the given operational context.

FDD is a major feature in the CBM strategy, as it is responsible for analyzing the monitoring data available for all physical assets, checking for any abnormal behavior, and indicating the reason for an eventual abnormality. Many authors have approached FDD methods, presenting review articles and comparative studies in several areas of application (Alauddin, Khan, Imtiaz, & Ahmed, 2018; Alzghoul, Backe, Löfstrand, Byström, & Liljedahl, 2014; Chen, Wang, Qiao, & Chen, 2018; Gao, Cecati, & Ding, 2015a, 2015b; Habibi, Howard, & Simani, 2019; Isermann, 2011; Krichen, Ben, Chaeib, & Neji, 2017; MacGregor & Cinar, 2012; Sartori, Amaro, De Souza, & Embiruçu, 2012; Tidriri, Chatti, Verron, & Tiplica, 2016; Yin, Ding, Haghani, Hao, & Zhang, 2012; Yin, Ding, Xie, & Luo, 2014; Yin, Li, Gao, & Kaynak, 2015).

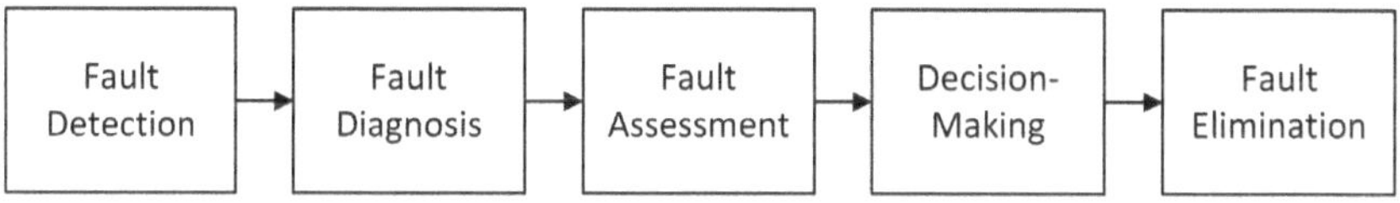

FIG. 6.1 Steps of the fault correction process.

In the CBM strategy, FDD is part of a system's fault correction process. This process, according to (Lampis, 2010), has five stages (Fig. 6.1):

1. Fault Detection: The occurrence of a fault can cause abnormal behaviors, called symptoms, in the physical asset or even in the system to which it belongs. The fault detection step is responsible for identifying such symptoms through the analysis of monitoring data;
2. Fault Diagnosis: The previously identified symptoms should be analyzed. The fault diagnosis step is responsible for identifying which physical asset, among all those monitored, is in a faulty state. Identifying the source of a fault can be done manually by inspecting the system components, but if it is large and complex or needs to be diagnosed quickly for safety reasons, the use of an automated fault diagnosis method is required;
3. Fault Assessment: The potential consequences that a fault may cause to the system, if not properly dealt with, should be analyzed. The fault assessment step is responsible for classifying the diagnosed fault according to its severity or the hazard it represents, in addition to eventually calculating the remaining useful life of the faulty asset;
4. Decision-making: After assessing the fault, a decision of what must be done has to be made. The decision-making step is responsible for defining whether the system will have its operation completely interrupted to correct the fault or will continue to operate and the correction is delayed;
5. Fault Elimination: The fault is corrected by repairing or replacing the component and the system returns to operation.

Of these five steps, generally only the first two, which comprise FDD, are currently automated. The other three are manually performed by the maintenance team. And even though the detection and diagnosis steps are highly interdependent, each has its function and can be performed by a different set of methods, under a different set of approaches.

Since the previous chapter presented how fault detection can be performed, this one will focus on fault diagnosis. In order to do that, a more formal definition of subject needs to be addressed. Fault diagnosis, according to Papadopoulos and McDermid (2001), represents the process of identifying the origin of a fault through a series of observable effects that it causes in the system to which it belongs.

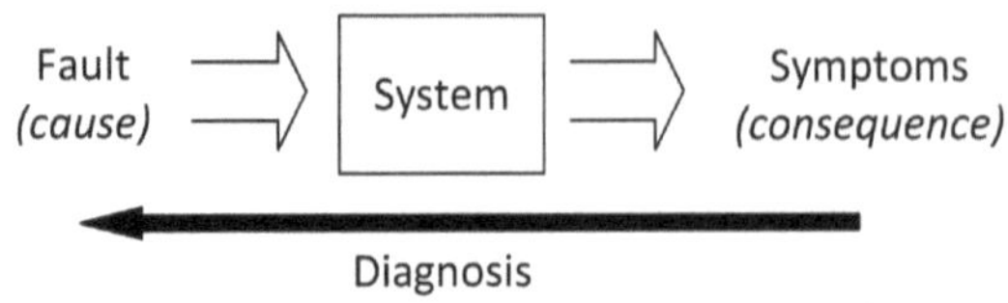

FIG. 6.2 Fault diagnosis: From consequences to causes.

It can be concluded, by interpreting this definition, as (Isermann, 2006) points out, that the fault diagnosis process goes inversely to the physical cause-consequence relationship. This inversion of causality, as shown in Fig. 6.2, although the essence of fault diagnosis, is also the source of its greatest challenges. This is because the reverse path, from symptoms to fault, is not easily reconstructed without the use of complete and structured knowledge about the behavior of a fault. Without this knowledge base, which is difficult to acquire, especially in complex engineering systems, the results obtained by fault diagnosis methods can be impaired or, at the very least, be uncertain. The next section will deal with the challenges, limitations, and desirable characteristics of a fault diagnosis method.

6.2 Challenges in fault diagnosis

There are many approaches that can be followed to implement a fault detection and diagnosis process, as the previously mentioned review articles show. Therefore, it can be difficult to decide which approach is best, that is, which one best meets a particular set of needs.

In order to make a conscious choice about which fault diagnosis method will contribute the most to decision-making regarding maintenance in a given operational context, three aspects need to be taken into account: particular characteristics of the system under study, the characteristics of the fault detection method, and the characteristics of decision-making to be made.

The system under study influences the choice of diagnosis process because the available knowledge about how it behaves when a fault occurs is used by the method. Therefore, the way in which this knowledge is available, either by a system expert or by a set of qualitative or quantitative data, or even if this knowledge does not exist or is not available, ends up influencing the choice of a diagnostic method.

The fault detection method also influences the choice of the diagnosis method because the first is the one who identifies and provides, as input information, the symptoms that the system is presenting for the second one, which then reasons about which fault is in progress. That is why it is very common, in the literature, to present the two processes, detection and diagnosis, combined.

Decision-making that is carried out based on the results of the diagnosis also influences the choice of the method. Decision-makers, i.e., maintenance managers, will decide the actions to be taken not only through the diagnostic result itself, but how much they trust that result. Accordingly, the uncertainty of diagnosis methods should be evaluated (or considered) during its selection.

Taking into account all these considerations, it is necessary to verify which characteristics are desirable in the fault diagnosis process. The next 11 subsections will present what are the 11 main challenges to be faced by a diagnosis method, i.e., what characteristics they should have to optimize their results. This list of challenges was based on work published by several authors (Melani, 2016; Isermann, 2006; Venkatasubramanian, Rengaswamy, & Kavuri, 2003; Venkatasubramanian, Rengaswamy, Kavuri, & Yin, 2003a, 2003b). It is worth mentioning that the methods used in the implementation of the fault diagnosis, in general, do not meet all these characteristics (Lampis, 2010). The choice of method to be used in a given system must be made according to the design needs.

6.2.1 Ability to represent and take advantage of prior knowledge

The diagnosis process can only accurately reason about which possible fault is responsible for the appearance of certain symptoms in the system if it knows the fault-symptom relationship. That is why it is essential that all types of knowledge about the behavior of the system are somehow inserted in and used by the diagnosis process.

This knowledge can present itself in several ways. It can come from the experience of a system expert, the historical data taken from the maintenance management system, the information obtained about similar systems, or even from the physical understanding of how the system works. The diagnosis process must take the advantage of the existence of this prior knowledge to make its results more accurate.

It is worth noting, however, that this knowledge is not always available. In this case, experimentally trained methods may be used to develop a diagnosis process, as will be described in the "Fault diagnosis methods" section.

6.2.2 Ability to take into account different types of symptoms

As there are different types of data that can be analyzed by the detection process, as described in Chapter 5, different types of symptoms can also be identified and supplied to the fault diagnosis process. Representation of a symptom can be done through binary, multivalued, interval-scaled, or continuous variables. Moreover, even a single symptom can be represented in different ways. The presence of a leakage in a pipeline, for example, can be represented by a binary (*yes* or *no*), a multivalued (*no, low leakage, high leakage,* and *very high leakage*), an interval-scaled (*0–10 mL/s, 10–100 mL/s,* and *100–1000 mL/s*) or by a continuous variable (the exact measured or inferred flow rate).

The diagnostic process, therefore, needs to be able to interpret different types of symptom representation and identify the fault or faults responsible for them. It is worth noting that greater precision about the real value of a symptom—precision, in this case, refers to the variable type used to represent the symptom, that is, a continuous variable would contain more accurate information about the symptom than a binary—can make the result of the diagnosis more accurate, but this results in greater difficulty in modeling the diagnosis method.

6.2.3 Agility

The sooner maintenance managers become aware of a fault in progress, the sooner they can plan to correct that fault. Reducing the time spent to diagnose faults is critical in systems that pose risks to operators and the environment, as well as contributing to the reduction of total maintenance time. In fact, agility must be considered not only in the fault diagnosis process but in the entire FDD cycle, including the acquisition of monitoring data.

A couple of factors can influence FDD agility, for example, the type of method being used and how efficient the algorithm that performs it is. The time for the execution of the FDD process is also dependent on the amount of data, whether quantitative or qualitative, that needs to be analyzed, in addition to the processing capacity of the computer system to which it is installed. On the one hand, a large amount of data can make the FDD results more accurate, but on the other hand, it may require more processing time, delaying decision-making.

The decision on how agile the FDD process should be can be made, for example, based on the behavior of the faults that the system presents. If the progression of such faults into failures is slow, perhaps agility is not one of the priority characteristics to be considered when choosing the method.

6.2.4 Ability to discriminate different faults

It is surprisingly common for different faults in a system to cause similar symptoms. Consider, for example, a simple electric motor. Misalignment on its axis or corrosion on one of its bearings can both cause excessive noise. Likewise, a defect in its copper winding insulation or insufficient lubrication in its bearings can cause a temperature increase in its housing.

Although such faults have similar symptoms, the maintenance actions required to correct them are completely different.

The diagnostic process must be able to identify and differentiate system faults, even if they have similar symptoms. For this to be possible, it must consider not only the symptoms themselves, but also their particular characteristics, such as intensity, location, or change rate. It is worth noting, nevertheless, that if the diagnostic process still fails to carry out such differentiation, this can be circumvented, for example, by adding more sensors to the system, which can provide more information/symptoms for a more accurate result.

6.2.5 Ease of elucidation

Another challenge for the fault diagnosis process is to allow decision-makers to understand its final result. The diagnostic results should make it clear, in a simple way, what conclusion he/she reached, i.e., which fault is in progress, and why he/she reached such a conclusion. Therefore, the fault diagnosis process should, ideally, also show which detected symptoms led to the conclusion that a certain fault is happening.

This transparency in the results makes the maintenance managers more confident in the conclusions presented. It also allows decision-makers to check whether any known noise or unexpected event may have interfered with the analysis.

6.2.6 Ability to identify multiple simultaneous faults

With the increasing complexity of engineering systems, it is inevitably becoming more common for multiple faults to happen simultaneously. In fact, it is normal, sometimes even expected, for one fault to cause another to arise, largely because of the interconnection and interdependence between the components of a system. The diagnosis process should, therefore, be able to identify whether a performance loss of a system is originated from a fault of a single piece of equipment or a combination of faults of several units.

Identification of simultaneous faults is particularly difficult to perform because a combination of faults can cause symptoms different from those observed when the same faults happen at different times.

6.2.7 Robustness

It is inevitable that the data acquired by the monitoring system will present a level of noise or uncertainty. This makes the FDD process responsible for minimizing the influence of such factors on the final result of its analysis. Ideally, the results presented by FDD should suffer little or no interference from noise present in the reading of sensors and of uncertainties regarding the monitoring system.

Usually, it is the detection process that is responsible for dealing with such noises and uncertainties, but it is possible that such factors cause the identification of false symptoms, propagating their effects to the diagnosis process. If this happens, a fault can be misdiagnosed.

To get around this problem, it is common to have the FDD results displayed only after a diagnosis confirmation, after several repetitions of the analysis using more recent data. Although this technique makes the result of the FDD more robust, it ends up compromising its performance, i.e., its agility. Because of this trade-off between robustness and performance, it is necessary to consider which of these characteristics is more desirable in the operational context under study.

6.2.8 Adaptability

All systems undergo changes. Change is an essential part of its life cycle. Several factors can contribute to a system's evolution, for example, the need to comply with novel environmental and safety regulations, a demand for a more efficient process, a technological update of components, a variation in the raw material used, and so on. But with each change, however small, the novel system may behave differently from what was previously expected. This behavioral change directly impacts the functioning of the diagnosis process; in fact, it impacts the entire FDD process.

The fault diagnosis process must therefore be able to adapt to the behavior of the updated system. It needs to be able to receive new information, knowledge, or even new monitoring data and update its reasoning process. Every FDD needs to be able to adapt to system updates and provide new outputs to better inform maintenance managers.

6.2.9 Novelty identifiability

Ideally, the FDD process should be able to identify all possible faults that could potentially occur in a system. In reality, however, the information, knowledge, and monitoring data available about a system are hardly capable of allowing the FDD, when it is developed and implemented, to be able to fulfill this requirement. This is even more complicated in complex systems, whose completeness of knowledge about all faults that can happen is generally nonexistent.

The diagnostic process must therefore recognize its limitations and be able to inform maintenance managers when any abnormal behavior is identified but the root cause is not. If such a feature is not implemented in the fault diagnosis method, it is likely that it will provide, in response to a novel abnormal behavior, a misclassification, i.e., the novel fault will be misidentified as one of the known ones. This incorrect diagnosis is likely to influence decision-makers to define incorrect maintenance strategies, impairing system efficiency and potentially increasing the risk of the operation.

6.2.10 Minimal development efforts

Because of the amount of information and knowledge handled by the diagnostic process, it is inevitable to imagine how much time and effort can be spent so that the developed diagnostic process can accurately diagnose the faults that the system may present. Especially in complex systems, the amount of data to be considered can be overwhelming.

Development of the diagnostic process, therefore, must take into account the time to be spent on its construction. It is desirable that the development effort is as small as possible. It is worth noting, however, that there is a trade-off between such effort and other of the characteristics previously described, since it is expected that a more robust, agile, and adaptive diagnosis of faults, for example, requires a more detailed modeling.

6.2.11 Owning uncertainty

It is inevitable that the diagnostic results are not 100% accurate. Uncertainties from different sources, such as inaccuracies arising from the detection process, can influence the final result. Limitations of the model in knowing and representing all fault-symptom relationships existing in the system, for example, can also impair diagnostic results.

It must be considered, therefore, that the results presented by the diagnostic process are probabilistic rather than deterministic, i.e., such results contain uncertainties that influence them and they may not represent reality. It is desirable, therefore, that the diagnostic process indicates and even estimates the uncertainty of its result, allowing decision-makers to ponder actions to be taken.

6.3 Fault diagnosis methods

There are many methods currently being used to perform FDD. There are so many possibilities that many authors have proposed different categorizations for them. It is very common, for example, when talking about the entire FDD process, the used approaches to be generally classified into

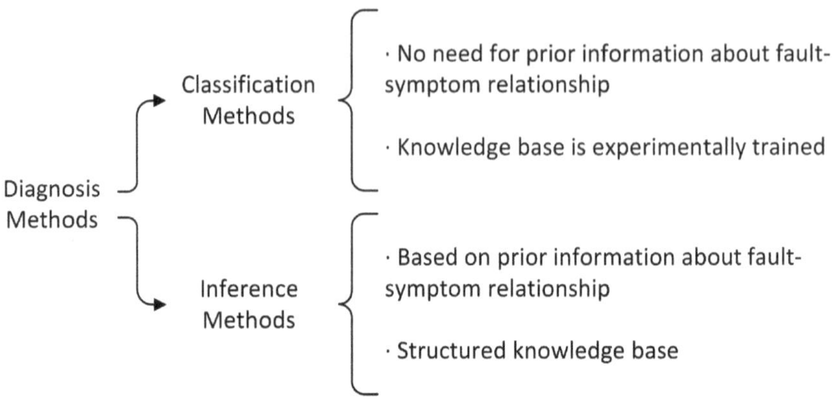

FIG. 6.3 Categories of fault diagnosis methods.

two categories: model-based and data-based approaches. This classification was largely discussed in Chapter 5, dedicated to the detection process.

Nevertheless, when dealing specifically with the fault diagnosis process, Isermann (2006) presented a more appropriate classification, which will be discussed in this section. According to him, diagnostic methods can be classified into two categories, as shown in Fig. 6.3: classification methods and inference methods. The next two subsections will discuss in detail the differences between these two approaches.

6.3.1 Classification methods

The fault diagnosis process relies on the knowledge about the behavior of the system when a fault occurs in order to be able to correctly indicate to the maintenance managers the reason for an eventual appearance of certain symptoms. It is common, however, that this type of knowledge is not available in a structured way, e.g., represented in the form of if-then rules. In this case, classification methods can learn about the relationship between faults and symptoms through experimentally stored data.

The classification methods, therefore, are highly dependent on the availability of symptom and fault records for its knowledge base to be trained. If such records, for example, do not comprise all possible faults that the system may present, the method will not be able to identify a novel fault. However, its knowledge base can be constantly updated, making the diagnosis process more robust over time.

Several techniques have been used as classification methods for fault diagnosis. In the next subsections, three of them will be briefly introduced: naïve Bayes classifier, support vector machine, and neural networks.

6.3.1.1 Naïve Bayes classifier

Naïve Bayes classifier is a machine learning model that applies the Bayes theorem, presented in Eq. (6.1), for probabilistic classification. By observing the values (input data) of a given set of features or parameters, represented as B in the equation, naïve Bayes classifier is able to calculate the probability of the input data belonging to a certain class, represented as A.

$$P(A|B) = \frac{P(B|A)P(A)}{P(B)} \tag{6.1}$$

For the classification of input data to take place, the probabilities of it belonging to each of the existing classes must be determined and the one with the highest probability will be the class to which the input data belongs. Therefore, class a with the highest probability must be found as expressed in Eq. (6.2), where b_i is one of the n features/predictors observed.

$$a = argmax_a P(a|b_1, \cdots, b_n) \tag{6.2}$$

Since a naïve Bayes classifier assumes that all variables are independent, only a small training data is necessary to estimate parameters necessary for classification. In the context of fault diagnosis, classes would represent faults or a set of faults that a system could develop and the predictors would represent the symptoms the system is presenting. Although naïve Bayes classifiers are easy and quick to implement, considering the predictors as independent variables can be seen as a disadvantage of the method, since in most real fault diagnosis cases, the symptoms can be dependent on each other (a high vibration can cause an increase in temperature, for example).

Naïve Bayes classifier has been used for fault diagnosis in several different systems, such as bearings (Yi, Chen, & Hou, 2017; Zhang, Wu, Yang, & Guan, 2018), analog circuits (He, He, Li, & Zhang, 2020), and fuel cells (Fan, Huang, & Yi, 2013).

6.3.1.2 Support vector machine

Support vector machine (SVM) is a machine learning technique widely used for data classification. It is a very efficient and robust prediction method used in several real-world applications (Steinwart & Christmann, 2008).

SVM's goal is to analyze a dataset and separate its data points into different classes. As it is common for the analyzed data to be multidimensional, such classification is made by determining a hyperplane, which separates the data into different areas (which determine the classes). As there are many hyperplanes that can be used to separate data into categories, SVM seeks to find the hyperplane that maximizes the margin or

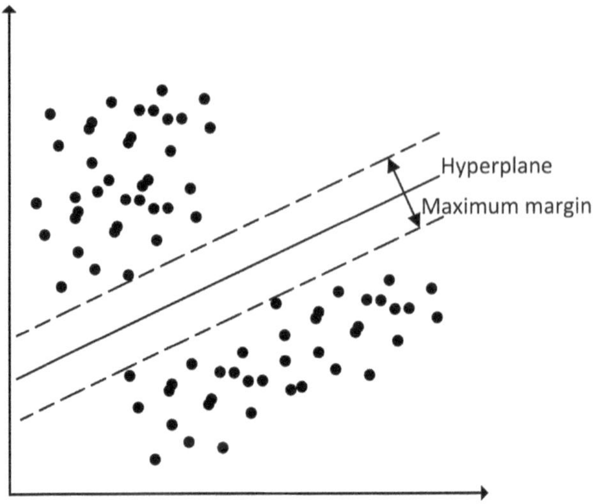

FIG. 6.4 Example of hyperplane determined by an SVM.

distance between classes of points. Such maximization allows future data points to be more accurately categorized. To better illustrate how classification via SVM works, Fig. 6.4 shows a two-dimensional representation of the optimal hyperplane for a given set of data.

It is important to mention that SVM is able to perform data classification considering a scenario with many classes. In the context of the diagnosis of faults, classes would represent faults or a set of faults that a system could develop and the input data (i.e., symptoms) would be analyzed and classified according to the hyperplanes determined by SVM. Depending on the result of the classification, it would be possible to estimate which fault the system may be presenting.

SVM has been shown to be a technique that presents consistent results even with a small database sample and has been used for fault diagnosis by several authors and in different systems, such as electrical motors (Li & Xue, 2018; Niu, Wang, Zhai, & Cheng, 2011), ball bearings (Lin, Chu, Wang, Chen, & Chen, 2019), and power transformers (Yin, Zhu, & Yu, 2011).

6.3.1.3 *Neural networks*

A neural network is a structure formed by processing units, called artificial neurons, which have the ability to perform calculation operations for data processing and knowledge representation. Such a concept was introduced in the 1940s but has recently become popular because of its ability to solve prediction problems. By mimicking the way a biological neuron operates, neural networks (NN) are able to recognize the hidden

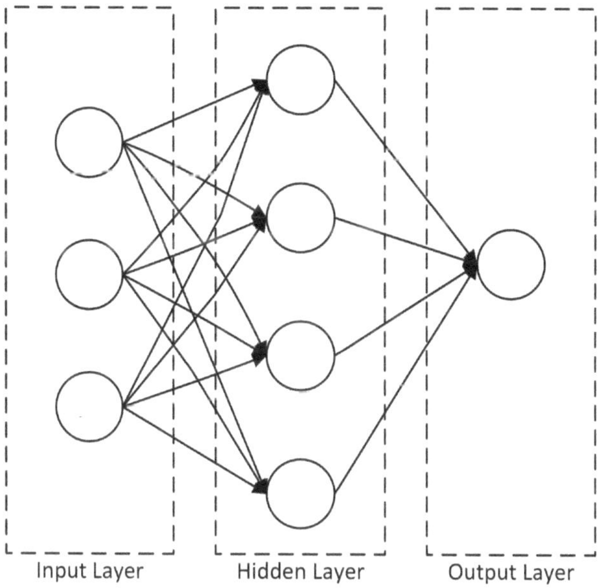

FIG. 6.5 Example of a neural network structure.

relationships among numbers or variables in a dataset. It has been largely applied in many disciplines because of its ability to predict and reproduce nonlinear phenomena.

As shown in Fig. 6.5, a neural network is usually formed by three layers: the input layer, where the input data are inserted; the output layer, where the results of the analysis are provided; and the hidden layer (which can be formed by a single or multiple layers of neurons), in which each neuron receives a linear combination of values from the previous layer so an activation function generates a new value for that neuron. As the name suggests, the linear combination of values of the previous layer, used as input to each of the neurons of the next layer, is a linear function in which its parameters will be the values of the previous layer and their combination will be done by a matrix of weights between layers.

The activation function, applied to each neuron, can be interpreted as a mathematical "gate" between the input and output data of the neurons. Such a function can be as simple as a binary step function, which is a threshold-based function that turns the neuron output on or off. It can also be something more complex as a nonlinear function (a sigmoid function, for example), capable of creating mappings between the input and output data. One of the advantages of using nonlinear functions as activation functions is the possibility of using backpropagation algorithms to train the neural network, i.e., to adjust the weights of the neurons so that the results are more accurate.

One of the advantages of neural networks is its ability to also use non-structured data as input. However, in order for the results obtained by the network to be more accurate, it is necessary to train it, which requires a quantity of input and output data previously available. In the context of fault diagnosis, this means that, ideally, data on the behavior of system faults would need to be previously recorded, which is not always the case.

In fault diagnosis, neural networks have been used in several different systems, such as power plants (Simani & Fantuzzi, 2000), production processes (Sorsa, Koivo, & Koivisto, 1991), induction motors (Asghar, Talha, & Kim, 2016), and electric power systems (Swarup & Chandrasekharaiah, 1991).

6.3.2 Inference methods

Unlike classification methods, the inference methods rely on prior knowledge about the faults-symptoms. This knowledge can even be represented in an analytical or heuristic way. Analytical knowledge comes from the laws of physics or quantitative observations, while heuristic knowledge comes from the experience acquired by operators and system engineers. Knowledge about the fault-symptom relationship can be represented in the form of rules, predicate logic, or directed graphs.

Inference methods take the advantage of the availability of this knowledge to reason and point out which faults may be responsible for a given set of symptoms. Since you do not need an experimental database, the inference methods can be robust and reliable from the beginning of their implementation—if, of course, the knowledge base used is comprehensive and accurate.

However, building this knowledge base can be very time-consuming, especially for complex systems. Accordingly, it is common to use knowledge already acquired for other purposes as a supply for inference methods. The failure modes and effects analysis (FMEA)—presented in Chapter 2—for example, has been used as a source of information for the diagnostic process (Price & Taylor, 1997; Yang, Lu, Liu, & Hao, 2009; Zhang, Zhou, Yang, & Zhang, 2014). It is important to keep in mind, however, that, when the knowledge base is not specifically developed for fault diagnosis, adaptations or additions to that base are generally necessary.

Several techniques have been used as inference methods for fault diagnosis. In the next subsections, three of them will be briefly introduced: fault tree analysis (FTA), expert systems, and Bayesian networks.

6.3.2.1 *Fault tree analysis*

FTA, as previously described in Chapter 2, is a deductive method in which an unwanted event, called a top event, is defined and, from that, the causes or combinations of causes that may lead to the occurrence of this event are systematically defined. Although it is a technique usually used for reliability analysis, many authors have shown that it can also be applied for fault diagnosis (Melani, 2016).

In general, when using FTA for fault diagnosis, the top event represents the observed symptom instead of representing the system fault as when it is used for reliability analysis. FTA is a relatively simple technique that can be used to extract knowledge from system experts about known system faults. Additionally, the fact that it can represent the occurrence of simultaneous faults is one of the advantages that the technique presents, especially when applied in complex systems.

For the fault diagnosis to be performed through FTA, it is necessary to compute the cut sets that lead to the top event, as already presented in Chapter 2. The most likely cut set represents the fault or set of faults that are most likely occurring in the system. It is worth mentioning that several authors have combined FTA with Bayesian networks, which will also be presented in this chapter, so that multiple symptoms are interpreted simultaneously, making the results of the diagnosis optimized (Melani, 2016).

Several authors have used FTA to diagnose faults in different systems, such as power transformers (Wang, Li, Ma, & Li, 2017), manufacturing processes (Peng, Shang, Shi, & Peng, 2007), and fuel cells (Yousfi Steiner et al., 2012).

6.3.2.2 Expert systems

The objective of the expert systems (ES) is to mimic the reasoning of an experienced professional or system expert, using a database and a reasoning algorithm (Jackson, 1998). In other words, it aims to solve problems quickly and intelligently and reproduce the decisions of a human specialist in a particular area of interest.

Fig. 6.6 presents a framework of how a typical ES works. Construction of an ES begins with filling of a knowledge base by an expert of the system or scenario under study. In the context of fault diagnosis, for example, the specialist must inform what symptoms he has already seen the system present when a given fault occurred. For filling of this knowledge base to be done in a structured way, it is common for a knowledge engineer to mediate this process.

Once the knowledge base is filled, an inference engine must be developed so that it is able to receive a query from a nonexpert user, interpret it based on the knowledge base, and provide advice as output. In the context of fault diagnosis, the user would inform the ES which symptoms the system is presenting and the ES, in turn, would compare them with those reported by the system expert and provide the user with which faults may be occurring in the system.

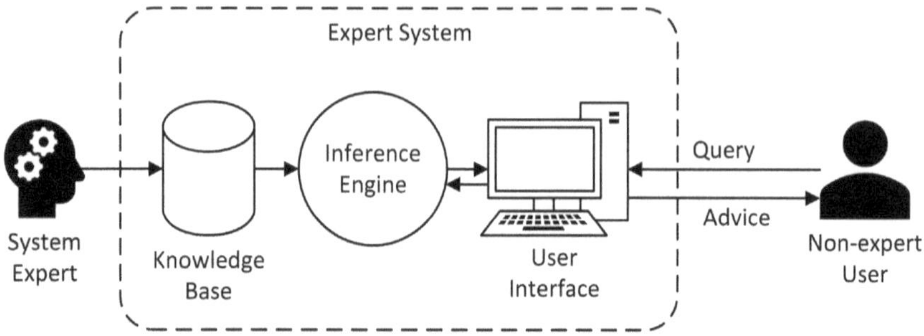

FIG. 6.6 Expert systems.

An inference engine can be developed using several strategies, but in general it is composed of if-then-else rules that follow a chain of conditions and derivations to finally deduce an outcome.

One of the advantages of ES for fault diagnosis is its fast and steady response, with less error rates. On the other hand, construction of the knowledge base is highly time-consuming and dependent on the availability of a system expert capable of providing the necessary information for fault diagnosis.

Several authors have used ES to diagnose faults in different systems, such as wind turbines (Deng, Gao, Zhang, Hu, & Yang, 2017), distribution substations (Lee, Ahn, & Park, 2000), and machines (Krishnamurthi & Phillips, 1992).

6.3.2.3 Bayesian networks

As already presented in Chapter 4, Bayesian network is a graphical model based on Bayes' theorem that represents variables of a domain and conditional dependencies between them. Because of its ability, as a representational tool, to represent an uncertainty problem in a compact, complete, and modular way, BN has been applied in many different domains, such as reliability engineering, risk analysis, and fault diagnosis. Special attention will be paid in this section to the diagnostic processes based on Bayesian Networks (BN) for two reasons: first, this reasoning tool is becoming increasingly popular in the area, as shown by Cai, Huang, and Xie (2017); second, such a tool will be used in the case studies to be presented in Chapter 9.

More formally, BN is a directed acyclic graph (DAG) composed of nodes and arcs. Each node represents a variable, discrete or continuous, of the universe under study. Each arc, or directed line, connects a pair of nodes and represents a conditional dependency between these variables. If an arc is connected from node A to node B, then node A is called the parent of node B. The presence of this arc implies that node A, the

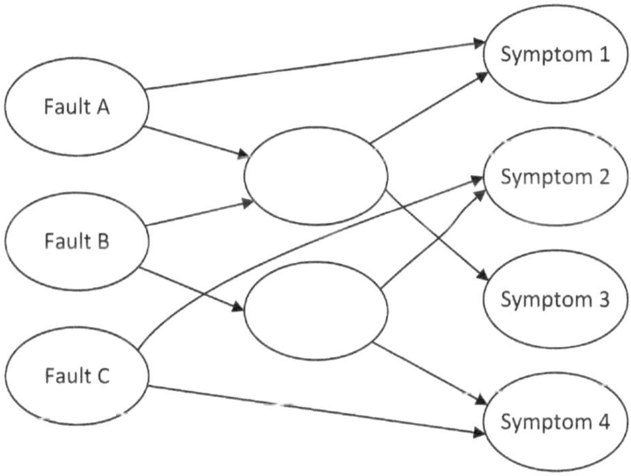

FIG. 6.7 Example of a BN graph.

parent, has a direct influence on node B, the child. This characteristic of BN makes parent nodes to be used to represent causes and child nodes consequences in the universe under study (Russell & Norvig, 2010). Fig. 6.7 shows an example of a BN graph, highlighting the "causes" (i.e., faults) and "consequences" (i.e., symptoms) nodes.

Besides, BN consists of quantitative and qualitative components (Neapolitan, 2004). The qualitative one is represented by DAG and the quantitative one by the conditional probabilities associated with each DAG node. That is because each node X has a conditional probability distribution, given by $P(X|Parents(X))$, which quantifies the effect of the parent node on the child node.

One of the most attractive features of BN is its ability to perform inference since it allows the probability of a certain variable to be updated by observing the state of a set of variables, that is, it makes possible the computation of the posterior probability. Inference can be performed using exact or approximate solutions. This is because exact inference, while providing exact posterior probability, can be very time-consuming, especially in very complex Bayesian networks.

For the execution of the exact inference, where X is the variable to be used to calculate the posterior probability, E is the set of evidence variables (e is the set of observed values for these variables) and Y is the set of unobserved variables (y is the set of possible values for these variables), Eq. (6.3) can be used, where $\alpha = 1/P(e)$:

$$P(X|e) = \frac{P(X,e)}{P(e)} = \alpha P(X,e) \tag{6.3}$$

Eq. (6.3) can be rewritten as Eq. (6.4) (Russell & Norvig, 2010):

$$P(X|e) = \alpha \sum_{y} P(X, e, y) \tag{6.4}$$

By analyzing Eq. (6.4), it is easy to see that, as the complexity of the Bayesian network increases, the summation present in the equation makes its use computationally very time-consuming. Some techniques may simplify this calculation, such as variable elimination, but in very complex networks, not even these techniques can make the exact inference feasible.

The approximate inference decreases the processing time of the posterior probability using, for example, stochastic simulation. Russell and Norvig (2010) present algorithms used to make the approximate inference.

Because of its ability to perform inference, BN is a highly recommended tool for modeling diagnostic systems, as it allows one to identify or at least indicate possible causes of observable symptoms. During the last decades, BNs have been highly used for fault diagnosis (Cai et al., 2017).

In addition to the Bayesian inference, previously presented, another possibility of obtaining important results from a Bayesian network is the maximum a posteriori probability (MAP). MAP's goal is to find the most likely configuration of a set of nodes given partial evidence on the remaining variables in the BN.

It is important to note the difference between the results obtained by the inference process and by MAP when a Bayesian network has observed evidence: while the first updates the probability of each node (which are not evidence nodes), the second finds the most likely configuration of nodes that would cause the observed evidence. That is, while the first analyzes nodes individually, the second analyzes them collectively.

In a more formal way, as explained by Yuan, Lu, and Druzdzel (2004), MAP is the instantiation that satisfies Eq. (6.5).

$$\hat{x} = max_X P(X|E) \tag{6.5}$$

where X is the set of nodes whose most probable configuration needs to be found and E is the set of evidence nodes.

When applying the chain rule and the law of total probability in Eq. (6.5), Eq. (6.6) is obtained.

$$\hat{x} = max_X \sum_{Y} \prod_{Z_i \in Z} P(Z_i | Parents(Z_i)) \tag{6.6}$$

where Y is the set of nodes that do not belong to X nor to E and $Z = X \cup E \cup Y$.

When comparing Eq. (6.6) with Eq. (6.4), it is noted that MAP problem is even more complex to solve than the inference process, especially with the increase in the number of variables present in the Bayesian network.

As it is computationally even more time-consuming, it is common to use approximate MAP techniques. Park and Darwiche (2006) investigated some of the best effort approximations and discussed the difficulty encountered by these methods in decreasing the computational time while guaranteeing a good approximation result. Although complex to solve, MAP can be very useful, as it offers a different perspective when analyzing the effect of evidence on the Bayesian network.

In addition to being excellent for representing the relationship between component faults and their effects on the system, Bayesian networks are highly recommended tools for modeling diagnostic systems, since the inference process, as well as the MAP algorithm, allows identification of possible causes of these effects.

6.4 Development of a fault diagnosis process

Development of a fault diagnosis process can offer several challenges to its designer. But, although several characteristics about the chosen method or the analyzed system can make the diagnosis process unique, the step-by-step process for its development is, in general, well known. Fig. 6.8 shows the seven main steps to be taken to develop a fault diagnosis process.

The first step in the development is to understand and define what requirements and specifications the fault diagnosis process must meet. Issues such as project cost, development time, or even computing power to be used should be acknowledged.

In the second step, all useful information about the system should be collected. The number and type of sensors that monitor it, for example, are fundamental information for the next development steps. In addition, it should be ascertained whether a knowledge base about the faults-symptoms relationship already exists or if at least it can be built, through the help of system experts, like operators and maintainers. Historical failure and monitoring records can also be critical, depending on the methods to be used for fault detection and diagnosis.

The third step is to choose the fault diagnosis method. It is important to remember that, although Fig. 6.5 focuses on the development of a diagnostic process, usually the choice is not limited to diagnosis, but encompasses the whole FDD process. The choice of the method that will carry out the diagnosis must be made based on the information obtained in the previous steps and on how this method will work alongside the detection method.

In the fourth step, the proposed method must be developed, alongside the detection method. In the case of complex systems, the possibility of

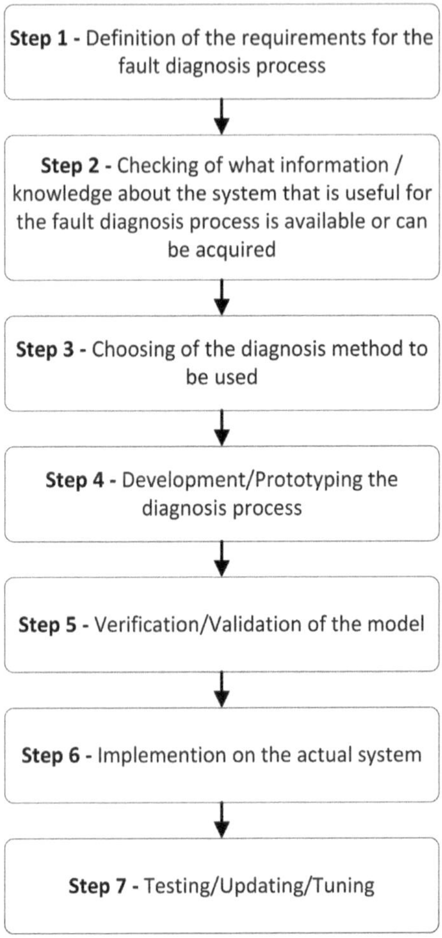

FIG. 6.8 Steps of the fault diagnosis development.

developing only a prototype will be considered. This possibility must be considered because, in the fifth step, verification/validation of the model must be done and, if it proves unsuccessful, the time and resources spent on development will be minimized. Validation of the model, in general, takes place through exhaustive tests, using simulated monitoring data or monitoring data taken from the system database as input.

In the last two steps of the development, the diagnostic method must be implemented in the real system and, from there, the results obtained by it must be tested. If necessary, an update or a slight tuning may be necessary to adjust the behavior of the FDD process to the operational context.

References

Alauddin, M., Khan, F., Imtiaz, S., & Ahmed, S. (2018). A bibliometric review and analysis of data-driven fault detection and diagnosis methods for process systems. *Industrial and Engineering Chemistry Research, 57*, 10719–10735.

Alzghoul, A., Backe, B., Löfstrand, M., Byström, A., & Liljedahl, B. (2014). Comparing a knowledge-based and a data-driven method in querying data streams for system fault detection: A hydraulic drive system application. *Computers in Industry, 65*, 1126–1135.

Asghar, F., Talha, M., & Kim, S. H. (2016). Neural network based fault detection and diagnosis system for three-phase inverter in variable speed drive with induction motor. *Journal of Control Science and Engineering, 2016*, 1–12.

Cai, B., Huang, L., & Xie, M. (2017). Bayesian networks in fault diagnosis. *IEEE Transactions on Industrial Informatics, 13*, 2227–2240.

Chen, X., Wang, S., Qiao, B., & Chen, Q. (2018). Basic research on machinery fault diagnostics: Past, present, and future trends. *Frontiers of Mechanical Engineering, 13*, 264–291.

Deng, X.-W., Gao, Q.-S., Zhang, C., Hu, D., & Yang, T. (2017). Rule - based fault diagnosis expert system for wind turbine. *ITM Web of Conferences, 11*, 07005.

Fan, L., Huang, X., & Yi, L. (2013). Fault diagnosis for fuel cell based on naive Bayesian classification. *TELKOMNIKA Indonesian Journal of Electrical Engineering, 11*(12), 7664–7670.

Gao, Z., Cecati, C., & Ding, S. X. (2015a). A survey of fault diagnosis and fault-tolerant techniques—Part I: Fault diagnosis with model-based and signal-based approaches. *IEEE Transactions on Industrial Electronics, 62*, 3757–3767.

Gao, Z., Cecati, C., & Ding, S. X. (2015b). A survey of fault diagnosis and fault-tolerant techniques-part II: Fault diagnosis with knowledge-based and hybrid/active approaches. *IEEE Transactions on Industrial Electronics, 62*, 3768–3774.

Habibi, H., Howard, I., & Simani, S. (2019). Reliability improvement of wind turbine power generation using model-based fault detection and fault tolerant control: A review. *Renewable Energy, 135*, 877–896.

He, W., He, Y., Li, B., & Zhang, C. (2020). A naive-Bayes-based fault diagnosis approach for analog circuit by using image-oriented feature extraction and selection technique. *IEEE Access, 8*, 5065–5079.

Isermann, R. (2006). *Fault-diagnosis systems*. Berlin, Heidelberg: Springer.

Isermann, R. (2011). *Fault-diagnosis applications model-based condition monitoring*. Berlin, Heidelberg: Springer.

Jackson, P. (1998). *Introduction to expert systems*. Boston, MA: Addison-Wesley Longman Publishing Co.

Krichen, M., Ben, N., Chaeib, M., & Neji, R. (2017). Fault detection and diagnosis methods in permanent magnet synchronous machines: A review. In *International conference on recent advances in electrical systems (ICRAES'17). Hammamet, Tunisia* (pp. 229–237).

Krishnamurthi, M., & Phillips, D. T. (1992). An expert system framework for machine fault diagnosis. *Computers and Industrial Engineering, 22*, 67–84.

Lampis, M. (2010). *Application of Bayesian belief networks to system fault diagnostics*. Loughborough University.

Lee, H.-J., Ahn, B.-S., & Park, Y.-M. (2000). A fault diagnosis expert system for distribution substations. *IEEE Transactions on Power Delivery, 15*, 92–97.

Li, S.-Y., & Xue, L. (2018). Motor's early fault diagnosis based on support vector machine. *IOP Conference Series Materials Science and Engineering, 382*, 32047.

Lin, C.-J., Chu, W.-L., Wang, C.-C., Chen, C.-K., & Chen, I.-T. (2019). Diagnosis of ball-bearing faults using support vector machine based on the artificial fish-swarm algorithm. *Journal of Low Frequency Noise Vibration and Active Control, 39*(4), 954–967. https://doi.org/10.1177/1461348419861822.

MacGregor, J., & Cinar, A. (2012). Monitoring, fault diagnosis, fault-tolerant control and optimization: Data driven methods. *Computers and Chemical Engineering, 47*, 111–120.

Melani, A. H. A. (2016). *Desenvolvimento de um método para diagnose de falhas na operação de navios transportadores de gás natural liquefeito através de redes bayesianas*. São Paulo: Universidade de São Paulo.

Melani, A. H. A., Murad, C. A., Caminada Netto, A., Souza, G. F. M., & Nabeta, S. I. (2018). Criticality-based maintenance of a coal-fired power plant. *Energy, 147*, 767–781.

Neapolitan, R. E. (2004). *Learning Bayesian networks* (1st ed.). New Jersey: Prentice Hall.

Niu, W., Wang, G. Q., Zhai, Z. J., & Cheng, J. (2011). Fault diagnosis model based on support vector machine and genetic algorithm. *Applied Mechanics and Materials, 130–134*, 2535–2539.

Papadopoulos, Y., & McDermid, J. (2001). Automated safety monitoring: A review and classification of methods. *International Journal of COMADEM, 4*, 1–32.

Park, J. D., & Darwiche, A. (2006). Complexity results and approximation strategies for MAP explanations. *Journal of Artificial Intelligence Research, 21*, 101–133.

Peng, H., Shang, W., Shi, H., & Peng, W. (2007). On-line monitoring and diagnosis of failures using control charts and fault tree analysis (FTA) based on digital production model. In *Knowledge science, engineering and management* (pp. 544–549). Berlin, Heidelberg: Springer Berlin Heidelberg.

Price, C., & Taylor, N. (1997). Multiple fault diagnosis from FMEA. In *Proceedings of the fourteenth national conference on artificial intelligence and ninth conference on innovative applications of artificial intelligence, AAAI'97/IAAI'97* (pp. 1052–1057). AAAI Press.

Russell, S., & Norvig, P. (2010). *Artificial intelligence a modern approach* (3rd ed.). New Jersey: Prentice Hall.

Sartori, I., Amaro, C. A., De Souza, M. B., & Embiruçu, M. (2012). Detecção, diagnóstico e correção de falhas: Uma proposição consistente de definições e terminologias. *Ciência & Engenharia, 21*, 41–53.

Simani, S., & Fantuzzi, C. (2000). Fault diagnosis in power plant using neural networks. *Information Sciences, 127*, 125–136.

Sorsa, T., Koivo, H. N., & Koivisto, H. (1991). Neural networks in process fault diagnosis. *IEEE Transactions on Systems, Man, and Cybernetics, 21*, 815–825.

Steinwart, I., & Christmann, A. (2008). *Support vector machines*. Springer.

Swarup, K. S., & Chandrasekharaiah, H. S. (1991). Power system fault diagnosis using artificial neural networks. In *ACE '90. Proceedings of [XVI annual convention and exhibition of the IEEE in India]* (pp. 81–85). IEEE.

Tidriri, K., Chatti, N., Verron, S., & Tiplica, T. (2016). Bridging data-driven and model-based approaches for process fault diagnosis and health monitoring: A review of researches and future challenges. *Annual Reviews in Control, 42*, 63–81.

Venkatasubramanian, V., Rengaswamy, R., & Kavuri, S. N. (2003). A review of process fault detection and diagnosis part II: Qualitative models and search strategies. *Computers and Chemical Engineering, 27*, 313–326.

Venkatasubramanian, V., Rengaswamy, R., Kavuri, S. N., & Yin, K. (2003a). A review of process fault detection and diagnosis part III: Process history based methods. *Computers and Chemical Engineering, 27*, 327–346.

Venkatasubramanian, V., Rengaswamy, R., Kavuri, S. N., & Yin, K. (2003b). A review of process fault detection and diagnosis part I: Quantitative model based methods. *Computers and Chemical Engineering, 27*, 293–311.

Wang, Y., Li, X., Ma, J., & Li, S. (2017). Fault diagnosis of power transformer based on fault-tree analysis (FTA). *IOP Conference Series: Earth and Environmental Science, 64*, 12099.

Yang, S., Lu, M., Liu, B., & Hao, B. (2009). A fault diagnosis model for embedded software based on FMEA/FTA and Bayesian network. In *2009 8th international conference on reliability, maintainability and safety* (pp. 778–782). IEEE.

Yi, X. J., Chen, Y. F., & Hou, P. (2017). Fault diagnosis of rolling element bearing using Naïve Bayes classifier. *Vibroengineering PROCEDIA, 14*, 64–69.

Yin, S., Ding, S. X., Haghani, A., Hao, H., & Zhang, P. (2012). A comparison study of basic data-driven fault diagnosis and process monitoring methods on the benchmark Tennessee Eastman process. *Journal of Process Control*, *22*, 1567–1581.

Yin, S., Ding, S. X., Xie, X., & Luo, H. (2014). A review on basic data-driven approaches for industrial process monitoring. *IEEE Transactions on Industrial Electronics*, *61*, 6414–6428.

Yin, S., Li, X., Gao, H., & Kaynak, O. (2015). Data-based techniques focused on modern industry: An overview. *IEEE Transactions on Industrial Electronics*, *62*, 657–667.

Yin, J., Zhu, Y., & Yu, G. (2011). Power transformer fault diagnosis based on support vector machine with cross validation and genetic algorithm. In *2011 international conference on advanced power system automation and protection* (pp. 309–313). IEEE.

Yousfi Steiner, N., Hissel, D., Moçotéguy, P., Candusso, D., Marra, D., Pianese, C., et al. (2012). Application of fault tree analysis to fuel cell diagnosis. *Fuel Cells*, *12*, 302–309.

Yuan, C., Lu, T. C., & Druzdzel, M. J. (2004). Annealed MAP. In *Proceedings of the 20th conference on uncertainty in artificial intelligence. Banff, Canada* (pp. 628–635).

Zhang, N., Wu, L., Yang, J., & Guan, Y. (2018). Naive Bayes bearing fault diagnosis based on enhanced Independence of data. *Sensors*, *18*, 463.

Zhang, Y. F., Zhou, R. B., Yang, J. M., & Zhang, Z. (2014). Application of FTA-FMEA method in fault diagnosis of tracked vehicle. *Advances in Materials Research*, *940*, 112–115.

CHAPTER

7

Framework for engineering systems health monitoring and fault diagnosis

An ounce of prevention is worth a pound of cure.

Benjamin Franklin (1706–1790) In: On Protection of Towns from Fire

When it comes to risk management, prevention is one of the most effective courses of action for risk treatment. However, it did not seem to be so evident in the past. For instance, Benjamin Franklin himself wrote an article for *The Pennsylvania Gazette* in 1735 concerned about how fires were not preventively handled in Philadelphia. On that occasion, the notorious quote "An ounce of prevention is worth a pound of cure" was used to introduce and promote a series of preventive measures against fire threats in the city.

Nowadays, regarding physical assets, failure prevention is also a strategy widely used to address the risks associated with the operations of complex engineering systems. Through a health monitoring and fault diagnosis process, failure prevention can be systematically implemented in the organization. It can provide maintenance planning with useful information for prioritizing maintenance tasks in order to anticipate failures and avoid all of their consequences.

Accordingly, this chapter intends to discuss a framework for engineering systems health monitoring and fault diagnosis to support maintenance management. Section 7.1 provides an overview of the Health Monitoring and Fault Diagnosis (HMFD) process. Then, a framework for it is presented and discussed in Section 7.2. Lastly, the main challenges that an organization may face about health monitoring and fault diagnosis and recommendations to address them are covered in Section 7.3.

https://doi.org/10.1016/B978-0-12-823521-8.00011-6

7.1 Fundamentals of health monitoring and fault diagnosis

It is essential for any business that its assets are always able to perform its functions and fulfil its objectives within the operational context. For this to happen, however, it is necessary to ensure that the assets are always in an optimal state of health, which means not only performing preventive maintenance actions, but also monitoring and mitigating any degradation they present.

Monitoring asset degradation is a complex process that should not be confused with the system supervision process. While the latter is responsible for ensuring that the system as a whole is operating safely, the former aims to anticipate the occurrence of failures, allowing for optimization in maintenance decision making and potentially increasing system availability. Fig. 7.1 highlights the difference between these processes, called, respectively, HMFD and system supervision.

From the point of view of preventing undesirable scenarios and assessing risks in any type of systems, both health monitoring and fault diagnosis and system supervision processes behave as Independent Protection Layers (IPLs). They are sets of actions capable of preventing a scenario from proceeding to an unwanted consequence without being adversely affected by the initial event or action of any other layer of protection associated with the scenario. As such, these processes have their own response time. Independent protection layer Response Time (IRT) is the time required for the IPL to detect the abnormal condition and complete the necessary actions to stop the progression of the condition that leads the system to an unsafe state (Center for Chemical Process Safety (CCPS), 2015).

In the context of Reliability Analysis and Asset Management of Engineering Systems, the IRT of the supervision and HMFD processes are quite distinct, as well as their actions to mitigate the condition that takes

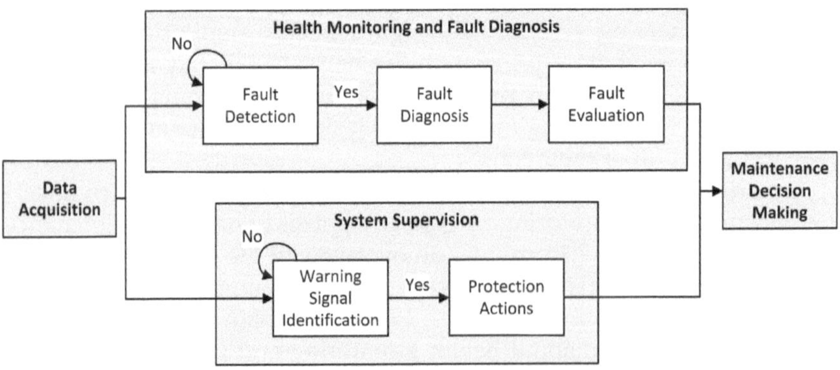

FIG. 7.1 Data-based HMFD and system supervision processes.

the system to an unsafe state and the process for assessing this condition. A Supervisory Control and Data Acquisition (SCADA) system, generally associated with the system supervision process and responsible for the automatic identification and handling of warning signals, determines when an alarm event has occurred by monitoring and checking certain threshold values. In addition, it can also take protection actions such as activating one or more alarm indicators, sending information to remote operators, and even executing the automatic shutdown of the monitored system to prevent major issues (such as higher risk conditions or the breakdown of any system component). Due to their function and specifications, the response of SCADA systems is generally immediate upon the detection of an event in progress.

On the other hand, as will be presented in Section 7.2 of this chapter, the HMFD process has a much more extensive and complex sequence of events, incorporating processes such as fault detection, fault diagnosis, and fault evaluation, besides data acquisition and maintenance decision making. Thus, the response time of an HMFD process, even when automated and executed with the aid of dedicated software, ends up being much longer than that of a SCADA system.

Although independent, the system supervision and HMFD processes are complementary. Since no process or system can be considered failure-proof, a good HMFD process is able to anticipate failures through the identification of early-stage faults. If there is an ongoing fault not detected by the HMFD process in the analyzed system, however, protection measures will eventually need to be taken, generally associated with the system supervision process.

Failures can expose employees to danger, impact the environment, interrupt the provision of essential services to the population, and cause various financial expenses to the organization. Without an HMFD process, the organization relies primarily on preventive maintenance tasks to avoid failures and the system supervisory control to protect it from the consequences of unexpected failures. However, this is not enough for maintenance optimization.

Although SCADA has an almost immediate IRT, it is not able to detect fault very early. When the data exceed threshold values, the fault degradation usually will be advanced and failure may be imminent. Maintenance planning ought to be able to detect, diagnose, and properly evaluate a fault in order to make the best decisions, and this is not possible by implementing only a system supervision process. An HMFD process is a recommended approach for this purpose, especially in organizations with complex engineering systems. As a result, the system's supervision process should become an important ally, especially when any supplementary protection is needed, but being no longer the lead agent in the prevention of risk scenarios for the system under consideration.

7.2 A framework for health monitoring and fault diagnosis

Organizations that have complex engineering systems are likely to have already implemented a system supervision process. It is usually integrated within the operation management system and routine, alerting operators, and triggering protection actions automatically when the behavior of engineering systems is not as expected. Nevertheless, it is not uncommon to find organizations that still struggle to implement a health monitoring and fault diagnosis process.

Many of the reasons for this disparity are related to the intrinsic challenges of the HMFD process that will be addressed in Section 7.3. These challenges are seen by organizations as barriers that delay or limit the implementation of maintenance strategies based on condition monitoring. Furthermore, there are also cases where the difficulty in implementing an HMFD process to support maintenance management is in the structuring of this process itself. Accordingly, Fig. 7.2 presents a conceptual framework that will be discussed in order to contribute to the implementation of the HMFD process in organizations.

The HMFD framework goes beyond the simplified HFMD structure shown in Fig. 7.1. It is composed of a more detailed set of activities, contributing to the understanding of the HMFD process and the relationships between them. In short, it all starts with the acquisition of the system's

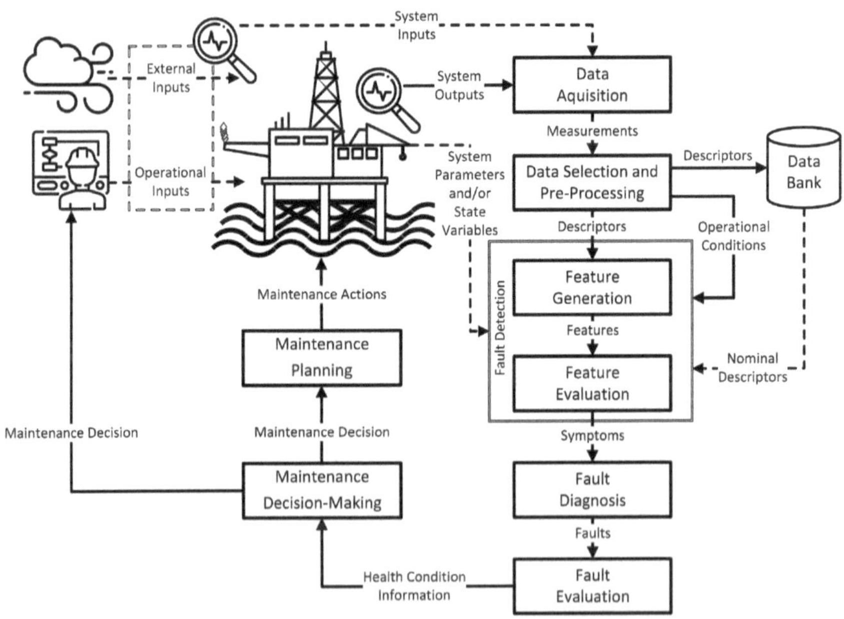

FIG. 7.2 HMFD conceptual framework.

inputs and outputs data for each engineering system covered by the HMFD process. These measurements are selected and preprocessed to perform the fault detection activity. When a fault is detected, fault diagnosis and fault evaluation activities are performed sequentially to support maintenance decision making. Finally, the maintenance decision guides the actions to be done in order to avoid a failure to occur.

It is worth mentioning that data acquisition is a continuous activity. Thus, when a failure is detected, diagnosed, and evaluated, new data are provided for these activities to be reviewed throughout the failure degradation process. Accordingly, maintenance leadership will be able to make new decisions as the analysis evolves with new data. Finally, the following subsections discuss in depth the activities of the conceptual framework for health monitoring and fault diagnosis.

7.2.1 Setting up the health monitoring and fault diagnosis process

As can be seen in Fig. 7.2, a framework for HMFD shall integrate several activities to effectively support maintenance management in the organization. However, before discussing these activities, it is appropriate to address the necessary preparation to implement an HMFD process in line with the presented framework. This setup usually comprises three activities as presented in Fig. 7.3.

First, the organization shall define the relevant engineering systems for health monitoring and fault diagnosis. From them, the engineering systems' failure modes which will be prevented with the support of the HMFD process shall be identified. Therefore, it should be noted that not all physical assets or their failure modes will be addressed through condition-based maintenance. These definitions are usually carried out by the maintenance planning process as it is responsible for defining the maintenance plans for the organization's physical asset portfolio, for instance, through the application of Reliability-Centered Maintenance (RCM).

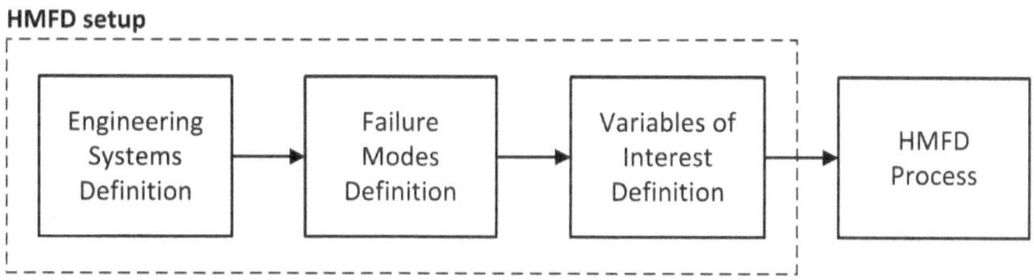

FIG. 7.3 The setup activities for an HMFD process.

Once defined the failure modes of the engineering systems that will be considered in the on-condition maintenance strategy, the maintenance planning manager shall define the variables of interest. These variables are identified through the relationship between each failure mode and its detectability. As this is the last activity for the setup of the HMFD process, after it the organization will be able to initiate the HMFD process properly by means of the data acquisition.

Finally, it is worth mentioning that these setup activities are expected to occur only once before the HMFD process is implemented. However, when the operational context of the organization is changed, it is pertinent to assess the need for updates in the scope of the HMFD process. For example, new engineering systems in the organization's physical asset portfolio may require the inclusion of new failure modes to be prevented in the HMFD process. On the other hand, the replacement of engineering systems by more modern engineering systems can eliminate the need to prevent a failure mode that no longer exists. Accordingly, if the operational context of the organization changes, it is recommended to assess how it can impact the HMFD process.

7.2.2 Data acquisition, selection, and preprocessing

As shown in Fig. 7.2, data acquisition is the basis of the entire process of detecting, diagnosing, and evaluating a system's failures. However, in addition to data acquisition, data selection and preprocessing are important activities for the creation of descriptors, that is, the set of monitored data that describe the system's condition at each instant of time. However, prior to this group of activities, it is necessary to define the variables of interest in the fault detection and diagnosis process.

Once the "*mise en place*" has been done, the physical assets to be considered in the HMFD process are known, as well as their failure modes and the variables of interest that must be used to observe the evolution of these failure modes in the selected assets.

The variables of interest are composed by, but are not limited to, a set of descriptors capable of describing the undergone changes by the components and subsystems of the studied system after the occurrence of the considered fault. However, from the point of view of fault detection and diagnosis requirements, descriptors may not be sufficient. In this way, the system's state variables and parameters may also be part of the set of variables of interest.

Table 7.1 presents an example of the correlation between subsystems, components, failure modes, and variables of interest. In this example, a single subsystem, two components, five failure modes, and nine variables of interest are considered.

TABLE 7.1 Correlation between subsystems, components, failure modes, and variables of interest.

Subsystem	Component	Failure mode	Variable
Subsystem 1	Component 1.1	Failure mode 1.1.A	Variable I
			Variable II
			Variable III
		Failure mode 1.1.B	Variable I
			Variable II
			Variable IV
			Variable V
	Component 1.2	Failure mode 1.2.A	Variable I
			Variable V
		Failure mode 1.2.B	Variable III
			Variable VI
		Failure mode 1.2.B	Variable VII
			Variable VIII
			Variable IX

Among the variables of interest, descriptors are the only ones that have been measured, going through the selection and preprocessing process. The other two kinds of variables of interest, i.e., state variables and parameters, are not measured, but estimated, being applied directly in the feature generation process. This particularity shows that the setup of the HMFD process has a profound influence not only on the data acquisition actions, but also on all other activities of fault detection, diagnosis, and evaluation, as will be seen in the next sections of this chapter.

The main purpose of preprocessing is to remove excessive noise, artifacts, and outliers from the measurements, that is, to make the collected signal free from variations that may eventually be misinterpreted as symptoms in the fault detection process. Several techniques can be used in this process, such as frequency filters in predefined bands, pattern-based analysis from preestablished models, statistical evaluations, among others.

After preprocessing, data selection is carried out, being probably the most relevant process between data acquisition and the Fault Detection and Diagnosis (FDD) process itself, especially in a realistic scenario in which multiple failure modes can develop simultaneously in an engineering system that operates under different conditions.

As discussed in Chapter 5, it is important to distinguish between measurements that are used as descriptors and those that define the different operational conditions under which the system operates. Defining the operational conditions that will be considered in the fault detection and diagnosis process is also part of the configuration step. The number of conditions, the variables considered in each case, and the limit values for such variables are the parameters that must be evaluated by the maintenance team for the implementation of a framework for HMFD. The distinction between different conditions is of great importance since an analysis that does not make such a distinction will inevitably lead to errors since variations in the operational condition of the system would be interpreted as disturbances or faults.

On the other hand, if multiple failure modes are analyzed at the same time, without the variables affected by each of them being grouped (or clustered), the detection and diagnosis process would require a greater correlation between detectability and observability, that is, a greater number of variables would have to be observed and analyzed simultaneously. This scenario, in addition to making the detection activity less sensitive, would hinder the diagnostic activity's ability to isolate and quantify ongoing faults. Thus, good practice in the elaboration of a framework for health monitoring and fault diagnosis for modern engineering systems is the definition of clusters with the variables of interest.

An intuitive way to define these clusters is to consider the variables affected by each failure mode considered. In this way, there would be at least many clusters as the number of failure modes. In the example case presented in Table 7.1, five clusters would be considered a priori. However, it may be interesting to create more comprehensive groupings, which observe the behavior of the components and even subsystems as a whole. In this way, three more clusters could be added to the list originally created from failure modes.

The use of clusters that group the variables of interest considering the different analyzed failure modes is based on the principle of the detectability of a system. From this principle, it is possible to state that only the variables of interest related to a particular failure mode need to be analyzed in the process of detecting this specific failure mode. In this case, even if multiple faults are ongoing in the system, clustering allows the failure modes to be isolated and analyzed one by one, even if there is an overlap in relation to the variables of interest affected by different failure modes.

It is interesting to note that with the adoption of different operation conditions and clusters in the fault detection and diagnosis process, different approaches can be considered depending on the different failure modes, components, and subsystems analyzed. In this way, the methods of feature generating and evaluating, for example, can be specific, theoretically, to each cluster and operational condition. In practice, such a selective approach could be very laborious or consume reasonable computational time, making the distinction between different operational conditions generally applied to the entire process. But it is interesting to keep in mind such a more surgical approach, which can be advantageous in specific cases.

Considering the same example presented in Table 7.1, the list of clusters that could be built is presented in Table 7.2.

Depending on the considered detection methods, one aspect that must be taken into account when clustering the variables of interest is their discretization sampling rate, that is, the time interval between two consecutive samples. Generally, the combination of descriptors and state variables or system parameters is not an issue, since the last two are obtained through models and can be discretized at any rate that is necessary.

On the other hand, the measurements that give rise to the descriptors may have different characteristics with regard to the method of collection and the rate of acquisition. For example, data that is collected online through monitoring or supervision systems and data collected offline through inspection routes have entirely different properties in relation to those aspects.

TABLE 7.2 Relationship between clusters and variables of interest.

Cluster	Variables	Cluster	Variables	Cluster	Variables
Cluster #1	Variable I	Cluster #6	Variable I	Cluster #8	Variable I
	Variable II		Variable II		Variable II
	Variable III		Variable III		Variable III
Cluster #2	Variable I		Variable IV		Variable IV
	Variable II		Variable V		Variable V
	Variable IV	Cluster #7	Variable I		Variable VI
	Variable V		Variable III		Variable VII
Cluster #3	Variable I		Variable V		Variable VIII
	Variable V		Variable VI		Variable IX
Cluster #4	Variable III		Variable VII		
	Variable VI		Variable VIII		
Cluster #5	Variable VII		Variable IX		
	Variable VIII				
	Variable IX				

In this case, if the compatibility of the discretization rate is necessary, one can choose to resample the data. Generally, data collected offline has a lower acquisition rate than data collected online, that is, at the same time interval, there will be fewer samples of offline data than of online data. A data resampling, in this way, can either reduce the number of samples collected online—discarding data that could be correlated with useful information—or increase the number of samples collected offline—generating data artificially, which could be interpreted as the introduction of false information in the database—mainly due to the fact that data collected offline is less accurate than data collected online. For both cases, the outcome of the proposed solution can be considered not ideal.

Thus, using clustering as a way to group data online and offline separately, without discarding useful information or generating artificial data, can be considered an interesting and efficient solution.

7.2.3 Fault detection and diagnosis

Once the clusters with the variables of interest are defined, as well as the different operational conditions of the analyzed system, the next step of a framework for health monitoring and fault diagnosis is the feature generation, starting the fault detection process.

Feature generation is probably at the heart of the detection process. The objective of this step is to generate variables that can characterize the system status, that is, in which condition the system is operating. Generally speaking, it is the generated features that are used in the detection process to determine how far the current behavior of the analyzed system is from its expected or nominal behavior.

As mentioned earlier, variables of interest can be descriptors, state variables, and parameters of a system. Descriptors are originally measurements that went through a preselection and preprocessing process as previously presented. Strictly speaking, descriptors are necessary for any fault detection process, regardless of the method selected.

State variables and parameters, on the other hand, are obtained from models during the feature generation phase, if any of these variables are considered necessary for the fault detection process. That is, if any state variable or system parameter is considered to be a variable of interest, some type of model-based method will be necessary to estimate these variables.

At this point, an interesting conclusion is reached regarding the feature generation according to the variables of interest considered in the fault detection process: If all variables of interest can be obtained from the data acquisition process, it will be more likely that a data-based method is best suited to detect any system fault. If state variables and/or system parameters are listed as variables of interest, a model-based method will most likely be more appropriate in the fault detection process. Taking into account the possibility of clustering, hybrid solutions that use data-based and model-based methods can be considered according to each analyzed failure mode.

To date, no defined procedure is known to assist in the selection of fault detection methods. However, in order to assist the reader in future endeavors, Table 7.3 presents the main characteristics (pros and cons) of the four principal families of fault detection methods, as presented in

TABLE 7.3 Main characteristics of fault detection methods.

	Data-based methods	Model-based methods
Quantitative methods	*Pros:* (1) Are based on large amounts of historical data; (2) Does not depend on any prior knowledge about the process; (3) Can handle high dimensional and correlated process variables; (4) Are simpler to implement than model-based approaches; (5) Are suitable for complex and large-scale systems; (6) Do not require first-principles models, reducing the analysis time and cost; (7) Are preferred when only measurements are available and a system model is not; (8) Have the ability to collect information and provide knowledge that is beyond the current knowledge of experts; (9) Are able to detect faults faster than model-based approaches.	*Pros:* (1) Are based on first-principles knowledge (or deep knowledge) to build mathematical models; (2) Integrates the physical understanding of the system and incorporates the interactions with the environment into the fault detection process; (3) Are more precise and specific about a system than other methods; (4) Have been widely studied, developed, and applied in many process systems; (5) Need small amount of collected data; (6) Can outperform data-driven approaches when an accurate model is available.
	Cons: (1) May require a preprocessing step to extract the useful information from data with a high computational cost; (2) Its performance is based on the quality and the quantity of the collected data; (3) May require a substantial number of labeled data samples for training; (4) Performance degradation may	*Cons:* (1) Require an explicit mathematical model; (2) Require a large effort in model construction especially if dynamical aspects are included; (3) Are not easily implementable for early fault detection of complex processes; (4) Require reliable a priori knowledge about the process; (5) Model update becomes more difficult as the system complexity

Continued

TABLE 7.3 Main characteristics of fault detection methods—cont'd

	Data-based methods	Model-based methods
	occur when the system is operating in an unknown condition.	increases; (6) Most models rely on a set of assumptions and approximations; (7) May require a model validation step; (8) Are generally suitable only for linear systems and some specific nonlinear problems; (9) In many cases, are not suitable for large-scale and complex systems; (10) Are performance-dependent on model accuracy and quality.
Qualitative methods	*Pros:* (1) Data are represented as primitives which establish a trend; (2) Are based on a representation of experiential knowledge in a domain; (3) Allow a syntactic way of representing knowledge; (4) Provides relatively accurate results.	*Pros:* (1) Are based on empirical associations or superficial knowledge about the system; (2) Yield predictions on complex systems behaviors; (3) Can incorporate important behavior patterns; (4) Produce descriptions that are closer to the level of human insights into the behavior of the system; (5) Represent reasoning about continuous aspects of entities and systems in a symbolic and human way; (6) Are suitable when the detailed mathematical model is not available.
	Cons: (1) May become computationally stiff for complex processes; (2) May suffer from the limitations common to rule-based methods.	*Cons:* (1) Depend on experience obtained in a specific context; (2) Do not cover behavior previously unknown or observed in the system; (3) Inferred rules are based on the historical failure cases and expert's knowledge; (4) Predictions may be often ambiguous in complex systems; (5) Are only suitable when the number of inputs, outputs, and states of a system is relatively small.

Chapter 5, and according to the current state of the art (Alauddin, Khan, Imtiaz, & Ahmed, 2018; Bahaloo, Ebrahimi, & Samadi, 2009; Bondavalli, Favilla, & Bodini, 2009; Tidriri, Chatti, Verron, & Tiplica, 2016; van Harmelen, Lifschitz, & Porter, 2008; Venkatasubramanian, Rengaswamy, & Kavuri, 2003; Venkatasubramanian, Rengaswamy, Kavuri, & Yin, 2003; Venkatasubramanian, Rengaswamy, Yin, & Kavuri, 2003).

After the feature generation phase, follows the feature evaluation phase. Some detection methods, however, can incorporate these two phases in a single process, with the response to the characteristics of the detection itself: whether or not there is an ongoing fault in the system, when it was detected and, if possible, a value that correlates with the probability of this detection being true or not.

It is during the feature evaluation phase that the definition of the operational condition in which the system is found is necessary in order to avoid false alarms or even the nondetection of a fault in progress. The more refined the definition of the different operating conditions of the system, the more precise the feature evaluation step will be.

Another aspect that must be taken into account during the feature evaluation phase is the possibility that disturbances in the system can be misinterpreted as faults in progress. Thus, methods that ratify the response of the feature evaluation phase must be considered and applied. As the FDD process is usually carried out systematically and periodically, considering that it is necessary to repeat the detection result a number of times before starting the diagnostic process can be a good solution.

Once the detection process has confirmed the presence of certain symptoms, the diagnostic process should check which faults may be responsible for them, and which one is most likely to be occurring. There are many methods capable of performing fault diagnosis as already presented in Chapter 6, which also offers a discussion on how to choose the best method for a given application. Table 7.4 presents the two categories of methods used when dealing with fault diagnosis.

It is worth mentioned that, in some cases, a single method can be considered for the entire FDD process, i.e., the detection and diagnosis are performed jointly. However, it would be interesting, especially when dealing with modern engineering systems, to use different methods for detection and diagnosis, and in these cases, an intermediate phase between the two processes may be necessary to translate or make the detection method output and the diagnosis method input compatible. For example, summarizing the detection result in a Boolean response or using fuzzy logic as a way to interpret the output of the detection process and create a graduated input in the diagnosis process can be interesting alternatives if it is desirable to simplify the flow of information between these two stages of the FDD process.

TABLE 7.4 Categories of fault diagnosis methods.

Category	Features	Examples of methods
Classification methods	• No need for prior information about fault-symptom relationship • Knowledge base is experimentally trained	Naïve Bayes classifier Support vector machine Neural networks
Inference methods	• Based on prior information about fault-symptom relationship • Structured knowledge base	Fault tree analysis Expert systems Bayesian networks

7.2.4 Fault evaluation

When the FDD process indicates the presence of a fault in the system being monitored, the maintenance managers will need to make a decision on what actions should be taken to mitigate, eliminate, or somehow deal with such a fault. However, before this decision is made, maintenance managers will need more information about the fault, i.e., they will need to evaluate the fault in order to make a more assertive decision. Fault evaluation should take into account three main aspects: its veracity, severity, and remaining useful life (RUL). Although it is not always possible to assess them, an effort shall be made to ensure that these aspects are always considered in the maintenance decision-making process.

The first aspect to be considered by decision makers about the diagnosed fault is its veracity. Keep in mind that, due to epistemic uncertainties, method limitations, and sensor reading errors, the FDD process may erroneously identify a fault. Hence, whenever possible, a verification regarding fault veracity should be made. This verification can be done through an inspection of the equipment, carried out by the maintenance team. Depending on the operational context and the type of equipment, such an inspection may be impossible if, for example, this results in a production stoppage that would have a major negative impact on revenue.

Another aspect to be considered when evaluating the fault is its severity. Knowing, in advance, the consequences that the progression of a fault into failure can bring to the system, the personnel, and the environment directly influence the decision making to be made by the maintenance managers. Several criteria shall be taken into account in order to categorize or classify the severity of a fault. Among them, the following stand out (Melani, Murad, Caminada Netto, de Souza, & Nabeta, 2018):

Environmental impact: represents the pollution or contamination of soil, water, or air that may potentially be caused by the fault or by the failure it may progress into. The greater the potential impact on the environment, the more critical the fault.

Impact on the community: represents the potential physical damage that the fault (or the failure it may progress into) can cause to workers or the community around it. The greater the potential impact on the community, the more critical the fault.

Impact on system availability: represents the potential damage that the fault (or the failure it may progress into) can cause in the system's production. The longer the system downtime due to a fault, the more critical it is.

Patrimonial impact: represents the total financial impact generated to the system due to a certain fault (or the failure it may progress into), including repair costs, loss of production, and damage to adjacent equipment.

The third aspect to be considered is the Remaining Useful Life (RUL) of the equipment presenting the fault. By definition, RUL is the time interval in which the equipment is likely to operate until a failure occurs and a repair is required. By knowing the RUL estimate, maintenance managers can optimize the scheduling of maintenance actions for the equipment, making the most of its production capacity, and, at the same time, minimizing the consequences of an unwanted failure.

Several authors propose the use of different techniques to predict RUL through prognosis analysis (Si, Wang, Hu, & Zhou, 2011; Wang, Zhao, & Addepalli, 2020; Zhang, Si, Hu, & Lei, 2018). In general, the technique to be used depends on the information previously available about the behavior of the equipment when it has a detected fault, in addition to the health monitoring data that is being collected. As the lack of monitoring data can make it impossible to predict the RUL, the health monitoring system shall be designed in order to make such analysis feasible.

Unlike fault detection and fault diagnosis that were covered in depth in two chapters of this book, fault evaluation will be only given an overview in this section, especially on prognosis. Prognosis is a challenging activity in the fault evaluation of the HMFD process, which has the potential to provide the greatest benefits in terms of both reduced Operational and Support (O&S) costs and Total Cost of Ownership (TOC) for the improved life cycle and safety of many types of complex machines and systems. The evolution of systems monitoring and diagnosis has led to the recognition that the predictive prognosis is desired and technically possible even for engineering systems. The aviation industry is probably the best example of this statement (Vachtsevanos, Lewis, Roemer, Hess, & Wu, 2006).

Among the main issues related to the prediction of RUL in prognosis, analysis is how to perform the prediction based on the monitoring information (Lei et al., 2018). In the literature, one finds that the approaches for calculating the RUL of a single variable are basically divided into two types: physics-based, also referred to as principle-based, and data-based (Hu, Liu, Lu, & Zhang, 2014; You, Li, Meng, & Ni, 2010). While the physics-based approach assumes an accurate and available mathematical model to describe the performance of the system, the data-based approach assumes that the statistical characteristics remain unchanged unless a defect occurs (You et al., 2010).

This classification into two categories, however, is not a consensus. Several studies have reviewed and classified the prognosis approaches but their nomenclature patterns are different from each other, causing some confusion (Lei et al., 2018). In order to unify the classifications, Lei et al. analyzed and compared the meanings of the nomenclatures and the scope in different works and reorganized the approaches into four categories taking into account the basic technique and associated methodologies: physics model-based approaches, statistical model-based approach, Artificial Intelligence (AI) approach, and hybrid approach (Lei et al., 2018), as shown in Table 7.5.

In order to specify precision requirements for the prognosis, the following aspects shall first be defined: the level of degradation that is considered unsatisfactory or undesirable for the engineering system in question; a minimum amount of warning time to provide operators and maintainers with the necessary information that can be used to define the actions to be taken before the failure occurs or the unwanted condition is reached; and a minimum level of confidence that the estimated remaining useful life is equal to or greater than the minimum level of warning time required (Vachtsevanos et al., 2006).

Finally, it is worth mentioning that there are still a few works that address the prognosis using multivariable approaches. In contrast to the methods developed for FDD, prognostic methods, despite their great importance for Industry 4.0, currently do not have many approaches capable of meeting the needs of working with increasingly complex systems. Accordingly, the RUL estimate is generally carried out based on single variable analysis, which can clearly be considered a field of future research in this context.

7.2.5 Maintenance decision making

Maintenance decision making is carried out after the detection, diagnosis, and evaluation of a fault, and it is supported by information provided by these three previous activities of the HMFD process. This is an

TABLE 7.5 Approaches to RUL estimation.

Approaches	Features	Main methods
Physics model-based approaches	It describes the degradation process by means of a mathematical model based on the failure mechanism or the damage principle. It corresponds to about 10% of the applications identified in the literature	The Paris-Erdogan (PE) model is one of the most widely used physical methods. Other models are Formam crack growth law and Norton law
Statistical model-based approaches	It estimates the equipment VUR through the establishment of statistical models based on empirical knowledge and, generally, it presents the prediction of the VUR conditioned to a probability density function (pdf) dependent on the observations. The category has become the most popular among the four and corresponds to about 56% of the applications identified in the literature	Wiener process; random coefficients; Markov models; proportional hazards; gamma process; inverse Gaussian process and Autoregressive (AR)
Artificial intelligence approaches	The approach attempts to learn the degradation pattern of the equipment using AI techniques in available observations rather than building physical or statistical models. It is capable of dealing with complex issues, making it more attractive with about 26% of the applications identified in the literature	Artificial neural network; support vector machine; fuzzy neural system and Gaussian process regression
Hybrid approaches	The hybrid approach tries to integrate the advantages of different approaches in its integration. Currently, it is the approach that covers the smallest part of the applications with about 8% of the applications identified in the literature	The methods combine the models described by the other three approaches, e.g., AI techniques with random coefficients or artificial neural network with AR

important activity that will guide how a particular fault will be managed. Depending on this decision, different courses of action may be taken for the physical asset.

In other words, it is during the maintenance decision-making activity that maintenance leaders decide what should be done in face of the detection, diagnosis, and evaluation information. As shown in Fig. 7.2, the maintenance planning process is commonly triggered to plan the maintenance actions for the fault correction, closing the loop in the HMFD framework. However, this is not the only possible decision in these situations.

Maintenance managers may also decide to immediately stop the operation when it is not possible to ensure a safe operation due to the fault evaluation. This decision is usually supported by the fault evaluation information such as trends of degradation and prediction of the remaining useful life as well as the consequences associated with the occurrence of the failure in the physical asset.

In some situations, maintenance decision making may combine different actions. For instance, maintenance planning is triggered to start the planning of the maintenance intervention for fault correction while the operation is informed to change the operational conditions to contribute to slowing down the degradation process or to mitigate the probability of failure.

Accordingly, it shall be clear to the reader that there is no general rule for making maintenance decisions in the context of the HMFD. Some organizations will decide to start maintenance planning even though the fault evaluation showed an early degradation, while others will decide to wait longer to start it. This can be associated with the criticality of the physical asset, with the failure assessment, and with the consequences of the failure among other factors. Finally, it is worth mentioning that maintenance planning is covered in detail in the next chapter that will provide the reader with further information about it.

7.3 Challenges of health monitoring and fault diagnosis

Each activity that makes up the HMFD framework in Fig. 7.2 can be implemented and performed in different ways. There are several methods, for example, that can be used in the detection and diagnosis activities, as presented in Chapters 5 and 6. Therefore, it can be difficult to decide which approach is best for each activity, that is, which one best meets a particular set of needs.

In order to make a conscious choice about how each of these activities can be performed in a given operational context, Sections 7.3.1–7.3.4 describe challenges faced when implementing and using an HMFD process.

7.3.1 Complexity of engineering systems

The complexity in engineering systems originates a difficulty in modeling its behavior because of the overwhelming number of interactions existing between its many components, operators, and environment.

This large number of relationships is the main source of nonlinearity and uncertainty in the behavior of the system (Webster, 2016).

With the arrival of the Fourth Industrial Revolution, the increasing automation of traditional production and manufacturing processes brought even more complexity to engineering systems. In order for the potential of industrial digitalization proposed by Industry 4.0 to be properly explored, several plants have undergone technological updates in order to increase their levels of automation, monitoring, and communication (Mourtzis, Papakostas, & Makris, 2019).

This increase in complexity has resulted in a greater amount of data generated and collected by these new engineering systems. Data have been generated in unprecedented volume, variety, and velocity. And it is in this Big Data scenario imposed by Industry 4.0 that the HMFD process is inserted. Since the HMFD process is extremely dependent on monitoring data, this scenario is, in fact, ideal for its implementation.

However, while this Big Data and high complexity scenario offer opportunities to develop an optimized HMFD process, it also poses challenges. Perhaps the greatest of these challenges is dealing with the fact that it is practically impossible to know all possible system operating states, that is, the number of components (and, consequently, the interaction between them) is so great that the behavior of the system when any set of unwanted or unpredictable events happens may be unknown. Fig. 7.4 illustrates the state-space diagram of a system.

When developing, implementing, and performing the HMFD process, it is important to know what its limitations are in identifying the current state of the system. Moreover, it is necessary to take into account that the

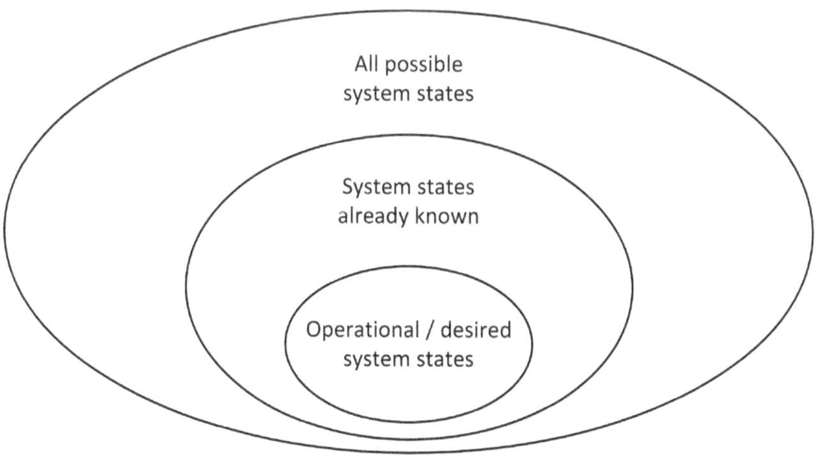

FIG. 7.4 Representation of system states.

HMFD needs to be constantly updated, in order to expand the set of known system states as they appear during the system operational life.

7.3.2 Data management

Data management is undoubtedly one of the most important topics of the Fourth Industrial Revolution, covering all disciplines related to data administration, control, and processing as a valuable resource.

In general, the term "data" can refer to raw data, information, and knowledge. Regarding HMFD processes, "data" may refer to all the information flows presented in the proposed framework of Fig. 7.2, that is, measurements, system parameters, state variables, descriptors, operational conditions, features, symptoms, faults, health condition information, maintenance decision, and maintenance actions.

It is recommended that any HMFD process creates a record that enables one to retrace the steps of any previous analysis. Obviously, such a practice tends to create a large mass of data that must be managed in a structured way. However, it can be said that the most delicate point of this management is how to handle the measurements, basic information for any HMFD process.

Currently, much commercial software that acts as monitoring and fault diagnosis systems tends to store only monitored data related to some fault detection; or, at most, create records of the data collected when any significant variation in the pattern of the monitored signals is detected. In both cases, two alternatives are generally found: or a synchronous data acquisition process is performed, analyzing the monitored data, and acting as a trigger for the data record; or all the data collected are stored in a temporary database, being analyzed asynchronously to the acquisition process and, then, only the data considered relevant are permanently stored.

The main issue with these data management approaches is the prior definition of which data is considered relevant or not, without knowing a priori which failure mode is underway in the system. In addition, many failure modes can progress very slowly, taking a long time for any deviation in the system behavior to be detectable.

Thus, correctly managing the data of an HMFD process becomes a major challenge. Assuming that all information is relevant, it is recommended that all data be stored in a structured way. As the measurement acquisition rate can be relatively high, as well as the number of points monitored, the resulting mass of data can become really colossal. Looking on the bright side, the cost of storing large amounts of data is currently low and technologies such as cloud storage are available, making data managers' jobs a little easier.

7.3.3 Alignment

In addition to these previous challenges, the organization may face other barriers regarding the alignment of models and interpretation of results in the HMFD process. As there are several methods for fault detection and diagnosis, there is not a standard way for the detection outputs to be incorporated as inputs into fault diagnosis. A general rule to keep in mind is that fault detection results shall be suitable as input to the chosen fault diagnosis method and, if necessary, these results shall be processed to ensure this.

Another concerned alignment is with the fault evaluation activity. After the fault diagnosis, the HMFD process performs the fault evaluation for the assessment of the detected fault that is in progress. Similarly, this activity also requests inputs from previous activities of the HMFD framework. Therefore, this information shall be suitable for fault evaluation methods as well.

The reader may find several works on the literature that addressed the fault detection, diagnosis, and evaluation under different approaches and combinations of methods (Alauddin et al., 2018; Lei et al., 2018; You et al., 2010). However, maintenance leaders may not find that same variety regarding computational tools to support the implementation of the HMFD process. Standard software is usually developed to address the HMFD in a specific way. In other words, the choice and integration of methods were previously defined by the software developer. Moreover, it is common to find commercial software that performs only part of the HMFD framework such as data acquisition until fault detection, for instance.

After an appropriate alignment of the activities within the HMFD process, the organization will have a set of information for maintenance decision making. By analyzing these results, maintenance managers shall be able to make better decisions in relation to the detected fault. Nevertheless, as this activity directly depends on the ability to interpret the results, it can also be a barrier for the organization.

This can be explained by the nondeterministic nature of the results of the activities of fault detection, fault diagnosis, and fault evaluation. In other words, these activities usually provide probabilistic results that make decision making even more complex for the decision makers. For instance, fault diagnosis may not ensure with 100% certainty that the detected fault is related to a specific failure mode as well as its prognosis may not ensure the predicted remaining useful life. Accordingly, the qualification of maintenance leaders and the participation of technical personnel in the maintenance decision making may contribute to mitigate the struggle regarding this barrier in the HMFD process.

7.3.4 Embracing HMFD

This section could not end without highlighting the importance of the human factor for the implementation of the HMFD process in organizations. Even if the previous technical challenges are overcome, maintenance may not obtain the benefits of HMFD if the personnel is not committed to the activities in the process. It can be said that the human factor shall be as strong as the technical setup in the HMFD process.

Organizations need to be aware of this challenge of empowering people to carry out the activities in the HMFD. In addition to being trained, personnel involved in the HMFD process need to be aware of their contribution to the effectiveness of the process and the benefits that can be achieved. Furthermore, the implications of not performing as expected in asset management and maintenance management shall be covered as well.

As motivation and leadership are one of five critical areas within human resources that significantly have a bearing on asset management (Kriege & Vlok, 2015), the organization may find it suitable to encourage interaction with the human resources (HR) department during the implementation of an important process for maintenance management. For instance, HR can support maintenance managers in identifying and assigning the activities for the employees and help maintenance managers and other leaders in promoting the implementation and demonstrating commitment.

Once the HMFD process has been implemented, it shall be given credibility within the organization. Top management should endorse the process by ensuring that leaders use its results in decision making, especially in maintenance and operation management. This could contribute to the changes in the organizational culture that an HMFD may require.

References

Alauddin, M., Khan, F., Imtiaz, S., & Ahmed, S. (2018). A bibliometric review and analysis of data-driven fault detection and diagnosis methods for process systems. *Industrial & Engineering Chemistry Research, 57*(32), 10719–10735. https://doi.org/10.1021/acs.iecr.8b00936.

Bahaloo, H., Ebrahimi, A., & Samadi, M. (2009). Misalignment modeling in rotating systems. In *Volume 6: Structures and dynamics, parts A and B* (pp. 973–979). ASMEDC. https://doi.org/10.1115/GT2009-60121.

Bondavalli, C., Favilla, S., & Bodini, A. (2009). Quantitative versus qualitative modeling: A complementary approach in ecosystem study. *Computational Biology and Chemistry, 33*(1), 22–28. https://doi.org/10.1016/j.compbiolchem.2008.07.012.

Center for Chemical Process Safety (CCPS). (2015). *Guidelines for initiating events and independent protection layers in layer of protection analysis* (p. 384). Wiley-Aiche.

Hu, Y., Liu, S., Lu, H., & Zhang, H. (2014). Remaining useful life assessment and its application in the decision for remanufacturing. *Procedia CIRP, 15*, 212–217. https://doi.org/10.1016/j.procir.2014.06.052.

Kriege, L., & Vlok, P. J. (2015). *Human resources within ISO 55000—The hidden backbone to the asset management system* (pp. 435–446). https://doi.org/10.1007/978-3-319-15536-4_34.

Lei, Y., Li, N., Guo, L., Li, N., Yan, T., & Lin, J. (2018). Machinery health prognostics: A systematic review from data acquisition to RUL prediction. *Mechanical Systems and Signal Processing, 104*, 799–834. https://doi.org/10.1016/j.ymssp.2017.11.016.

Melani, A. H. A., Murad, C. A., Caminada Netto, A., de Souza, G. F. M., & Nabeta, S. I. (2018). Criticality-based maintenance of a coal-fired power plant. *Energy, 147*, 767–781. https://doi.org/10.1016/j.energy.2018.01.048.

Mourtzis, D., Papakostas, N., & Makris, S. (2019). Complexity in Industry 4.0 systems and networks. *Complexity, 2019*, 1–2. https://doi.org/10.1155/2019/7817046.

Si, X.-S., Wang, W., Hu, C.-H., & Zhou, D.-H. (2011). Remaining useful life estimation—A review on the statistical data driven approaches. *European Journal of Operational Research, 213*(1), 1–14. https://doi.org/10.1016/j.ejor.2010.11.018.

Tidriri, K., Chatti, N., Verron, S., & Tiplica, T. (2016). Bridging data-driven and model-based approaches for process fault diagnosis and health monitoring: A review of researches and future challenges. *Annual Reviews in Control, 42*, 63–81. https://doi.org/10.1016/j.arcontrol.2016.09.008.

Vachtsevanos, G., Lewis, F., Roemer, M., Hess, A., & Wu, B. (2006). *Intelligent fault diagnosis and prognosis for engineering systems*. John Wiley & Sons, Inc. https://doi.org/10.1002/9780470117842.

van Harmelen, F., Lifschitz, V., & Porter, B. (2008). *Handbook of knowledge representation* (p. 1034). Elsevier.

Venkatasubramanian, V., Rengaswamy, R., & Kavuri, S. N. (2003). A review of process fault detection and diagnosis. *Computers & Chemical Engineering, 27*(3), 313–326. https://doi.org/10.1016/S0098-1354(02)00161-8.

Venkatasubramanian, V., Rengaswamy, R., Kavuri, S. N., & Yin, K. (2003). A review of process fault detection and diagnosis. *Computers & Chemical Engineering, 27*(3), 327–346. https://doi.org/10.1016/S0098-1354(02)00162-X.

Venkatasubramanian, V., Rengaswamy, R., Yin, K., & Kavuri, S. N. (2003). A review of process fault detection and diagnosis. *Computers & Chemical Engineering, 27*(3), 293–311. https://doi.org/10.1016/S0098-1354(02)00160-6.

Wang, Y., Zhao, Y., & Addepalli, S. (2020). Remaining useful life prediction using deep learning approaches: A review. *Procedia Manufacturing, 49*, 81–88. https://doi.org/10.1016/j.promfg.2020.06.015.

Webster, C. S. (2016). Safety in unpredictable complex systems—A framework for the analysis of safety derived from the nuclear power industry. *Prometheus, 34*(2), 115–132. https://doi.org/10.1080/08109028.2017.1279873.

You, M.-Y., Li, L., Meng, G., & Ni, J. (2010). Two-zone proportional hazard model for equipment remaining useful life prediction. *Journal of Manufacturing Science and Engineering, 132*(4). https://doi.org/10.1115/1.4001580.

Zhang, Z., Si, X., Hu, C., & Lei, Y. (2018). Degradation data analysis and remaining useful life estimation: A review on Wiener-process-based methods. *European Journal of Operational Research, 271*(3), 775–796. https://doi.org/10.1016/j.ejor.2018.02.033.

CHAPTER

8

Engineering systems asset management

A man who doesnot think and plan long ahead will find trouble right by his door.
Confucius (551–479 BCE) in: Analects

It is not by chance that the Analects of the Chinese philosopher Confucius is one of the most widely read and influential books in China. With a collection of ancestral sayings that seem immune to time, it still serves as an inspiration for modern problems. By encouraging people to cultivate planning in order to obtain a journey without unexpected problems, this selected saying fits well into this chapter's context of engineering systems asset management.

Maximizing the realization of value from physical assets through asset management is a contemporary approach to support the achievement of organizational goals. Nevertheless, in view of the increase in complexity and pressure for better performance of engineering systems, asset management depends on appropriate planning and implementation of support techniques, such as those covered in previous chapters.

Accordingly, this chapter intends to discuss asset management of modern engineering systems with a particular focus on the maintenance stage of their life cycle. Sections 8.1 and 8.2 provide an overview of the asset management discipline through its fundamentals and its interfaces with maintenance management. Then, a maintenance planning framework for asset management of complex engineering systems is presented in Section 8.3. Finally, the main issues related to decision-making and improvement in maintenance planning are discussed, as well as available support methods.

Reliability Analysis and Asset Management of Engineering Systems

https://doi.org/10.1016/B978-0-12-823521-8.00008-6

8.1 Asset management fundamentals

All organizations have assets that are used to generate value in their organizational context. Regarding physical assets such as machinery, equipment, utilities, vehicles, buildings, and other infrastructure, they play key roles in supporting the organization's core activities. Nevertheless, like the reader, leaders of organizations may wonder if this has been done effectively and efficiently.

According to ISO 55000, asset management is the coordinated activity of an organization to realize value from assets (ISO, 2014a). In other words, it is an effort on the part of the organization to plan, implement, review, and improve asset management processes and activities in order to maximize the achievement of value from its asset base. Therefore, asset management does not focus on the assets themselves, but on the value that the organization can derive from them (GFMAN, 2014).

This realization of value normally involves a balancing of costs, risks, and opportunities against the performance benefits in order to achieve organizational objectives (ISO, 2014a). Moreover, it is so related to the time perspective that the value of asset management is not only a question of how much value an organization can realize from its assets but also when this value can be realized (GFMAN, 2016a). Therefore, asset management comprises different stages of the asset's life cycle, which can be considered from its acquisition, through operation and maintenance, to disposal and replacement, as shown in Fig. 8.1. Although this chapter initially provides general concepts on asset management, its primary focus, like of the book, is on realizing value from assets during the maintenance stage of their life cycle.

In fact, this concept of value is so important for asset management that it is the first of its four fundamentals in the ISO 55000 series for asset management. Assets exist to provide value for the organization and its stakeholders (ISO, 2014a). However, this term can be confusing when sometimes used as a direct substitute for "benefit" while ignoring any disadvantages. An appropriate representation of value is a gain being greater than any associated pain, and, therefore, good value implies an attractive cost-benefit ratio (Woodhouse, 2018).

Accordingly, what constitutes value will depend on the organizational objectives, the nature and the purpose of the organization, and the needs and expectations of its stakeholders (GFMAN, 2016a). The understanding of the value obtained from assets through asset management is more evident with the alignment between asset management objectives and organizational objectives.

This alignment is known as another fundamental of asset management. According to ISO 55000, asset management translates the organizational

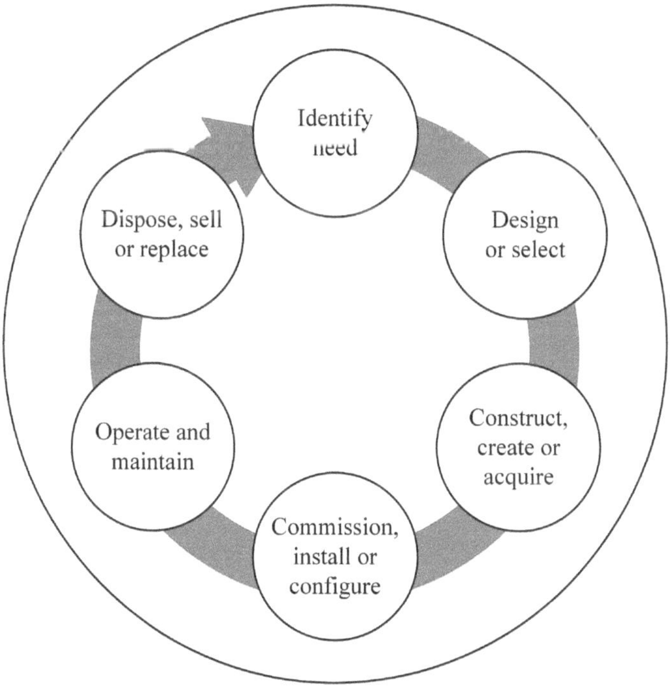

FIG. 8.1 Physical assets' life cycle stages.

objectives into technical and financial decisions, plans, and activities (ISO, 2014a). By connecting an organization's strategic plan and the asset management activities delivered by staff, one enables everybody to understand how they contribute to achieving success (IAM, 2015).

The asset management alignment makes it possible to ensure that all the employees, suppliers, and contractors that might influence the life cycle of the assets are aware of what the organization requires of them to achieve the expected objectives (Blackmore & Leeprechanon, 2015). In other words, this alignment contributes to the effectiveness of the realization of value from the assets. Lastly, it is worth mentioning that during the useful life of the assets organizational objectives are likely to change, requiring this alignment to be continuously reviewed (Ruitenburg & Braaksma, 2017).

This brings us to leadership, the third and determinant fundamental of asset management for the realization of value. Leadership and commitment from all managerial levels are essential for successfully establishing, operating, and improving asset management within the organization (ISO, 2014a). In addition, leadership represents one of the most important human factors for asset management, because the performance of employees can be significantly influenced by the quality of leading (Kriege & Vlok, 2015). Without it, asset management plans can fail due to insufficient or inappropriate asset provision or support (Hastings, 2015).

Finally, in order to realize value from assets efficiently, a final fundamental supports the discipline of asset management: assurance. As asset management encompasses monitoring and auditing of processes and outcomes to confirm that assets and the asset management process are operating as intended (IAM, 2015), it gives assurance that assets will fulfill their required purpose (ISO, 2014a).

For a better understanding, Fig. 8.2 represented the influence of these four fundamentals in a simplified conceptual framework for realizing value from the assets through asset management.

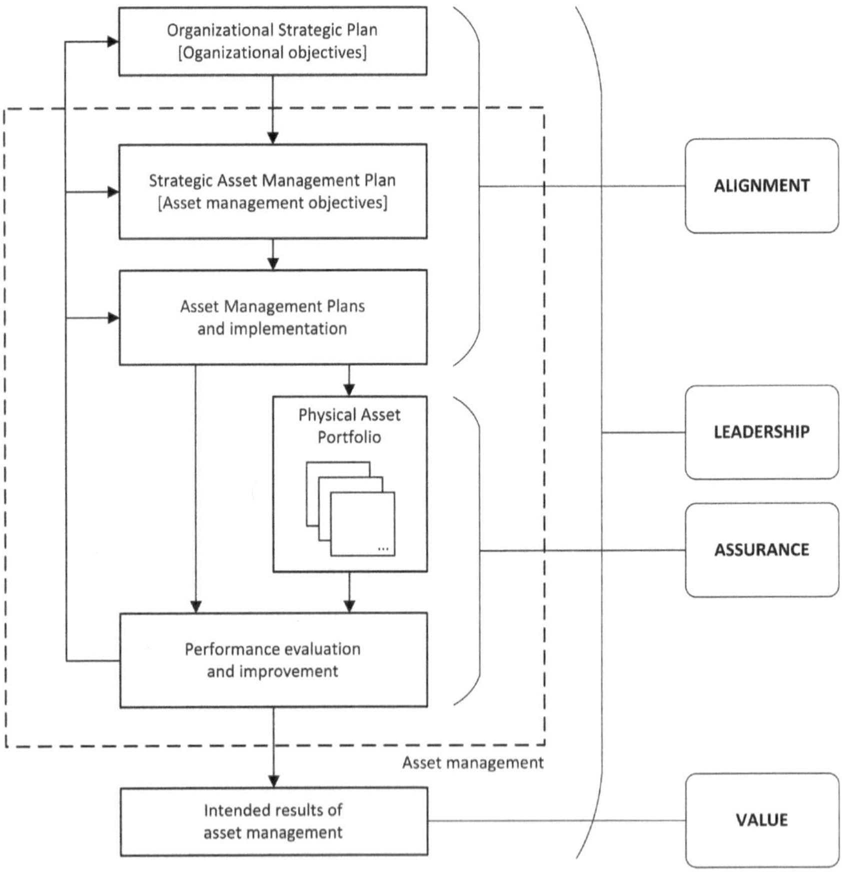

FIG. 8.2 Asset management and its four fundamentals.

Although the development of physical assets has been part of human activity since early times, asset management as a recognized discipline arises from the complex technical nature of modern systems (Hastings, 2015). For a long time, the term "asset management" has been related only to the business and financial areas. According to The Institute of Asset Management, it was only in the 1980s that this term began to be used in the context of physical assets by private and public companies around the world (IAM, 2015).

This was a period marked by major incidents such as the Bhopal industrial disaster in 1984, the Chernobyl nuclear accident in 1986, and the explosion of the Piper Alpha oil platform in 1988. In addition, both in Australia and New Zealand the public sector was struggling with low levels of service, increasing costs, and poor planning. This was also observed in the United States, where the US National Council on Public Works adopted federal asset management policies with an emphasis on achieving a desired level of service at the lowest life cycle cost in areas such as transportation, water, and wastewater (IAM, 2015).

Accordingly, in the '90s, businesses became aware of the asset management concept and, although the subject was not well defined at the time, there seemed already to exist a consensus about its interdisciplinary character (Wijnia & de Croon, 2015). Due to the expansion of infrastructure around the world, the paradigm began to migrate from construction to maintenance, repair, and management of facilities (Park, Park, & Lee, 2016).

This growing demand on the subject led to the development of several asset management guidelines such as the "Total Asset Management Manual" in 1993 and later the "International Infrastructure Management Manual" in 2000. In addition, these developments were accompanied by the formation of professional organizations in many countries which acted as catalysts for the recognition of the key role of asset management (Hastings, 2015).

Nevertheless, asset management as a discipline was only fully conceptualized with the formal requirements of the Publicly Available Specification 55 (PAS 55) in 2004 whose development was conducted by the IAM, in collaboration with the British Standards Institution (BSI) and 49 organizations from 10 countries (Woodhouse, 2014). This publication can be considered the first international specification for structured and standardized asset management. Due to its success and international acceptance, PAS 55 was revised in 2008 and included two parts: PAS 55–1 which specifies the requirements for the optimal management of physical assets and PAS 55–2 which presents guidelines for the application of PAS 55-1.

In 2009, PAS 55 was put forward to ISO, the International Organization for Standardization, as the basis for a new international standard for asset management (Woodhouse, 2014). As one of its main differences, the ISO 55000 series was built to be applied to any organization and type of asset as long as the assets are an important key factor for the achievement of organizational objectives (Konstantakos, Chountalas, & Magoutas, 2019), while PAS 55 focused primarily on physical assets. The development of the ISO 55000 series for asset management was approved in 2010, and three international standards (ISO 55000, ISO 55001, and ISO 55002) were published in 2014 (Hodkiewicz, 2015). After allowing one year of overlap with ISO 55000 for organizations to update their certifications, PAS 55 was withdrawn as a formal specification in 2015 (Konstantakos et al., 2019).

The main milestones in the development of asset management over the years are summarized in Table 8.1.

According to the IAM, modern society is heavily reliant on physical assets in order to function effectively (IAM, 2015). As these assets allow organizations to provide services and products, their performances impact the achievement of organizational objectives. Furthermore, their performances have been increasingly important to maintain and improve competitive advantages in asset-intensive industries (Parlikad, Wang, Russell, & Attwater, 2014) such as electricity generation and supply, water

TABLE 8.1 Milestones in the development of asset management.

Before 1989	1990 to 1999	2000 to 2009	2010 to 2019
In the '70s: Terotechnology emerges with practices applicable to physical assets in the maintenance phase (installation, commissioning, maintenance, and replacement)	**In the '90s:** Asset management began to be associated with planning and maintaining the condition and performance of physical assets and infrastructure	**2004 to 2008:** Publication of PAS 55 by IAM and BSI (first version in 2004 and second in 2008) for asset life cycle management	**2010 to 2013:** Development of the international ISO 55000 series of standards for asset management by the ISO technical committee (TC251)
Until the end of the '80s: The term asset management was associated with the financial sector (business economics)	Growing demand for good practices and guidelines in the approach to asset management throughout the entire life cycle	**In 2009:** Definition of PAS 55 as a reference for the development of an ISO international standard	**In 2014:** Launch of the international ISO 55000 series of standards (ISO 55000, 55001, and 55002)

supply and treatment, transportation, oil and gas extraction and processing, mining and minerals processing, chemicals, manufacturing, aviation, and defense (Hastings, 2015).

In addition to pressure on asset performance, these organizations face the inherent challenges of maintaining complex engineering systems. Their portfolios comprise several industrial assets that are composed of different interrelated components whose interactions, together with the state of each component, affect the state and performance of the system itself (Roda & Macchi, 2016). This corroborates the need for a systematic approach to the realization of value from physical assets in these organizations.

8.2 Maintenance in asset management landscape

Exposure to the global economy and international competition constantly requires organizations to adapt and create competitive advantages in order to guarantee their survival. This context has placed enormous pressure on maintenance management in the pursuit of better performance (Ihemegbulem, Baglee, & Wheatley, 2017). For a long time, the maintenance of physical assets was seen only as an operating expense and the maintenance management was not seen as strategic for organizations. However, this paradigm was overcome after years of maintenance evolution and the rise of physical asset management.

Facilities and equipment, however well designed, do not remain safe or reliable if there is no maintenance (Arunraj & Maiti, 2007). Managing maintenance comprises the planning, organizing, implementing, and controlling of the maintenance activities supported by maintenance leaders (Haroun & Duffuaa, 2009). In addition, it aligns maintenance delivery activities with organizational objectives and strategies (GFMAN, 2016b).

Maintenance is one of the key levers to deliver business outcomes as about two thirds of the total life cycle cost is consumed in the operate and maintain phase of a physical asset (GFMAN, 2016b). Moreover, evidence shows that poor maintenance can result in process failures that can put people at risk, cause loss of revenues, and impact operational availability (Ihemegbulem & Baglee, 2017).

Accordingly, the maintenance process is not just about the restoration of the physical assets to their operational state after a failure. At least it shouldn't be. Through asset management, the maintenance process contributes to the achievement of organizational objectives by supporting the strategic asset management plan. In other words, the maintenance objectives are aligned with the asset management objectives and are deployed to decision-making and maintenance activities on physical assets.

Nevertheless, the reality of organizations regarding maintenance is often quite different from the discussions in this book. The presence of reactive strategies for maintenance management is still a reality today, despite all the technological advances and methodologies applicable to maintenance management. As a consequence, these organizations struggle with a large number of unplanned maintenance interventions due to unexpected failures that negatively impact business results.

At this point, it is expected that the attentive reader has realized that "asset management" is not a synonym for "maintenance management." On the contrary, it is in his interest to understand how maintenance management can be improved in the asset management landscape. Therefore, an overview of the main subjects and requirements covered in asset management is discussed before proceeding in depth to the maintenance planning of complex engineering systems.

A global consensus on the main subjects covered in asset management was achieved through the nonprofit organization known as Global Forum on Maintenance and Asset Management (GFMAN), founded in 2010. According to its vision, GFMAN intends to be "a worldwide community, providing leadership for maintenance & asset management communities." Therefore, the forum is composed of important nonprofit organizations of different regions of the world that share the same values and objectives with the GFMAN. The international organizations that are currently members of GFMAN are listed in Table 8.2.

All GFMAN members share a common core of the asset management discipline that comprises 39 asset management subjects divided into six subject groups. In addition, this core also integrates the four fundamentals of asset management previously discussed. As these asset management subjects are not independent, it is worth mentioning that they need to be addressed as a holistic integrated body of knowledge in order to properly understand asset management discipline (GFMAN, 2014).

Accordingly, asset management was represented surrounded by its subject groups in Fig. 8.3 which are "Strategy and Planning," "Asset management decision-making," "Life cycle delivery," "Asset information," "Organization and People," and "Risk and Review." Thus, each of these groups contains a set of asset management subjects that relate to each other. It is an approach that contributes to the understanding of the broad scope of asset management by summarizing its main subjects into groups.

Since maintenance is a stage in the life cycle of physical assets and is covered by asset management, these asset management subject groups can be deployed and discussed to the level of maintenance management. However, as this conceptual model was conceived under a holistic view of

TABLE 8.2 International organizations that are members of the GFMAN.

Organization	Country/region
Associação Brasileira de Manutenção e Gestão de Ativos (ABRAMAN)	Brazil
Asset Management Council (AM Council)	Australia
Belgian Maintenance Association (BEMAS)	Belgium
European Federation of National Maintenance Societies (EFNMS)	Europe
Gulf Society for Maintenance and Reliability (GSMR)	Arabian Gulf
Institut Français d'Asset Management Industriel et Infrastructures (IFRAMI)	France
The Institute of Asset Management (IAM)	United Kingdom
Japan Association for Asset Management (JAAM)	Japan
Japan Institute for Plant Maintenance (JIPM)	Japan
PEMAC Asset Management Association of Canada	Canada
Southern African Asset Management Association (SAAMA)	South Africa
Society for Maintenance and Reliability Professionals (SMRP)	USA

FIG. 8.3 Asset management subject groups.

the asset management discipline, the subjects that make up each of these groups may relate to all stages of the asset's life cycle or be specific to some of them. Thus, these asset management subjects are shown in Table 8.3 so that the reader understands the breadth and complexity of the asset management discipline.

TABLE 8.3 The 39 asset management subjects.

G1. Strategy and planning	**G4. Asset information**
1. Asset management policy	22. Asset information strategy
2. Asset management strategy and objectives	23. Asset information standards
3. Demand analysis	24. Asset information systems
4. Strategic planning	25. Data and information management
5. Asset management planning	**G5. Organization and people**
G2. Asset management decision-making	26. Procurement and supply chain management
6. Capital investment decision-making	27. Asset management leadership
7. Operations and maintenance decision-making	28. Organizational structure
8. Lifecycle value realization	29. Organizational culture
9. Resourcing strategy	30. Competence management
10. Shutdowns and outage strategy	**G6. Risk and review**
G3. Life cycle delivery	31. Risk assessment and management
11. Technical standards and legislation	32. Contingency planning and resilience analysis
12. Asset creation and acquisition	33. Sustainable development
13. Systems engineering	34. Management of change
14. Configuration management	35. Asset performance and health monitoring
15. Maintenance delivery	36. Asset management system monitoring
16. Reliability engineering	37. Management review, audit and assurance
17. Asset operations	38. Asset costing and valuation
18. Resource management	39. Stakeholder engagement
19. Shutdown and outage management	
20. Fault and incident response	
21. Asset decommissioning and disposal	

The asset management subjects set out in the group "Strategy and Planning" address the alignment between organizational and asset management objectives. Subsequently, this alignment is deployed for maintenance as its performance contributes to the success of asset management. Therefore, the objectives and plans for maintenance management are derived from the asset management line of sight.

According to the IAM, effective asset decision-making is essential for organizations in order to maximize the value realization from assets over their lives (IAM, 2015). Thus, the group "Asset management decision-making" addresses relevant subjects to decision-making at different stages of the asset's life cycle. Regarding maintenance decision-making, it should incorporate processes and activities to determine maintenance requirements and guidelines in support of the asset management objectives.

The third group "Life cycle delivery" incorporates asset management subjects that are related to the implementation of the plans derived from the asset management strategy. Among these subjects, "Maintenance delivery" and "Reliability engineering" are directly related to the maintenance stage of the life cycle. The former covers the organization of maintenance in order to successfully implement its plans, and the latter supports physical assets' maintenance and operation through the integration of reliability engineering activities across the organization.

Since asset management processes and activities require, modify, or generate information throughout the realization of value from assets, all asset management subjects related to asset information are represented by the "Asset information" group. It includes topics such as strategy, standards, systems, and management of the asset information which support the processes, activities, and decision-making in maintenance management.

Like the previous group, "Organization and People" asset management subject group also addresses transversal subjects to the stages of the assets' life cycle. Through asset management, organizations can rethink the traditional way of working by reviewing organizational structures, roles and responsibilities, and contractual relationships (IAM, 2015). Regarding maintenance management, topics such as outsourced activities and spare parts can be addressed in the "Procurement and supply chain management" subject. In addition, the technical skills required in maintenance can be monitored through "Competence management."

Finally, the conceptual model of GFMAN takes us to the last asset management subject group: Risk and Review. In general, it can be said this group incorporates the subjects related to the identification, understanding, and management of risks to achieve the objectives of asset management as well as the activities and processes to ensure the improvement of asset management in the organization. Particularly, the "Asset performance and health monitoring" subject was addressed in the scope of this book given its importance for asset management of complex engineering systems, as set out in Chapter 7.

In face of the increase in complexity of equipment and processes, the number of physical assets, the speed of technological change, the pressure for cost reduction, and the concern for the safety of workers and the environment, asset management has been seen as an important source of benefits and competitive advantages for organizations (de la Fuente, González-Prida, Crespo, Gómez, & Guillén, 2018). However, in order to make it feasible due to its complexity, it is common for organizations to seek formal specifications of what needs to be implemented in international standards. Thus, this is where the recent ISO 55000 series for asset management can support maintenance management in the organizations.

It is worth mentioning the ISO 55000 series does not cover all the subjects proposed in the conceptual model of GFMAN, as presented in Table 8.3 (IAM, 2015). Moreover, its specified requirements only express what needs to be done, but not how, and this can be a barrier for some organizations (Wijnia & de Croon, 2015). In fact, this lack of support for implementation has been critical since PAS 55 (Jooste & Vlok, 2015). This corroborates the complementary character of the GFMAN model for the understanding and implementation of the asset management discipline. In fact, it should be noted that compliance with all requirements of the ISO 55000 series should be seen as a minimum level for an effective asset management system and should not be seen as the final goal (ISO, 2018).

The ISO 55000 series for asset management comprises three international standards that can be applied to all types of assets and by organizations of all types and sizes, although they are particularly intended to be used for physical asset management (ISO, 2014a, b, 2018). It is in ISO 55001, "Asset Management—Management Systems—Requirements," where the requirements for establishing, implementing, maintaining, and improving a management system for asset management within the context of the organization are specified. Then, ISO 55000 and ISO 55002 provide an overview of asset management, its principles and terminologies, and guidelines for the application of ISO 55001, respectively (da Silva & de Souza, 2020).

The concept of the PDCA cycle (Plan-Do-Check-Act), which is a management method used to support continuous improvement in organizations, is the basis for the ISO 55000 series. Accordingly, the elements of an asset management system can be represented in a conceptual framework, as shown in Fig. 8.4.

ISO 55001 states that the order in which the requirements are presented in the standard does not reflect their importance or implies the order in which they must be implemented (ISO, 2014b). As can be seen in Table 8.4, there are 10 sections in the ISO 55001 structure where the first

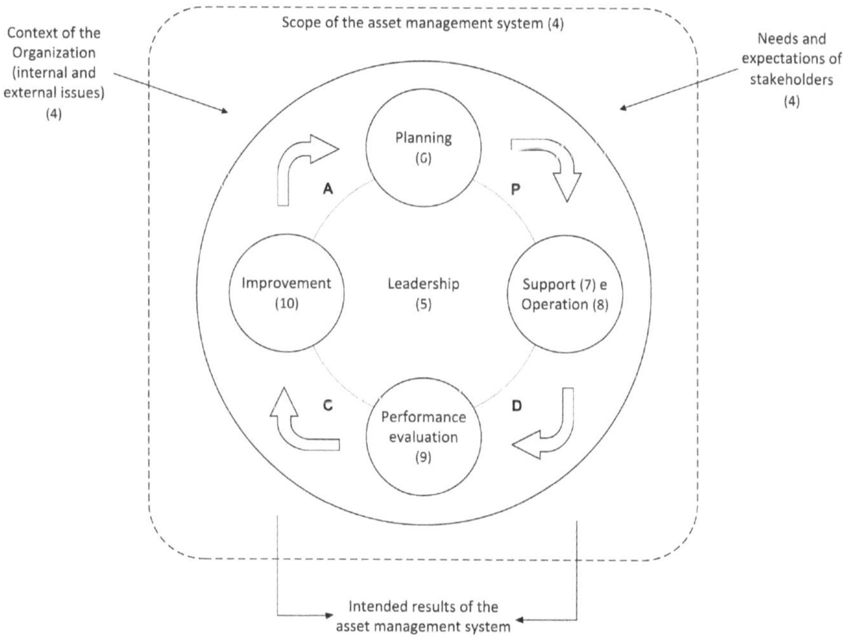

FIG. 8.4 Representation of ISO 55001 structure in a PDCA cycle.

TABLE 8.4 Topics of requirements for an asset management system.

Sec.	Elements	Topics
4	Context of the organization	4.1. Understanding the organization and its context;
		4.2. Understanding the needs and expectations of stakeholders;
		4.3. Determining the scope of the asset management system; and
		4.4. Asset management system
5	Leadership	5.1. Leadership and commitment; 5.2. Policy; and
		5.3. Organizational roles, responsibilities, and authorities
6	Planning	6.1. Actions to address risks and opportunities for the asset management system; 6.2. Asset management objectives and planning to achieve them
7	Support	7.1. Resources; 7.2. Competence; 7.3. Awareness;
		7.4. Communication; 7.5. Information requirements; and
		7.6. Documented information

Continued

TABLE 8.4 Topics of requirements for an asset management system—cont'd

Sec.	Elements	Topics
8	Operation	8.1. Operational planning and control; 8.2. Management of change; and 8.3. Outsourcing
9	Performance evaluation	9.1. Monitoring, measurement, analysis, and evaluation
		9.2. Internal audit; and 9.3. Management review
10	Improvement	10.1. Nonconformity and corrective action;
		10.2. Preventive action; and 10.3. Continual improvement

three are introductory and the others specify the requirements for the asset management system according to different topics. Furthermore, it is worth mentioning that an asset management system based on ISO 55001 can be integrated with systems based on other management standards, such as ISO 9001 for quality management and ISO 14001 for environmental management, as they all share the same base and elements.

Accordingly, the asset management landscape brings maintenance management to a more strategic position within organizations. Moreover, it also provides different conceptual models and international standards in order to guide practitioners to be more successful during their implementation of the discipline. Both the 39 asset management subjects from GFMAN conceptual model and the ISO 55000 series for asset management, as briefly presented in this section, can support organizations to realize values from its physical assets through maintenance management. Finally, it is worth mentioning that although there may be other sources of reference on asset management, these stand out due to the global participation in their preparation.

8.3 Maintenance planning in asset management

Asset management has generated significant interest across several industries as it emerges and matures as a "mainstream" expectation for organizations (Komljenovic, Gaha, Abdul-Nour, Langheit, & Bourgeois, 2016). As previously presented, it has a challenging and extent scope to be addressed across the context of the organization and throughout the life cycles of the physical assets. Due to the purpose of this book, maintenance planning and its associated activities are discussed and detailed under the asset management guidelines for maintenance management.

As an asset management system is primarily designed to support the delivery of an organizational strategic plan (IAM, 2008a), planning is a very important element of this system. This is where physical asset management begins or at least it should. From the organizational objectives, the organization aligns the objectives and strategic plans of asset management, as shown previously in Fig. 8.2. Subsequently, these plans are deployed to the asset portfolio throughout the stages of their life cycle.

This alignment with the objectives of asset management contributes to the understanding of what is expected from maintenance management and how it contributes to the achievement of organizational objectives. Considering this, maintenance planning shall plan the activities to be carried out in order to achieve the objectives of maintenance management. This is not at all simple, and it is even more challenging for organizations that rely heavily on complex engineering systems.

According to ISO 55001, when planning how to achieve these objectives, the maintenance management shall determine what will be done, what resources will be required, who will be responsible, when it will be completed, and how the results will be evaluated (ISO, 2014b). Accordingly, maintenance planning requires coordination between many departments in an organization such as material, operations, and engineering. In fact, in many situations, it is a major cause of delays and bottlenecks (Duffuaa & Raouf, 2015).

In addition, maintenance planning also shall determine the processes and method to be employed in managing its assets, the financial and nonfinancial implications of the maintenance plans, the review period for these plans, and the method and criteria for decision-making and prioritizing of the activities and resources (ISO, 2014b). These demands corroborate the importance of maintenance planning as a fundamental element of maintenance management.

Since a business process comprises a set of activities that are performed in coordination in an organizational and technical environment (Weske, 2019), maintenance planning can be addressed as a subprocess of maintenance management. It is worth mentioning that in business process management (BPM), the business process can be classified into operational, management, or supporting processes (von Rosing, Scheer, & von Scheel, 2015). Therefore, according to this classification, maintenance management is a support process since it supports the core operational processes of an organization.

Through business process modeling, it is possible to use graphical notations to represent the activities and the relationship between them within the process. In addition, this is an interesting approach as it provides an easier understanding since the first contact with the process. Thus, in Fig. 8.5, a maintenance planning process is modeled for maintenance management under the asset management guidelines.

Maintenance planning contemplates three possible start events that are represented by the circles on the left side of Fig. 8.5. The empty circle represents the default start event while the circle with the envelope represents

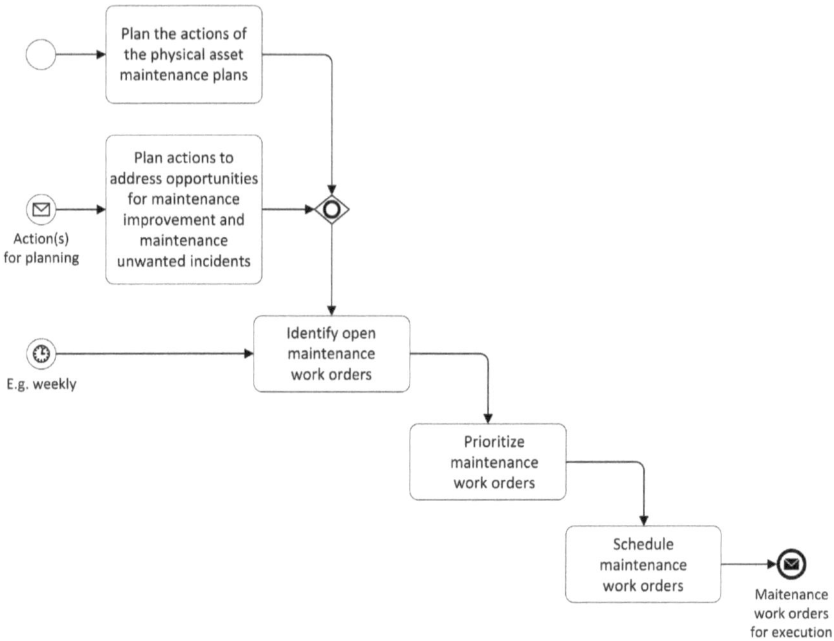

FIG. 8.5 A model of maintenance planning process.

the message type start event and the circle with a clock represents the timer start event. In other words, these represent different mechanisms for triggering the start of the process.

The default start event flow leads to the planning of actions for the maintenance plans of the organization's physical asset portfolio. This event is usually triggered when the maintenance planning process is being implemented as there is the initial demand for planning maintenance actions for the physical assets. Upon completion of this activity, the maintenance planning process is initiated by the other two start events.

As the second start event is a message start event, it requires the reception of some kind of communication (e.g., a document, an email, and a phone call) in order to trigger the beginning of the maintenance planning process (OMG, 2011). When this happens, it leads to the planning of actions to address the opportunities for maintenance improvement and/or unwanted maintenance incidents. In general, these demands arise from health monitoring and fault diagnosis, as presented in Chapter 7, and from the performance indicators as will be discussed in Section 8.5.

Finally, the timer start event flow leads directly to the activity of identifying open maintenance work orders waiting for execution. As an example, this start event was modeled so that it would be triggered weekly. It should be noted that this sequence flow is the only permanent one since the other two start events are only initiated on demand.

Afterward, maintenance planning is required to check if there is a need to prioritize planned maintenance work orders. Since maintenance resources are limited, prioritizing maintenance activities is not only essential to allocate them properly in order to achieve maintenance management objectives, but also a challenge as it requires information about assets for decision-making. This corroborates the importance of a framework for engineering systems health monitoring and fault diagnosis that, in addition to identifying demands for maintenance actions, provides useful information for decision-making.

Finally, maintenance planning prioritizes maintenance work orders and forwards them to execution. Unwanted incidents that are identified during execution can be forwarded to a maintenance improvement process and then returned to maintenance planning or can be forwarded directly to maintenance planning without the need to investigate causes and propose improvement actions. In both cases, the maintenance planning process is restarted by the message start event flow. Common unwanted incidents include the need to plan a maintenance work order that has been partially executed or to reschedule a maintenance order that has not been executed.

Several tools are available to support the activities of the maintenance planning process. In fact, some of them the reader has already been introduced to in previous chapters. As to the term "tool," it is applied here in a somewhat broader meaning, which may include approaches, techniques, and methods. As presented in Fig. 8.6, these tools can be classified into three groups according to their nature and considering their benefits. They are highly recommended during maintenance planning for organizations that have complex engineering systems.

It is difficult to imagine maintenance planning today without the support of computational tools such as Computerized Maintenance Management System (CMMS) and Maintenance, Repair, and Operations Management System (MROMS). CMMS is mainly used for the

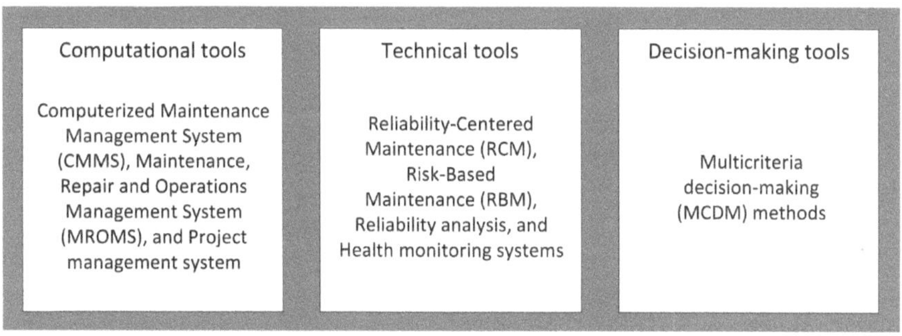

FIG. 8.6 Maintenance planning support tools.

management of maintenance work orders and all aspects associated with them, such as cost, labor, procedures, and scheduling. On the other hand, MROMS can address the management of the maintenance inventory such as consumables items, industrial equipment and spare parts, and plant upkeep supplies.

In addition, for organizations that regularly plan long maintenance stops, project management tools are often incorporated into maintenance planning. As an example, the Gantt chart is a widely known and used tool for tracking the progress of a project schedule that is commonly applied to maintenance stops (Baykasoğlu & Ozsoydan, 2018; Oke & Charles-Owaba, 2006; Radwan, 2000). It is very useful for tracking the planned maintenance work orders versus accomplishments on the same time scale. Moreover, through Gantt charts interdependencies among maintenance tasks and maintenance teams are evidenced (Al-Turki, Ayar, Yilbas, & Sahin, 2014).

In general, these computational tools are transversal to the maintenance planning and support all the tasks modeled in the process of Fig. 8.5. It is also worth mentioning that in organizations with fewer resources support for maintenance planning is usually done through simple spreadsheets and manual controls. Although more limited, those are encouraged in cases where specialized systems cannot be provided. In fact, the worst scenario for maintenance planning would be one that is not aided.

Among the technical tools, Reliability-Centered Maintenance (RCM) and Risk-Based Maintenance (RBM) support the execution of two main tasks of the maintenance planning process: "Plan actions for physical asset maintenance plans" and "Prioritize maintenance work orders." Both tools were presented in Chapter 2 where the reader may find a detailed description of them. RCM is applied to determine maintenance actions when planning maintenance plans for the physical asset portfolio (SAE International, 2009), while RBM is normally used to prioritize maintenance work orders based on a risk assessment of the physical assets (Arunraj & Maiti, 2007; Khan & Haddara, 2003; Ruparathna, Hewage, & Sadiq, 2018).

In addition to these, reliability analysis and health monitoring systems are also used as technical tools to support maintenance planning. Both can support the "Prioritize maintenance work orders" by providing useful information. While the former can estimate reliability in a certain time period or deliver reliability indicators such as mean time between failures, the latter can inform the health condition of the asset. Note that these contents were also covered in the previous Chapters 4 and 7, respectively.

Although these technical tools are systematic, their application in complex engineering systems contexts is often time-consuming and requires specialized technical knowledge. For example, through RCM complex systems are individually analyzed, identifying all their functions, functional failures, potential failure modes, failure effects, failure consequence categories, and finally selected maintenance actions (SAE International, 2009). Accordingly, the organizations need to be aware of this and ensure appropriate support and resources for applying them.

The combination of these computational and technical tools enables the maintenance planning process in organizations. However, when the needs and expectations of the stakeholders or the context of the organization reflect greater pressure on the results of maintenance management to achieve the asset management objectives and, consequently, the organizational objectives, organizations can also incorporate decision-making tools to support the maintenance planning process. Therefore, the concepts and tools for decision-making applications are properly explored in the following section due to their importance in the discipline of asset management.

8.4 Maintenance planning decision-making

Maintenance planning involves decision-making all the time. Questions such as "what activities need to be planned for physical assets?" or "What equipment or systems should be prioritized based on risk?" are usually answered through the application of technical tools like RCM and RBM. Nevertheless, these tools are not capable of handling complex decision problems that consider different attributes at the same time. That is where Multicriteria Decision-Making (MCDM) methods support organizations to make better decisions in maintenance planning.

As MCDM methods are based on a decision analysis, they involve the decomposition of a decision problem into a set of smaller problems. After each of them has been dealt with separately, decision analysis provides a formal mechanism for integrating the results so that a solution can be provisionally selected (Goodwin & Wright, 2004). In other words, MCDM comprises mathematical and computational approaches to support the subjective evaluation of a finite number of decision alternatives under a finite number of performance criteria by a single decision maker or by a group (Loostma, 1999).

Through decision analysis, decision makers are enabled to structure and clarify their thinking, explore trade-offs between attributes, and deliver a documented and defensible rationale for a given decision (Goodwin & Wright, 2004). These benefits are relevant to maintenance planning, as it faces complex decisions such as which maintenance work orders should be prioritized to best optimize maintenance results? It is no longer a simple and technical decision, and it needs to be supported as different criteria need to be considered.

Nevertheless, some key factors appear to influence decision makers when selecting support for decision-making. Among them are the time available to make a decision, the effort that a given strategy will involve, the decision maker's knowledge regarding the subject, the importance of making an accurate decision, and whether or not the decision maker has to justify his or her choice to others (Goodwin & Wright, 2004). In other words, despite all the benefits, decision-making tools are not always applied to a given decision problem for some of these reasons.

Larger organizations, in particular, are more aware of the importance of developing, implementing, maintaining, monitoring, analyzing, and improving asset management practices to support higher overall performance (Maletič, Maletič, Al-Najjar, & Gomišček, 2018). In this context, it is expected that decision makers throughout their asset management decision-making processes take into account different aspects in order to achieve the objectives of asset management and consider the support and structure provided by MCDM methods instead of simple heuristics.

In general, multicriteria decision-making can be carried out in eight steps, as shown in Fig. 8.7.

In the first two steps, decision makers need to deepen the understanding of the current problem and establish the goal for the decision-making (Bhushan & Rai, 2004). For the third step, criteria that are important to decision makers regarding the decision problem are determined in order

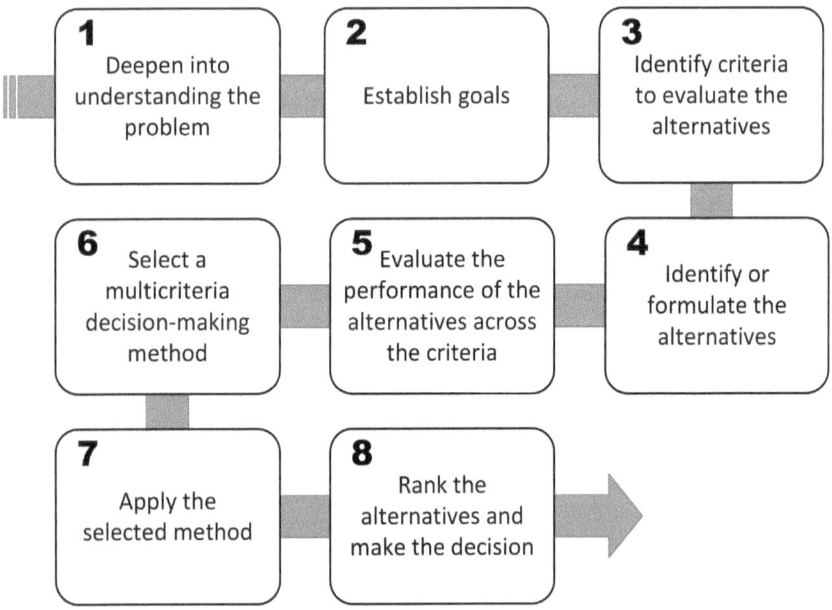

FIG. 8.7 Steps of multicriteria decision-making.

to support the decision-making. From the perspective of asset management, these criteria must be aligned with the objectives of asset management in order to ensure that decisions taken at lower levels contribute to the achievement of organizational objectives.

Then, decision makers identify or formulate a set of alternatives and evaluate their performance across each of the criteria in the fourth and fifth steps. These steps are important to differentiate the alternatives regarding the decision maker's preferences (Sabaei, Erkoyuncu, & Roy, 2015). It is worth mentioning that this order, first the identification of criteria and then the selection of alternatives, is a decision-making approach known as "value-focused thinking" as it prioritizes defining the criteria that are important to the decision before defining the alternatives. Some researchers prefer the "alternative-focused thinking" approach where the order of steps 3 and 4 are reversed (Goodwin & Wright, 2004). In other words, the criteria are defined after the identification of the set of alternatives and not derived from the goal of the decision-making.

In the sixth step, decision makers need to select an MCDM method to apply to their decision problem. There are several developed methods with different mathematical approaches for elaborating the results. Nevertheless, it is reported that several methods used in parallel with the same problem reveal that recommended alternatives do not vary significantly with the method applied (Huang, Keisler, & Linkov, 2011). Accordingly, this definition of an MCDM method should not be a main issue for the reader since the focus is on the benefits of deepening and systematizing decision-making rather than unaided decision-making.

Then, the method is applied and the results of the decision analysis are obtained in the last two steps. Usually, through them, it is possible to rank the alternatives from the best to the worst in relation to the goal of the decision analysis. Before making the final decision, it is recommended that the decision makers compare the best alternatives in relation to their final scores. In some cases, the difference between them may be so small that the decision makers may consider them to be equivalent and the final decision is not made only by the best position in the ranking.

In fact, some authors consider the results of the MCDM method as a provisional decision and recommend that a sensitivity analysis be performed before the final decision. This extra step is intended to see how robust the decision is to changes in the figures supplied by the decision makers, e.g., performance evaluation or criteria weights (Goodwin & Wright, 2004). This corroborates the fact that the result of the MCDM method should be seen as a recommendation and that it does not need to be strictly followed. Accordingly, decision makers may have some freedom to make decisions based on ranking the best alternatives according to other preferences that were not used in the decision analysis.

An extensive list of MCDM methods is currently available in the literature. This number is even greater when considering hybrid applications or the combination of these methods with the Fuzzy Set Theory. While hybrid methods combine different MCDM methods to overcome weaknesses, the Fuzzy MCDM considers imprecision, subjectivity, and vagueness in the performance evaluations of the decision-making analysis (Shafiee, 2015).

Regarding maintenance management applications, it was identified through a search in the SCOPUS database that five MCDM methods are predominant (Sabaei et al., 2015). These methods are presented in Table 8.5 with a brief description of their mathematical approach.

As can be seen in Table 8.5, these methods can be classified into compensatory and noncompensatory, also known as outranking. In a compensatory strategy, an option's poor performance on one attribute is

TABLE 8.5 Common MCDM methods applied to maintenance.

MCDM method	Mathematical approach	Classification
Analytic Hierarchy Process (AHP)	The criteria are pairwise compared in order to estimate the vector of priorities while all the alternatives in relation to each criterion are pairwise compared to estimate the local priorities. They are then synthesized to establish global priorities	Compensatory
ELimination Et Choix Tradusaint la REalité (ELECTRE)	The concordance and nondiscordance concepts are used to make pairwise comparisons of the alternatives. Then, one or more outranking relation is built and explored in order to present the result	Outranking
Preference Ranking Organization METHod for Enrichment Evaluation (PROMETHEE)	The alternatives are pairwise compared across the criteria to determine performance deviations across criteria. Then, a preference function is applied to translate these deviations and calculate the net outranking flow for each alternative	Outranking
Simple Multi-Attribute Rating Technique (SMART)	The alternatives are evaluated in terms of their performance in each criterion. These performances are then aggregated using a simple weighted average after determining the weight of each criterion	Compensatory
Technique for Order of Preference by Similarity to Ideal Solution (TOPSIS)	The alternatives are ordered using a closeness coefficient that relates their Euclidean distances from the positive ideal solution and the negative ideal solution	Compensatory

compensated by good performance on others during the score aggregation (Goodwin & Wright, 2004). On the other hand, the outranking methods compare pairwise the alternatives in terms of each criterion in order to identify the extent to which a preference for one over the other can be asserted and then aggregate such preference information across the criteria (Belton & Stewart, 2002).

All these MCDM methods can be applied as maintenance planning support tools. Maintenance resources are limited, and it is expected that they are appropriately allocated according to the priorities and the maintenance objectives during maintenance planning. Although tools such as RBM, health monitoring systems, prognosis analysis, and reliability analysis provide information that can be used for this purpose, they are generally used individually.

Accordingly, it is common for organizations to use asset criticality as a single criterion for prioritizing maintenance resources. This concept is the recognition that assets and asset systems have differing importance, or represent different vulnerabilities, to the organization (IAM, 2008b). Through this approach, the portfolio of physical assets is classified considering different aspects such as reliability, maintainability, risk, health index, and impact of their failures on the operation that are aggregated to compose the criticality. Thus, it is an example of how MCDM methods support maintenance planning decision-making.

Asset criticality can be used in the maintenance planning process, for example, to establish an order for defining or reviewing maintenance plans. Since these activities take time and dedication, their execution can be planned by prioritizing the most critical physical assets first. Moreover, it can also be applied for prioritizing maintenance work orders.

Although asset criticality is an interesting approach to maintenance work orders' prioritization, it alone may not express the needs of maintenance planning in order to achieve maintenance objectives. Consequently, asset criticality can be combined with other criteria such as maintenance work order cost, maintenance work order type, and waiting time as an input of a new decision-making analysis. In other words, MCDM methods can also be applied to prioritize maintenance work orders using criticality as one of their criteria.

Finally, it is worth mentioning that although MCDM methods are not recent techniques, their applications for maintenance started to gain prominence in the literature from the mid-2000s. In many industries such as power generating, oil and gas, and chemical, MCDM is a fairly young subject (Shafiee, 2015).

Organizations that have complex engineering systems are likely to have already implemented or intend to implement the computational and technical tools to support maintenance planning. On the other hand, tools for decision-making are still not as common despite their benefits. This needs to be changed. Since computational and technical tools provide valuable information to be used as input to decision analysis, aided decision-making should be a natural process. If maintenance planning decisions are not yet supported by decision-making tools, they can be made in a biased and nonoptimized way, impacting what is expected from maintenance.

8.5 Maintenance planning performance evaluation and improvement

Although the implementation of different maintenance planning support tools contributes to boosting maintenance planning, it is pertinent that the organization formally evaluates its outcomes based on the performance of maintenance planning and maintenance management. In fact, performance evaluation is one of the elements required for an asset management system in the ISO 55000 series. Through it, the organization can monitor, measure, analyze, and evaluate asset management regularly (ISO, 2018).

As stated in ISO 55001, an organization shall evaluate and report the asset management performance in different domains such as assets, asset management, and asset management system (ISO, 2014b). Considering only the maintenance stage of the assets' life cycle, these domains can be interpreted as the domains of physical assets, maintenance management, and maintenance management system. Then, these three performance evaluation domains can be broken down into subdomains according to the organization's needs.

Regarding the physical assets' domain, the performance evaluation may consider the segmentation of assets by type, location, criticality, and operational condition, for example, assessing different subdomains. Similarly, the maintenance management domain may conduct a performance evaluation for each of its processes in addition to the overall performance. Finally, the performance evaluation of the maintenance management system can also be assessed for each implemented element of the system in addition to the overall system.

The performance evaluation domains are commonly tracked through performance indicators. By definition, performance indicators are specifically defined variables that are characterized by performance measures and are used for performance measurement of any system or process (Parida, Kumar, Galar, & Stenström, 2015). According to ISO 55002, these indicators should provide useful information to determine areas of success as well as areas requiring corrective action or improvement (ISO, 2018). Therefore, a proper performance evaluation is an essential process for improving maintenance management (da Silva et al., 2020).

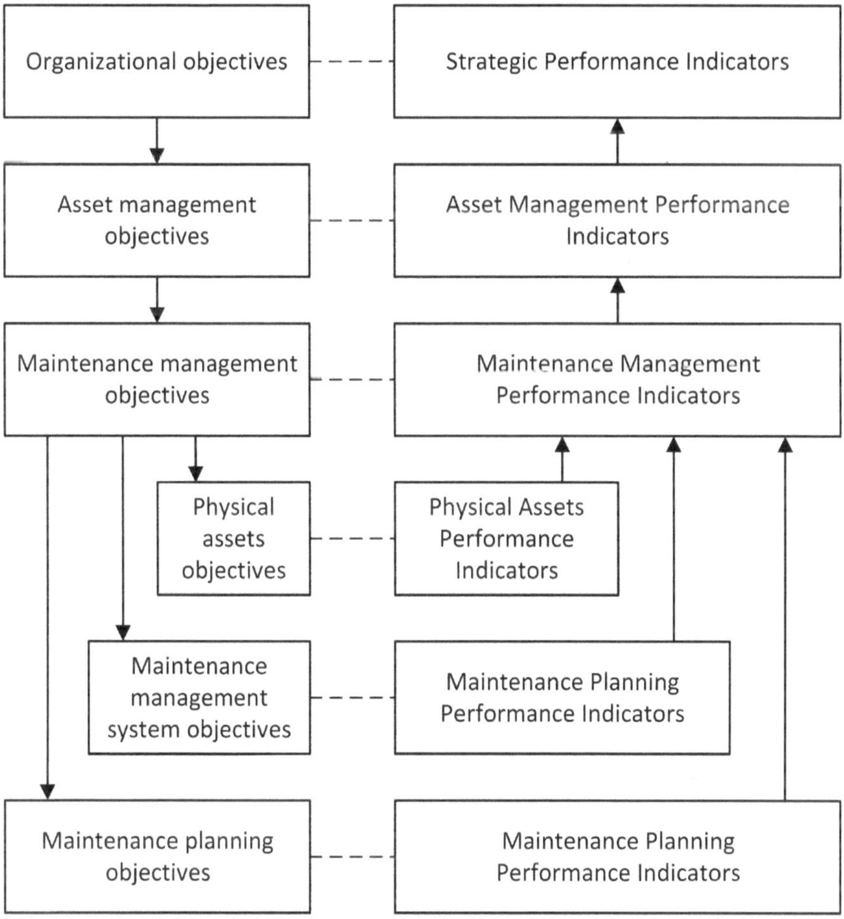

FIG. 8.8 Relationship among performance evaluation domains.

There is a hierarchical relationship between the performance indicators of different evaluation levels, showing how the lower-level indicators contribute to the higher-level indicators, as shown in Fig. 8.8. For example, the maintenance management performance evaluation domains are hierarchically related in order to achieve the organizational objectives. Physical assets and maintenance management system performance can positively impact maintenance management results. Then, maintenance management performance contributes to the achievement of the asset management objectives and, consequently, organizational objectives.

Concerning the maintenance planning performance indicators, there should be an alignment with the maintenance management performance indicators. This can be justified by the fact that the objectives of the maintenance planning process are derived from the objectives of maintenance management, as shown in Fig. 8.8. Accordingly, the maintenance planning performance indicators should evaluate how effectively maintenance planning is achieving its objectives in order to achieve maintenance management objectives.

Maintenance planning performance indicators may vary depending on the organization as they depend on the successive breakdown of organizational objectives into maintenance planning objectives. These performance indicators should not be defined arbitrarily, but through the translation of maintenance planning objectives into performance indicators. In other words, maintenance planning should monitor and evaluate what can impact the achievement of the maintenance planning objectives.

The main performance indicators used to evaluate maintenance planning are presented in Table 8.6. It is worth mentioning that although this chapter prioritizes maintenance planning indicators, the reader can find

TABLE 8.6 Maintenance planning performance indicators.

Indicator	Description	Unit
Planned work	Planned hours worked in relation to the total hours worked	%
Unplanned work	Unplanned hours worked in relation to the total hours worked	%
Schedule adherence	Scheduled hours worked in relation to the total scheduled hours	%
Reactive work	Hours worked in corrective maintenance in relation to the total hours worked	%
Proactive work	Hours worked in preventive and predictive maintenance in relation to the total hours worked	%
Maintenance distribution	Hours worked in each type of maintenance activity	%
Work documentation	Hours reported by the maintenance team in relation to the total hours worked	%
Work order cycle time	Average time from opening to closing a maintenance work order	days
Maintenance backlog	Hours of maintenance work awaiting execution over total available hours of maintenance crew	%
Awaiting maintenance work orders	Number of maintenance work orders awaiting execution	qty
Outsourced hours	The percentage of hours worked outsourced in relation to total hours worked	%

support for defining performance indicators for other maintenance management performance evaluation domains in the references (Al-Turki et al., 2014; da Silva et al., 2020; Parida & Kumar, 2009).

These maintenance planning performance indicators listed in Table 8.6 evaluate different aspects of maintenance planning. "Planned work" and "Unplanned work" assess the percentage of hours worked that were planned. In other words, they indicate how much the unplanned activities are impacting the weekly schedule. A low level of planned activity or high level of unplanned activity impacts compliance with the maintenance schedule that concentrates maintenance activities that were previously prioritized, planned, and scheduled for execution. As a consequence, the "schedule adherence" indicator may not reach the target indicated by the organization.

"Reactive work," "Proactive work," and "Maintenance distribution" are different ways of assessing the distribution of work by type of maintenance activity. Through them, it is possible to observe in which type of activity the maintenance team has concentrated its efforts. Regarding organizations with complex engineering systems, maintenance activities are expected to be mostly preventive and predictive, contributing to the prevention and anticipation of failures and risk management.

As important as the planning and execution of maintenance activities is the proper recording of them for maintenance control and analysis. To this end, it is common for organizations to use the "Work documentation" or equivalent indicator to assess how this routine maintenance information is being documented. For example, corrective maintenance activities that took place during the workweek need to be documented in the CMMS with the appropriate hours of associated work and spare parts used. In addition to justifying the impact on compliance with the weekly schedule, this information is the basis for reliability analysis and improvement of maintenance management.

"Work order cycle time," "Maintenance backlog," and "Awaiting maintenance work orders" can be used to give an overview of the maintenance planning efficiency. While the first assesses the average cycle time of maintenance service orders, the second assesses the maintenance demand in relation to the available workforce and the third assesses the total amount of maintenance work. A trend of increase over time in these indicators may be a consequence of a small maintenance team or a large number of unplanned and corrective maintenance activities. Finally, "Outsourced hours" assesses how much maintenance planning depends on external labor to meet internal maintenance demands.

After choosing the maintenance planning performance indicators, the organization will be able to monitor the performance of this process and how it has contributed to the achievement of maintenance management objectives. This is essential so that opportunities for improvement can be identified and addressed within the scope of maintenance planning. It is worth mentioning that this applies similarly to performance evaluation of other maintenance management processes and the physical assets and maintenance management system performance evaluation domains.

Accordingly, when the performance achieved on a maintenance planning performance indicator is less than expected, the organization shall react to this unwanted incident by taking action to control and correct it and/or dealing with its consequences. In addition, the organization shall evaluate the need for actions to eliminate the causes of this low performance so that it does not recur (ISO, 2014b). These demands are usually addressed by a process of maintenance management improvement that receives and analyzes them for the elaboration of corrective actions.

Depending on the type of these actions, they may need to be forwarded to maintenance planning to plan their execution. In these cases, the maintenance planning is communicated, triggering the maintenance planning process as previously presented in Fig. 8.2. This loop allows maintenance planning to be continuously improved, resulting in revisions of maintenance plans, criteria for prioritizing service orders, implementations, or replacements of maintenance planning support tools, along with others.

Even though this chapter has mainly focused on maintenance planning and maintenance planning decision-making due to the purpose of the book, the reader is encouraged to explore other maintenance management processes and even other stages in the life cycle of assets in the same way later on. Asset management is a complex and broad discipline, and it covers many subjects that are relevant to organizations that have complex engineering systems.

References

Al-Turki, U. M., Ayar, T., Yilbas, B. S., & Sahin, A. Z. (2014). *Integrated maintenance planning* (pp. 25–57). https://doi.org/10.1007/978-3-319-06290-7_3.

Arunraj, N. S., & Maiti, J. (2007). Risk-based maintenance—Techniques and applications. *Journal of Hazardous Materials, 142*(3), 653–661. https://doi.org/10.1016/j.jhazmat.2006.06.069.

Baykasoğlu, A., & Ozsoydan, F. B. (2018). Dynamic scheduling of parallel heat treatment furnaces: A case study at a manufacturing system. *Journal of Manufacturing Systems, 46,* 152–162. https://doi.org/10.1016/j.jmsy.2017.12.005.

Belton, V., & Stewart, T. J. (2002). *Multiple criteria decision analysis*. USA: Springer. https://doi.org/10.1007/978-1-4615-1495-4.

Bhushan, N., & Rai, K. (2004). *Strategic decision making applying the analytic hierarchy process* (p. 183). Springer.

Blackmore, P., & Leeprechanon, N. (2015). Improving network reliability through effective asset management. In *2015 IEEE innovative smart grid technologies—Asia (ISGT ASIA)* (pp. 1–9). IEEE. https://doi.org/10.1109/ISGT-Asia.2015.7387188.

da Silva, R. F., & de Souza, G. F. M. (2020). Asset management system (ISO 55001) and Total Productive Maintenance (TPM): A discussion of interfaces for maintenance management. *Revista Gestão Da Produção Operações e Sistemas, 15*(2), 288–313. https://doi.org/10.15675/gepros.v15i2.2528.

da Silva, R. F., Melani, A. H. D. A., Michalski, M. A. D. C., de Souza, G. F. M., Nabeta, S. I., & Hamaji, F. H. (2020). Defining maintenance performance indicator for asset management based on ISO 55000 and Balanced Scorecard: A hydropower plant case study. In *Proceedings of the 30th European safety and reliability conference and the 15th probabilistic safety assessment and management conference.*

de la Fuente, A., González-Prida, V., Crespo, A., Gómez, J. F., & Guillén, A. (2018). Advanced techniques for assets maintenance management. *IFAC-PapersOnLine, 51*(11), 205–210. https://doi.org/10.1016/j.ifacol.2018.08.260.

Duffuaa, S. O., & Raouf, A. (2015). Maintenance planning and scheduling. In *Planning and Control of Maintenance Systems* (pp. 155–186). Springer International Publishing. https://doi.org/10.1007/978-3-319-19803-3_7.

GFMAN. (2014). *The asset management landscape* (2nd ed., p. 57). GFMAN.

GFMAN. (2016a). *The maintenance framework* (1st ed., p. 24). GFMAN.

GFMAN. (2016b). *The value of asset management to an organisation* (1st ed., p. 11). GFMAN.

Goodwin, P., & Wright, G. (2004). *Decision analysis for management judgements* (3rd ed., p. 493). John Wiley and Sons Ltd.

Haroun, A. E., & Duffuaa, S. O. (2009). Maintenance organization. In *Handbook of maintenance management and engineering* (pp. 3–15). London: Springer. https://doi.org/10.1007/978-1-84882-472-0_1.

Hastings, N. A. J. (2015). *Physical asset management.* Springer International Publishing. https://doi.org/10.1007/978-3-319-14777-2.

Hodkiewicz, M. R. (2015). *The development of ISO 55000 series standards* (pp. 427–438). https://doi.org/10.1007/978-3-319-09507-3_37.

Huang, I. B., Keisler, J., & Linkov, I. (2011). Multi-criteria decision analysis in environmental sciences: Ten years of applications and trends. *Science of the Total Environment, 409*(19), 3578–3594. https://doi.org/10.1016/j.scitotenv.2011.06.022.

IAM. (2008a). *PAS 55-1:2008 Asset management—Part 1: Specification for the optimized management of physical assets* (2nd ed., p. 40). IAM.

IAM. (2008b). *PAS 55-2: 2008 Asset Management—Part 2: Guidelines for application of PAS 55-1* (2nd ed., p. 72). IAM.

IAM. (2015). *Asset management—An anatomy* (3rd ed., p. 84). IAM.

Ihemegbulem, I., & Baglee, D. (2017). ISO55000 standard as a driver for effective maintenance budgeting. In *2nd international conference on maintenance engineering (IncoME)* (p. 16).

Ihemegbulem, I., Baglee, D., & Wheatley, A. (2017). The role of ISO 55000 standard in asset integrity. In *30th international conference on condition monitoring and diagnostic engineering management (COMADEM)* (pp. 172–178).

ISO. (2014a). *ISO 55000—Asset management—Management systems—Overview, principles and terminology.* ISO.

ISO. (2014b). *ISO 55001 - Asset management - Management systems - Requirements* (p. 20). ISO.

ISO. (2018). *ISO 55002 — Asset management—Management systems—Guidelines for the application of ISO 55001.* International Organization for Standardization.

Jooste, J. L., & Vlok, P. J. (2015). *Identifying the critical success factors for engineering asset management services—An empirical study* (pp. 397–413). https://doi.org/10.1007/978-3-319-15536-4_32.

Khan, F. I., & Haddara, M. M. (2003). Risk-based maintenance (RBM): A quantitative approach for maintenance/inspection scheduling and planning. *Journal of Loss Prevention in the Process Industries, 16*(6), 561–573. https://doi.org/10.1016/j.jlp.2003.08.011.

Komljenovic, D., Gaha, M., Abdul-Nour, G., Langheit, C., & Bourgeois, M. (2016). Risks of extreme and rare events in asset management. *Safety Science, 88*, 129–145. https://doi.org/10.1016/j.ssci.2016.05.004.

Konstantakos, P., Chountalas, P., & Magoutas, A. (2019). The contemporary landscape of asset management systems. *Quality—Access to Success, 20*(169), 10–17.

Kriege, L., & Vlok, P. J. (2015). *Human resources within ISO 55000—The hidden backbone to the asset management system* (pp. 435–446). https://doi.org/10.1007/978-3-319-15536-4_34.

Loostma, F. A. (1999). *Multi-criteria decision analysis via ratio and difference judgement* (p. 292). Kluwer Academic Publishers.

Maletič, D., Maletič, M., Al-Najjar, B., & Gomišček, B. (2018). Development of a model linking physical asset management to sustainability performance: An empirical research. *Sustainability, 10*(12), 4759. https://doi.org/10.3390/su10124759.

Oke, S. A., & Charles-Owaba, O. E. (2006). A sensitivity analysis of an optimal Gantt charting maintenance scheduling model. *International Journal of Quality & Reliability Management, 23*(2), 197–229. https://doi.org/10.1108/02656710610640952.

OMG. (2011). *Business process model and notation (BPMN)* (2nd ed., p. 538). OMG.

Parida, A., & Kumar, U. (2009). Maintenance productivity and performance measurement. In *Handbook of maintenance management and engineering* (pp. 17–41). London: Springer. https://doi.org/10.1007/978-1-84882-472-0_2.

Parida, A., Kumar, U., Galar, D., & Stenström, C. (2015). Performance measurement and management for maintenance: A literature review. *Journal of Quality in Maintenance Engineering, 21*(1), 2–33. https://doi.org/10.1108/JQME-10-2013-0067.

Park, S., Park, S. I., & Lee, S.-H. (2016). Strategy on sustainable infrastructure asset management: Focus on Korea's future policy directivity. *Renewable and Sustainable Energy Reviews, 62*, 710–722. https://doi.org/10.1016/j.rser.2016.04.073.

Parlikad, A., Wang, J. Q., Russell, P., & Attwater, A. (2014). Measuring the performance of asset management systems. In *Asset management conference 2014* (p. 304). Institution of Engineering and Technology. https://doi.org/10.1049/cp.2014.1046.

Radwan, A. (2000). Computerised dynamic Gantt-chart for maintenance scheduling and management. *International Journal of Computer Applications in Technology, 13*(3/4/5), 251. https://doi.org/10.1504/IJCAT.2000.000246.

Roda, I., & Macchi, M. (2016). Studying the funding principles for integrating asset management in operations: An empirical research in production companies. *IFAC-PapersOnLine, 49*(28), 1–6. https://doi.org/10.1016/j.ifacol.2016.11.001.

Ruitenburg, R. J., & Braaksma, A. J. J. (2017). Mitigating change in the goals and context of capital assets: Design of the Lifetime Impact Identification Analysis. *CIRP Journal of Manufacturing Science and Technology, 17*, 50–59. https://doi.org/10.1016/j.cirpj.2016.05.008.

Ruparathna, R., Hewage, K., & Sadiq, R. (2018). Multi-period maintenance planning for public buildings: A risk based approach for climate conscious operation. *Journal of Cleaner Production, 170*, 1338–1353. https://doi.org/10.1016/j.jclepro.2017.09.178.

Sabaei, D., Erkoyuncu, J., & Roy, R. (2015). A review of multi-criteria decision making methods for enhanced maintenance delivery. *Procedia CIRP, 37*, 30–35. https://doi.org/10.1016/j.procir.2015.08.086.

SAE International. (2009). *JA1011 evaluation criteria for reliability-Centered Maintenance (RCM) processes* (2nd ed., p. 12). SAE International.

Shafiee, M. (2015). Maintenance strategy selection problem: An MCDM overview. *Journal of Quality in Maintenance Engineering, 21*(4), 378–402. https://doi.org/10.1108/JQME-09-2013-0063.

von Rosing, M., Scheer, A., & von Scheel, H. (2015). *The complete business process handbook—Body of knowledge from process modeling to BPM* (1st ed., p. 763). Elsevier.

Weske, M. (2019). *Business process management*. Springer Berlin Heidelberg. https://doi.org/10.1007/978-3-662-59432-2.

Wijnia, Y., & de Croon, J. (2015). *The asset management process reference model for infrastructures* (pp. 447–457). https://doi.org/10.1007/978-3-319-15536-4_35.

Woodhouse, J. (2014). Briefing: Standards in asset management: PAS 55 to ISO 55000. *Infrastructure Asset Management*, *1*(3), 57–59. https://doi.org/10.1680/iasma.14.00013.

Woodhouse, J. (2018). What is the value of asset management? *Infrastructure Asset Management*, *6*(2), 102–108. https://doi.org/10.1680/jinam.17.00040.

CHAPTER

9

Examples of application

For the things we have to learn before we can do them, we learn by doing them.
Aristotle (384–322 BCE) In: The Nicomachean Ethics

As some things do not come to us by nature, people may get them by first exercising them. At least, that is what the ancient philosopher Aristotle claimed about the moral virtues of humanity. For him, virtues are derived from the practice and go beyond knowing what is virtuous. His notorious quote that introduces this chapter expresses that the general concept of learning through experience is ancient.

That inspired the conception of two examples of application to be presented in this chapter. It is worth mentioning that even if the applications were not carried out by the reader, one is still able to analyze the results in a reflective observation ofthese applications and connect the ideas to the knowledge previously shared in this book. In other words, this practical approach covers the reliability analysis and asset management of modern engineering system subjects through a complementary perspective.

Accordingly, this chapter intends to discuss the subjects covered in the book through examples of application. Section 9.1 provides an overview of the methods and techniques of the previous chapters that will be used in these applications. Then, the following sections present their application to two modern engineering systems. Section 9.2 analyzes a Combined Diesel or Gas (CODOG) propulsion system for military ships, while Section 9.3 examines a power generation system of Floating Production Storage and Offloading (FPSO) vessels. Finally, Section 9.4 provides the final considerations of these examples of application.

https://doi.org/10.1016/B978-0-12-823521-8.00010-4

9.1 Methods and techniques for the examples of application

At this point, the reader has already been introduced to several concepts, methods, and techniques related to reliability analysis and asset management of modern engineering systems. Although previous chapters discussed these topics in detail and supported an understanding of the state of the art of the covered subjects, the presented approach is still theoretical. Therefore, this chapter presents two examples of practical applications in engineering systems to demonstrate to the reader how the content of this book can be brought into our daily lives.

The analysis of these examples included three main stages: the study of the engineering systems, the Reliability, Availability, and Maintainability (RAM) analysis, and the Health Monitoring and Fault Diagnosis (HMFD) process. The procedure for carrying out these stages will be discussed in the next sections as well as the presentation of the methods and techniques used in each application example.

9.1.1 Study of the engineering systems

The examples start with the study of the engineering system in both applications. In practice, this step is quite interesting to be applied first since it supports the subsequent analyses with information and characteristics of the systems. Besides, conducting the system study allows those involved to establish common technical knowledge regarding the engineering system under analysis.

The study of an engineering system contemplates its description, the elaboration of a functional tree of the system, the identification of its potential Failure Modes (FMs) and their detectability, and the collection of failure and repair data. These activities were addressed in the previous chapters of this book such as Chapter 2 for the description, the system's functional tree, and the identification of the potential FM; Chapter 5 for the system's observability and the failure modes' detectability of the FM; and Chapter 4 for the collection of failure and repair data for RAM analysis.

The description of the engineering systems provides the general information and operational context of the systems of interest in the study for the entire team involved in this stage. Thus, through the functional tree, one is able to hierarchically structure the subsystems and components of the systems in order to understand its levels. In addition, the functional tree may also help the team to define the scope of the analysis for each considered system. These trees were developed after a comprehensive study of each system, taking into account the guidelines found in ISO 14224 (ISO, 2016).

From the scope, the potential FM of the subsystems and/or equipment can be identified with methods such as FMEA. In order to establish an HMFD afterward, a detectability analysis can be attached to the list of identified potential FM. Usually, the measures associated with the FM and their observation window are defined in this activity. Finally, the study of the system also gathers the data regarding failures and repairs of the subsystems and/or equipment of interest in the analysis in order to support the performance of the RAM and HMFD analyses. It is worth mentioning that these data can be obtained from the organization's failure and repair records or based on technical databases, such as the Offshore and Onshore Reliability Data (OREDA, 2009) and the Nonelectronic Parts Reliability Data Publication (Quanterion Solutions Incorporated, 2016), in the lack of such data history.

9.1.2 Reliability, availability and maintainability (RAM) analysis

After the study of the engineering systems, the second stage of the examples of applications is to perform a Reliability, Availability, and Maintainability (RAM) analysis. RAM analysis is a well-known type of analysis for repairable systems to examine the overall reliability, availability, and maintainability of modern engineering systems. It is worth mentioning that the concepts and support for this stage were previously addressed in Chapters 2 and 4.

In the analysis of the CODOG propulsion system, the traditional Reliability Block Diagram (RBD) technique was applied to perform the RAM analysis. Through it, the overall result of the system depends on the arrangement of the elements of the system in terms of reliability configuration. In this example of application, three main RBDs were modeled: the port-side propulsion system, the starboard-side propulsion system, and the auxiliary systems. They were individually simulated as well as combined for the overall analysis of the military ship CODOG propulsion system.

In the FPSO power generation system, which constitutes the second application example, it was applied a Generalized Stochastic Petri net (GSPN) model to perform the RAM analysis. The GSPN was used instead of the RBD because of its ability to consider dynamic aspects of the system, such as the variation in the energy demand to be generated over time. Accordingly, the developed Petri net was able to estimate the reliability and availability of the studied system.

The results of both analyses were discussed and interpreted considering the operational contexts in which the engineering systems are inserted. RAM analysis can provide useful information for improvement and maintenance planning in organizations. Weaknesses in the engineering system can be identified in order to increase operational availability as well as recommendations for maintenance plan reviews.

9.1.3 Health monitoring and fault diagnosis

The last stage of both examples of application refers to the implementation of the HMFD framework, presented earlier in Chapter 7. In this stage, the analysis starts by picking up information provided in the study of the engineering systems (Section 9.1.1), such as the identification of potential FM and their detectability.

This prior information is essential since the definition of FM and the way to observe their symptoms defines both the measurements that shall be taken and the parameters and state variables of the system that shall be estimated to detect the system faults. Likewise, the definition of methods for fault detection and diagnosis is directly influenced by these pieces of information and definitions. Although there is still no defined procedure to assist in the selection of methods that should be applied to detect, diagnose, and evaluate faults, the reader will be able to find some useful information on which path to take in Chapters 5 and 6.

Particularly concerning the measurements' selection, when the analyzed system already has some type of protocol for performing periodic measurements or even dedicated software, such as Supervisory Control and Data Acquisition (SCADA) systems or even Health Monitoring Systems (HMS), it is interesting that the implementation of the HMFD framework makes the most of the information already available. Although the cost of installing new sensors, probes, and data acquisition hardware is not an issue nowadays, in general, such care helps employees and maintenance teams of a given organization to embrace the HMFD process that is being implemented.

In both examples, the preexistence of a SCADA system and its respective database (in which the collected data is stored) is considered available. Thus, the measurement acquisition process is only about reading the variables of interest from the SCADA database. As it includes only measurements already processed and directly related to the dynamic behavior of the monitored system, the data stored in the SCADA system database can already be considered descriptors for these examples, which allows to bypass the process of data selection and preprocessing described in the framework previously presented in Chapter 7.

Regarding the Fault Detection and Diagnosis (FDD) methods, as will be noted by the reader, for the ascertained FM a model-based approach was not considered necessary for the fault detection process. In this case, for both application examples, an adaptive variant of the Principal Component

Analysis (PCA) (Michalski, 2021), a linear Gaussian multivariate quantitative technique, was the chosen method mainly due to its ease of implementation, robustness, and ability to deal with multiple variables in a nonsupervised manner. Additionally, as a measure for evaluating the results of the PCA and define whether fault detection occurs or not, the T^2-*statistic* (Mujica, Rodellar, Fernández, & Güemes, 2011) was applied to the PCA eigenvector matrix. Acting in a way to complement the PCA technique, Bayesian Network (BN) was the approach adopted for the fault diagnosis process, especially due to its capability of representing an uncertainty problem in a compact, complete, and modular way (Melani, Michalski, da Silva, & de Souza, 2021).

Considering the principle of detectability, presented in Chapter 5, the monitored variables were grouped into clusters. Different criteria were used in each example to create these clusters, as will be presented in each case study individually. However, for both cases, the output of the detection process was simplified to a Boolean indicator (true or false type) for each cluster. These values determine the state of each cluster over time, which is defined as the system symptoms and, also, the input data for the fault diagnosis process. The correlation between parent nodes (failure modes) and child nodes (clusters) of the Bayesian network is made taking into account the measurements related to the effects of a given failure mode on the system and also in which clusters such measurements are included.

Finally, based on the failure analysis process, the calculation of the Remaining Useful Life (RUL) was made taking into account the behavior of the variables associated with the diagnosed failure mode, as well as their threshold values from which the system could have compromised functionality or even integrity. As it is an analysis of multiple variables, the resulting RUL is summarized in a range of values whose limits (the maximum and minimum values) are, respectively, associated with the variables closest and farthest from reaching their respective operational threshold.

9.2 Analysis of a CODOG propulsion system of military ships

This section presents a CODOG naval propulsion system for military ships as the first example of application. It is worth mentioning that it is presented according to the three stages of the analysis presented in Section 9.1. Therefore, Sections 9.2.1–9.2.3 address the study, the RAM analysis, and the HMFD for the CODOG propulsion system, respectively.

9.2.1 Study of the CODOG propulsion system

A CODOG is a type of naval propulsion system where for every propeller shaft of the ship, there is at least one diesel engine for cruising speed and one geared gas turbine for high-speed dashes. Both types of engines are connected to the shaft with clutches, but only one system is driving the ship at any time. A controllable-reversible pitch propeller provides the means of reversal. The CODOG concept has been proven to provide great improvements over the characteristics available with either all diesel or all gas turbine drives.

Fig. 9.1 presents a diagrammatic representation of the standard double-shaft ship with CODOG propulsion systems that were studied in this example of application.

As shown in Fig. 9.1, there are two diesel engines and one gas turbine for each propulsion system of each side of the ship, port, and starboard. Similarly, each propulsion system has a Speed Increaser Gearbox (SIG) that is coupled with the diesel engines and a Speed Reducer Gearbox (SRG) that is coupled with the propeller shaft. Therefore, as shown in Fig. 9.1, the reader is expected to have an overview of the CODOG propulsion system for this example of application.

After this brief description of the CODOG propulsion system, Figs. 9.2–9.4 present the functional trees for the delimited scope of this first

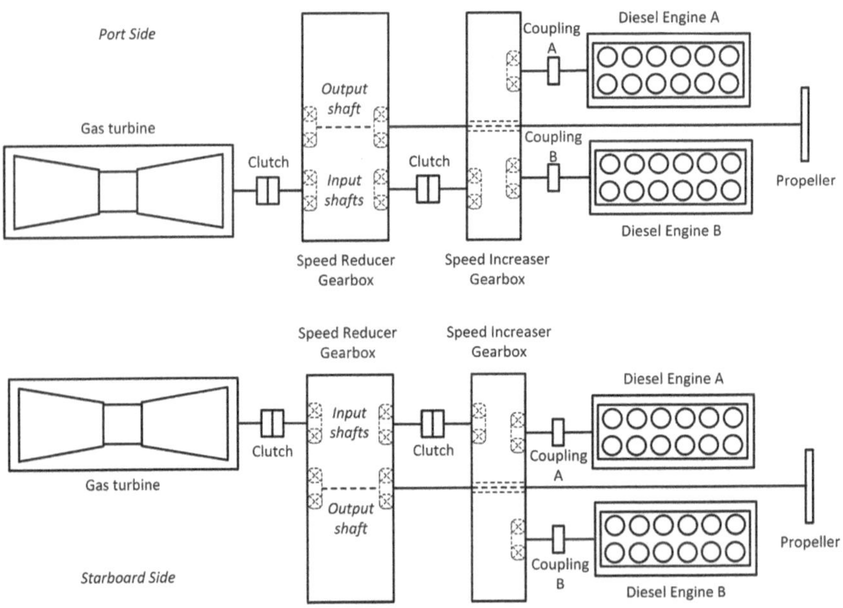

FIG. 9.1 Representation of the CODOG propulsion systems.

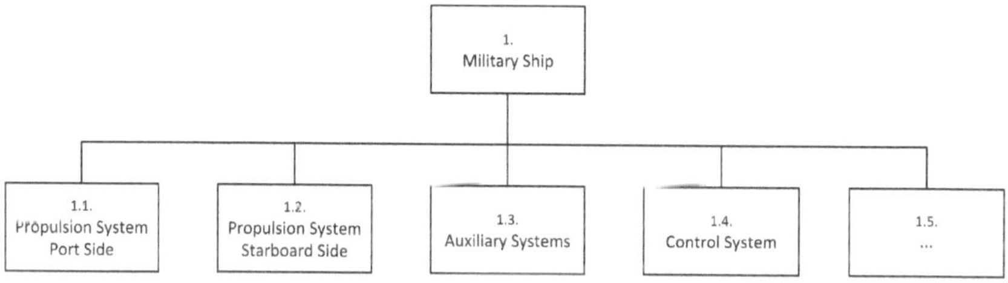

FIG. 9.2 Systems of the military ship unit.

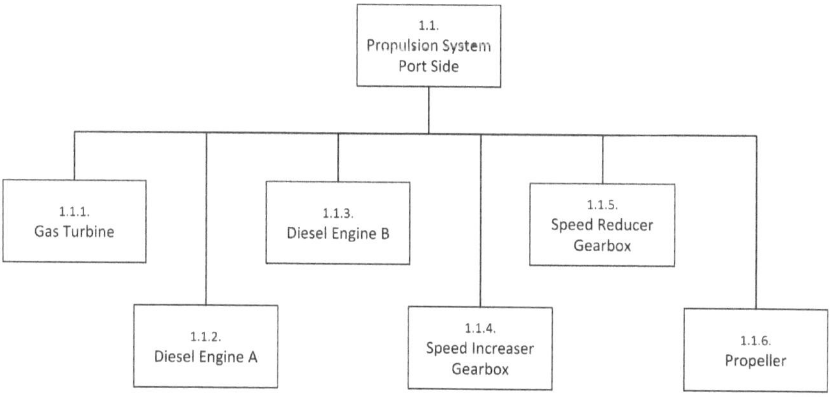

FIG. 9.3 Equipment units of the military ship's propulsion system.

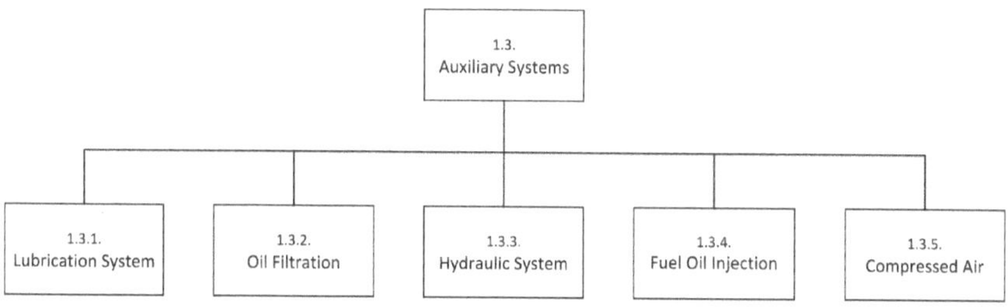

FIG. 9.4 Subsystem units of the military ship's auxiliary system.

example. First, Fig. 9.2 represents the functional tree for the military ship unit with its main systems. In other words, it shows the first level of the hierarchical structure of the military ship. It is worth mentioning that, in this example of application, the scope includes only the two CODOG propulsion systems (1.1. and 1.2.) and the auxiliary system (1.3.).

As both propulsion systems are identical, Fig. 9.3 represents the equipment units of only one propulsion system (1.1.). Similarly, Fig. 9.4 presents the subsystem units of the auxiliary systems (1.3.). These figures provide an overview of the second level of the functional tree for the corresponding systems of the military ship.

After the elaboration of the functional trees, the next activity in the study of the engineering systems is to identify the potential FM and their detectability for the items of interest for the application. This activity is fundamental for the establishment of an HMFD as it allows to define which are the FMs that will be prevented by a condition monitoring strategy. In this example, the study of potential FM, as well as their detectability, was limited to the equipment of the CODOG propulsion system due to the scope of the HMFD example of application. However, it is worth mentioning that the identification of the potential FM is recommended for all engineering systems in the organization as it supports the definition of the maintenance plans, as previously presented in Chapter 8.

Accordingly, Table 9.1 presents a simplified analysis of the potential FM and effects for the equipment in the CODOG propulsion systems as presented in Fig. 9.3. From the identified potential FM of Table 9.1, a

TABLE 9.1 Analysis of potential failure modes and effects of the CODOG propulsion system.

ID item	Item	Function	Potential failure mode	FM #	Potential failure effects
1.1.1.	Gas turbine	Provides rotation to the turbine output shaft from the expansion of the gas after the combustion chamber	Obstructed air outlet	FM01	Decreased output power
			Compressor rubbing	FM02	Vibration, propulsion impairment
			Obstruction in fuel injection	FM03	Power loss, generation failure, propulsion impairment
			Power turbine rubbing	FM04	Vibration, propulsion impairment
			Cutch slip	FM05	Vibration, speed coupling failure, propulsion impairment
			Lubricating oil insufficient cooling	FM06	Overheating, potential turbine trip
			Insufficient bearing lubrication	FM07	Overheating, increased wear, potential locking of mechanical parts

TABLE 9.1 Analysis of potential failure modes and effects of the CODOG propulsion system—cont'd

ID item	Item	Function	Potential failure mode	FM #	Potential failure effects
1.1.2.	Diesel engine A	Provides rotation to the output shaft from the internal combustion process	Obstructed air outlet	FM08	Decreased output power
			Clogged fuel filter	FM09	Decreased output power
			Insufficient engine cooling	FM10	Decreased output power, engine overheating
			Insufficient engine lubrication	FM11	Overheating, increased wear, potential stuck of mechanical parts, excessive vibration
1.1.3.	Diesel engine B	Provides rotation to the output shaft from the internal combustion process	Obstructed air outlet	FM12	Decreased output power
			Clogged fuel filter	FM13	Decreased output power
			Insufficient engine cooling	FM14	Decreased output power, engine overheating
			Insufficient engine lubrication	FM15	Overheating, increased wear, potential stuck of mechanical parts, excessive vibration
1.1.4.	Speed increaser gearbox	Increases the output shaft rotation according to the gear ratio between the gears	Cutch slip	FM16	Vibration, speed coupling failure, propulsion impairment
1.1.5.	Speed reducer gearbox	Reduces the output shaft rotation according to the gear ratio between the gears	Rolling-contact fatigue, input shaft, gas turbine side bearing	FM17	Vibration, potential propulsion impairment
			Rolling-contact fatigue, propeller shaft, gas turbine side bearing	FM18	Vibration; Potential propulsion impairment
			Rolling-contact fatigue, propeller shaft, diesel side bearing	FM19	Vibration; Potential propulsion impairment

Continued

TABLE 9.1 Analysis of potential failure modes and effects of the CODOG propulsion system—cont'd

ID item	Item	Function	Potential failure mode	FM #	Potential failure effects
			Rolling-contact fatigue, input shaft, diesel side bearing	FM20	Vibration, potential propulsion impairment
			Damaged gear teeth	FM21	Vibration, contamination of lubricating oil by metallic particles, potential slippery transmission
1.1.6.	Propeller	Transforms rotational power into thrust for propulsion	Coupling clearance	FM22	Vibration on the propeller shaft, impact on thrust generated by the propeller
			Cracked shaft	FM23	Vibration, decreased propulsion

detectability analysis was elaborated and presented in Table 9.2, considering the measurements affected by each FM, the clusters that group these measurements, and the failure rate (expressed in h^{-1}). These analyses combined are of importance for the provision of necessary information later in the HMFD application.

Finally, the study of the system gathered the failure and repair data of the subsystems and equipment of interest for the RAM analysis. Therefore, Table 9.3 presents the failure and repair data assumed in the CODOG propulsion system example of application. It is worth mentioning that, in this application, it was considered that the temporal unit is measured in hours; therefore, β is dimensionless, η is expressed in hours, and λ is expressed in h^{-1}.

With the results of the activities in the study of engineering systems, it was possible to carry out the next two stages of this example of application. Accordingly, Section 9.2.2 presents the RAM analysis, while Section 9.2.3 presents the HMFD for the CODOG propulsion system.

9.2.2 RAM analysis of the CODOG propulsion system

As previously mentioned in Section 9.1.2, the RAM analysis for the CODOG propulsion system was based on Reliability Block Diagram (RBD) technique. Thus, the functional trees were modeled into RBD

TABLE 9.2 Failure modes detectability analysis for the CODOG propulsion system.

FM #	Failure rate (h^{-1})	Measurements	Clusters	
FM01	0.00017080	Compressor inlet air pressure	C01	C24
		Power turbine exhaust temperature		
		Power turbine inlet temperature		
FM02	0.00009109	Gas turbine rotation speed	C02	
		Propeller shaft rotation speed		
		Compressor case vibration		
FM03	0.00007970	Gas turbine rotation speed	C03	
		Propeller shaft rotation speed		
FM04	0.00011380	Gas turbine rotation speed	C04	
		Propeller shaft rotation speed		
		Power turbine case vibration		
		Speed reducer gearbox input shaft bearing vibration, turbine side		
FM05	0.00013660	Gas turbine clutch temperature	C05	
		Speed reducer gearbox input shaft bearing vibration, turbine side		
		Propeller shaft rotation speed		
FM06	0.00022770	Gas turbine cooling water temperature	C06	
		Gas turbine cooling water pressure		
FM07	0.00034160	Compressor case vibration	C07	
		Turbine bearing noncoupling side temperature		
		Diffuser case vibration		
		Compressor turbine case vibration		
		Power turbine case vibration		
		Turbine bearing coupling side temperature		
FM08	0.00009769	Charge air pressure at cylinder inlet, Diesel Engine A	C08	C25
		Exhaust gas temperature, Diesel Engine A		

Continued

TABLE 9.2 Failure modes detectability analysis for the CODOG propulsion system—cont'd

FM #	Failure rate (h^{-1})	Measurements	Clusters	
FM09	0.00012480	Fuel oil pressure at engine inlet, Diesel Engine A	C09	
FM10	0.00010860	Cooling cylinder water pressure at engine inlet, Diesel Engine A	C10	
		Cooling cylinder water temperature at engine outlet, Diesel Engine A		
		Exhaust gas temperature, Diesel Engine A		
FM11	0.00016280	Lubricating oil pressure at engine inlet, Diesel Engine A	C11	
FM12	0.00009769	Charge air pressure at cylinder inlet, Diesel Engine B	C12	C26
		Exhaust gas temperature, Diesel Engine B		
FM13	0.00012480	Fuel oil pressure at engine inlet, Diesel Engine B	C13	
FM14	0.00010860	Cooling cylinder water pressure at engine inlet, Diesel Engine B	C14	
		Cooling cylinder water temperature at engine outlet, Diesel Engine B		
		Exhaust gas temperature, Diesel Engine B		
FM15	0.00016280	Lubricating oil pressure at engine inlet, Diesel Engine B	C15	
FM16	0.00002168	Lifting gearbox clutch temperature	C16	C27
		Speed reducer gearbox input shaft bearing vibration, diesel side		
		Propeller shaft rotation speed		
FM17	0.00000304	Speed reducer gearbox input shaft bearing vibration, turbine side	C17	
		Speed reducer gearbox input shaft bearing temperature, turbine side		
FM18	0.00000282	Speed reducer gearbox output shaft bearing vibration, turbine side	C18	
		Speed reducer gearbox output shaft bearing temperature, turbine side		

TABLE 9.2 Failure modes detectability analysis for the CODOG propulsion system—cont'd

FM #	Failure rate (h^{-1})	Measurements	Clusters	
FM19	0.00000260	Speed reducer gearbox output shaft bearing vibration, diesel side	C19	
		Speed reducer gearbox output shaft bearing temperature, diesel side		
FM20	0.00001084	Speed reducer gearbox input shaft bearing vibration, diesel side	C20	
		Speed reducer gearbox input shaft bearing temperature, diesel side		
FM21	0.00000217	Speed reducer gearbox input shaft bearing vibration, turbine side	C21	
		Speed reducer gearbox input shaft bearing vibration, diesel side		
		Speed reducer gearbox output shaft bearing vibration, turbine side		
		Speed reducer gearbox output shaft bearing vibration, diesel side		
FM22	0.00053590	Speed reducer gearbox output shaft bearing vibration, turbine side	C22	C28
		Speed reducer gearbox output shaft bearing vibration, diesel side		
		Propeller shaft rotation speed		
FM23	0.00035730	Speed reducer gearbox output shaft bearing vibration, turbine side	C23	
		Speed reducer gearbox output shaft bearing vibration, diesel side		

considering their reliability configuration. Fig. 9.5 presents the overall level of the scope of this RAM analysis with three main blocks: the CODOG propulsion system port-side (1.1.), the CODOG propulsion system starboard-side (1.2.), and the auxiliary systems (1.3.).

As shown in Fig. 9.5, these three reliability blocks are arranged in a series configuration which means that a failure in any of the results in the failure of the entire model. In other words, in this analysis of the propulsion for the military ship, all these three systems shall be working for the mission to succeed.

TABLE 9.3 Failure and repair data for the CODOG propulsion and auxiliary systems.

	Reliability ($R(t)$)			Maintainability ($M(t)$)		
Item	**Distribution**	**Parameter(s)**		**Distribution**	**Parameter(s)**	
1.1.1. Gas turbine	Weibull	$\beta = 2.0$	$\eta = 991.0$	Normal	$\mu = 350$	$\sigma = 75$
1.1.2./1.1.3. Diesel Engine (A/B)	Weibull	$\beta = 2.5$	$\eta = 2076.6$	Normal	$\mu = 150$	$\sigma = 25$
1.1.3. Speed increaser gearbox	Weibull	$\beta = 3.5$	$\eta = 51{,}257.8$	Normal	$\mu = 50$	$\sigma = 20$
1.1.4. Speed reducer gearbox	Weibull	$\beta = 3.1$	$\eta = 51{,}569.7$	Normal	$\mu = 50$	$\sigma = 20$
1.1.6. Propeller	Weibull	$\beta = 1.5$	$\eta = 1240.1$	Normal	$\mu = 100$	$\sigma = 25$
1.3.1. Lubrication system	Exponential	$\lambda = 0.000053058$		Normal	$\mu = 10$	$\sigma = 5$
1.3.2. Oil filtration	Exponential	$\lambda = 0.000082795$		Normal	$\mu = 3$	$\sigma = 1$
1.3.3. Hydraulic system	Exponential	$\lambda = 0.000044318$		Normal	$\mu = 10$	$\sigma = 3$
1.3.4. Fuel oil injection	Exponential	$\lambda = 0.000046730$		Normal	$\mu = 10$	$\sigma = 3$
1.3.5. Compressed air	Exponential	$\lambda = 0.00014000$		Normal	$\mu = 15$	$\sigma = 3$

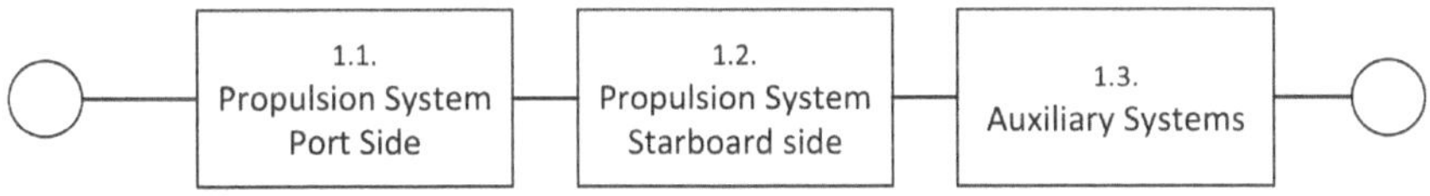

FIG. 9.5 RBD for the overall CODOG propulsion system level.

TABLE 9.4 Operational configurations of the CODOG propulsion systems.

Operational configuration	Output power
Operating with one diesel engine	Until 60% of maximum power
Operating with both diesel engines	Until 75% of maximum power
Operating with the gas turbine	Until 100% of maximum power

This overall level was broken down into subdiagrams that represented the lower levels of the functional hierarchy as previously presented in Figs. 9.3 and 9.4. The equipment units of both CODOG propulsion systems can have different reliability configurations depending on the power demand of the military ship. The reason for this is that each of the propulsion systems can operate with only one diesel engine, with both diesel engines at the same time, or with the gas turbine as the power source. These three operational configurations were represented in Table 9.4 for better comprehension.

Through Table 9.4, three ranges of power demands were defined for the military ship: 0 to 60%, greater than 60% to 75%, and greater than 75% to maximum power (100%). For each of them, a reliability configuration was modeled to represent the CODOG propulsion systems (A, B, and C). This was necessary since different operational configurations are available depending on the requested power demand. Accordingly, these RBDs were presented in Figs. 9.6–9.8, respectively.

As shown in Fig. 9.6, when the demand output power is less than 60% of the maximum power, the CODOG propulsion system of the military ship can be operated by only one of the diesel engines or by only the gas turbine. For instance, if Diesel Engine A is running, Diesel Engine B is on standby for it, as well as the gas turbine. Although they are both in a passive redundancy, the RBD represented them in simple parallel configuration since the failure rates of the standby components were considered the same in the quiescent mode as it is in the active mode (hot standby).

For the CODOG propulsion system configuration B, it can be operated by both diesel engines running at the same time or by only the gas turbine. Thus, one option acts as a standby for the other. Similarly, the RBD also represented them in a simple parallel configuration due to the fact that

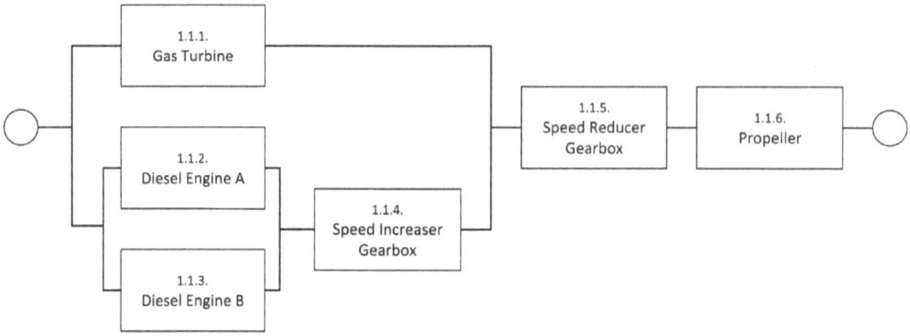

FIG. 9.6 RBD for the CODOG propulsion system A (up to 60% of maximum power.

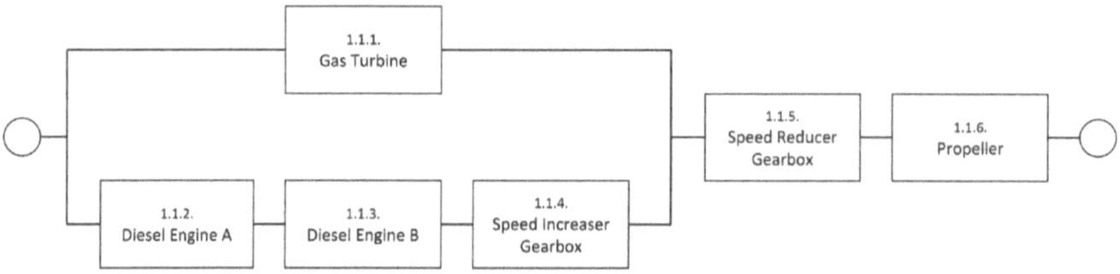

FIG. 9.7 RBD for the CODOG propulsion system B (from 60% to 75% of maximum power).

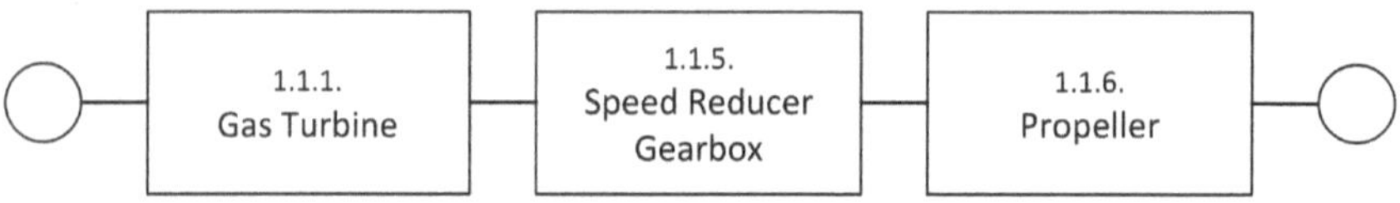

FIG. 9.8 RBD for the CODOG propulsion system C (more than 75% of maximum power).

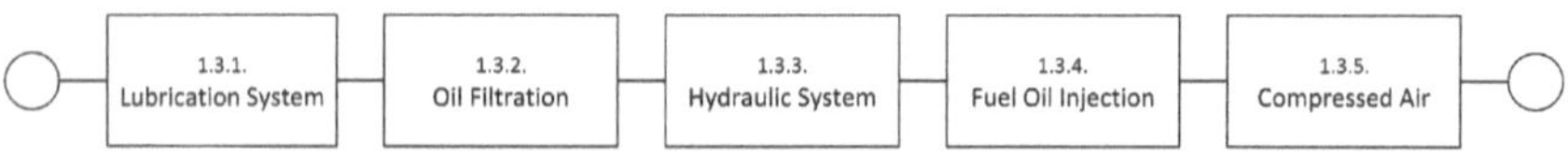

FIG. 9.9 RBD for the auxiliary systems.

it was assumed the same failure rate for the active and quiescent modes of the components in this example of application. Finally, when the output demand is higher than 75% of the maximum power, the CODOG propulsion system can be only operated by the use of the gas turbine with no redundancy available.

Finally, the auxiliary systems (1.3.) were represented in Fig. 9.9 in a series configuration. All the subsystems of the hierarchy are necessary for the proper operation of the military ship since they are centralized and serve both CODOG propulsion systems.

With these models in RBD and the failure and repair data, it was possible to perform a set of RAM analyses at different levels. The RAM analysis through RBD is based on Monte Carlo simulation that relies on repeated random sampling for each block according to their reliability and maintainability distribution (Table 9.3). The numerical results are aggregated according to the reliability configuration and several simulations are performed for the convergence of the main results. In this example of application, the results are first presented for the subdiagrams and then for the overall results. Among the results are the reliability curve, the operational availability, and the number of expected failures.

Fig. 9.10 presents the reliability curve for the auxiliary systems RBD, while Fig. 9.11 presents the reliability curves for the three configurations of the CODOG propulsion systems (A, B, and C). In both analyses, 1000 simulations with punctual results every 3 h were considered for the convergence of results for 1 year of operation (8760 h) with the support of Reliasoft BlockSim software.

According to the simulations, the reliability of the auxiliary systems after 1 year of operation is about 5%. In other words, there is only about a 5% probability that the system represented by the RBD in Fig. 9.9 will not fail with 8760 h of operation.

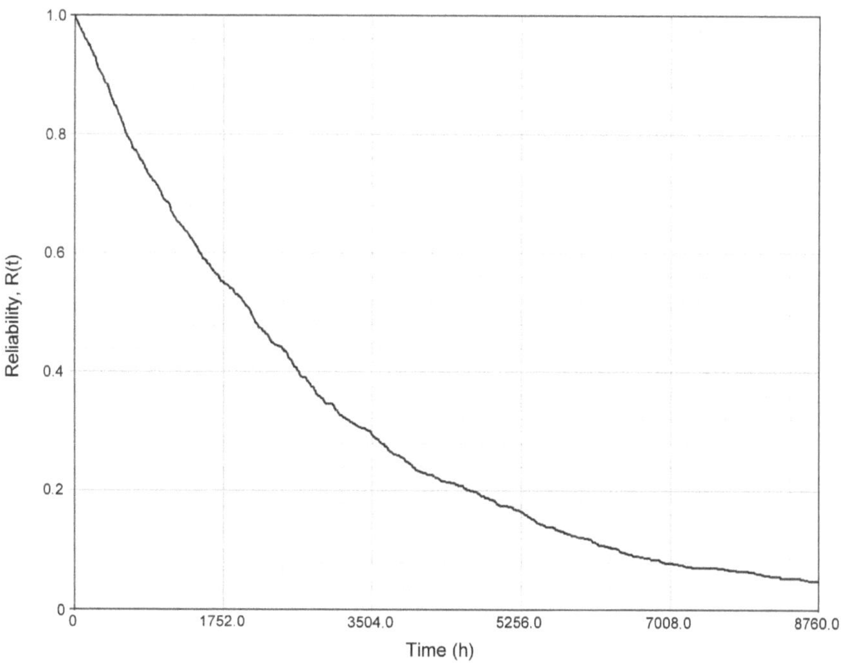

FIG. 9.10 Reliability versus time for the CODOG auxiliary systems.

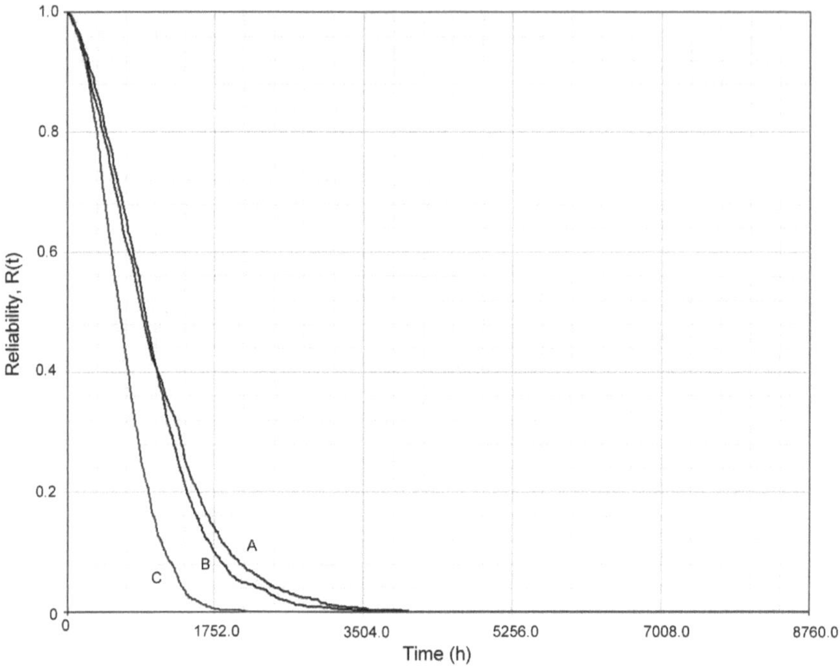

FIG. 9.11 Reliability versus time for the CODOG propulsion systems A, B, and C.

As shown in Fig. 9.11, the estimation of the reliability for the CODOG propulsion systems depends on their operational configuration. The configuration C required when the output power demand is greater than 75% to the maximum power is the least reliable configuration of the three available for the CODOG propulsion system. On the contrary, configuration A is the most reliable configuration. However, in none of the configurations is a flawless operation expected for 8760 h of operation. For instance, for 1752 h of operation, around two and half months, the reliability for the CODOG propulsion system is around 13%, 9%, and 0%, respectively, in configurations A, B, and C.

As this example of application considered all items as repairable systems, it was also possible through the simulations to estimate the availability of the RBD based on the combined reliability and maintainability distributions (Table 9.3). Accordingly, punctual availability for the auxiliary system over time is presented in Fig. 9.12 and for the three CODOG propulsion system configurations (A, B, and C) in Fig. 9.13.

The convergence of the simulations showed that the availability of auxiliary systems for 1 year of operation is high, greater than 99%. On the contrary, the three configurations of the CODOG propulsion system have less availability for the same period. In face of the configurations A and B, it is

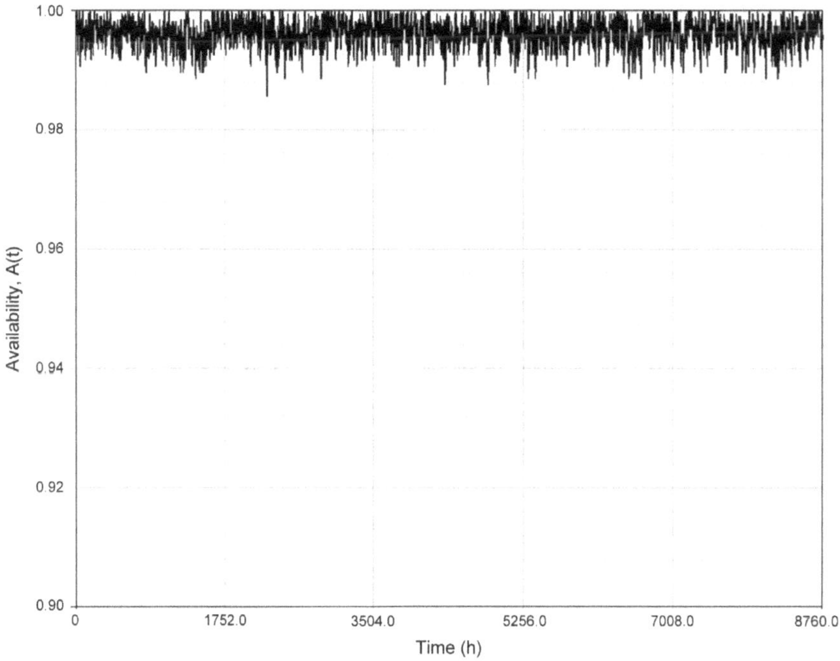

FIG. 9.12 Availability versus time for the CODOG auxiliary systems.

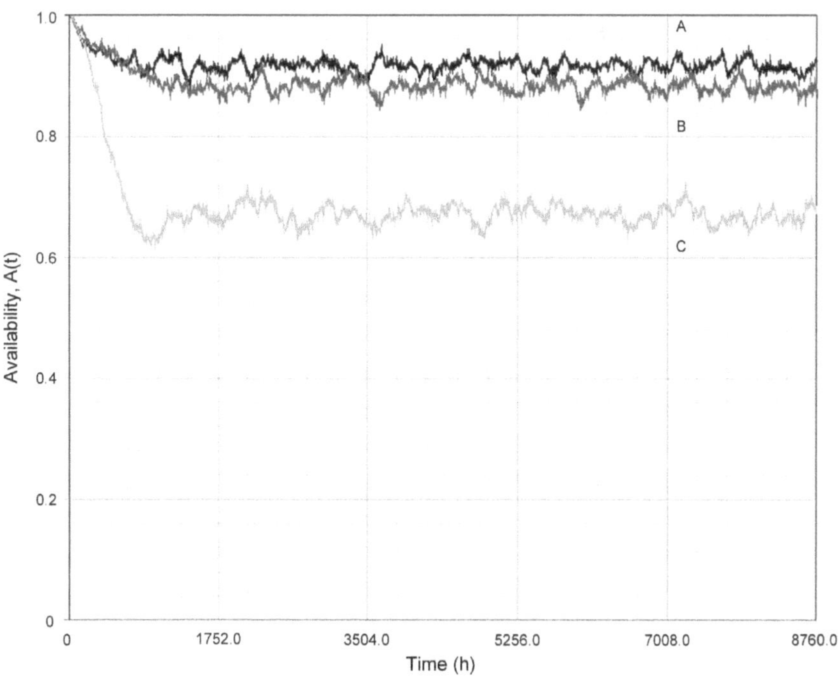

FIG. 9.13 Availability versus time for the CODOG propulsion systems A, B, and C.

expected an average availability around 92% and 88% for them, respectively. The CODOG propulsion system in configuration C, when the output power demand is greater than 75% to the maximum power, has an even lower operational availability, around 68%. Although this availability is the lowest, it is worth mentioning that the military ship remains in operation for a short time in this configuration since it is usually specific for situations of attack and evasive maneuvers.

Finally, the RAM analysis simulated the number of expected failures for the subdiagram levels for 1 year of operation (8760 h). Fig. 9.14 presents the results for the auxiliary systems, while the results of the three configurations of the CODOG propulsion system are presented in Table 9.5. Moreover, a comparison of the total expected failures in the three configurations is shown in Fig. 9.15.

In this example, it is worth noting that the equipment units of the auxiliary system have fewer expected failures than the CODOG propulsion system equipment units. This reiterates the lower reliability and availability presented for the three configurations of the CODOG propulsion system when compared to the auxiliary system. The exception is for both gearboxes that, due to their high reliability, do not present expected failures for 1 year of operation.

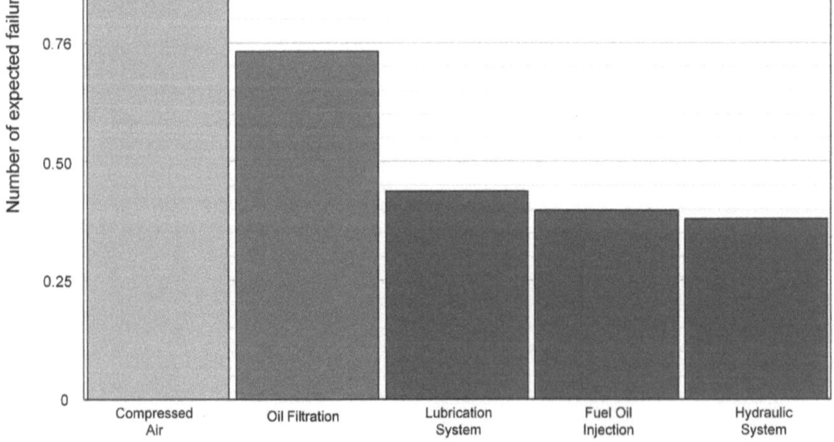

FIG. 9.14 Number of expected failures for the CODOG auxiliary systems.

TABLE 9.5 Number of expected failures for the CODOG systems A, B, and C.

ID item	Config. A 0%–60% of max. Power	Config. B >60%–75% of max. Power	Config. C >75% of max. Power
1.1.1. Gas turbine	6.66	6.55	6.57
1.1.2. Diesel Engine A	3.65	3.68	–
1.1.3. Diesel Engine B	3.65	3.68	–
1.1.4. Speed increaser gearbox (SIG)	0.00	0.00	–
1.1.5. Speed reducer gearbox (SRG)	0.05	0.05	0.05
1.1.6. Propeller	6.98	6.83	5.25

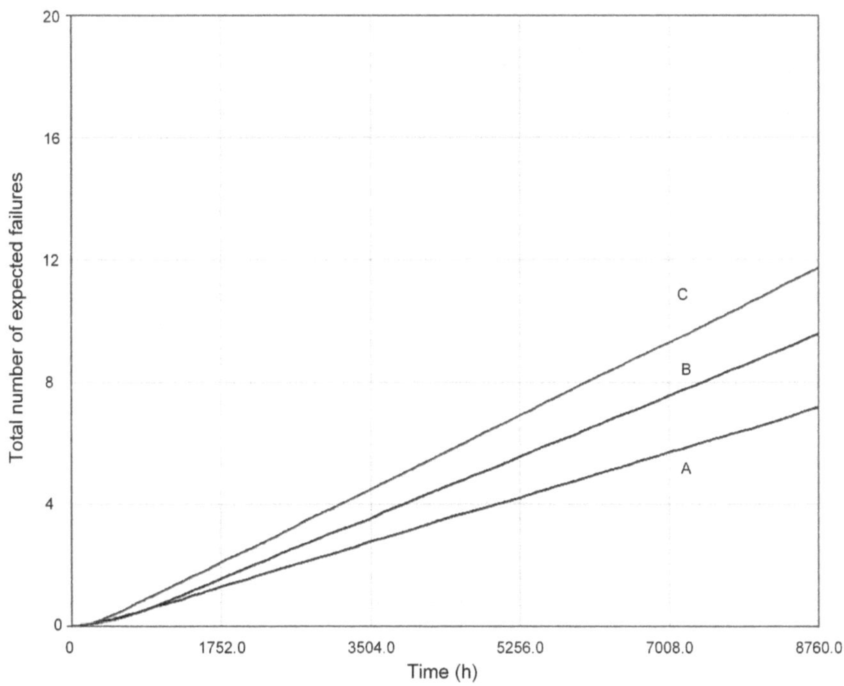

FIG. 9.15 Total expected failures for the CODOG propulsion systems A, B, and C.

After the simulations of these subdiagrams individually, the RAM analysis performed the simulations for the overall CODOG propulsion system level. In this analysis, three main reliability blocks are considered, as presented in Fig. 9.5: the CODOG propulsion system port-side, CODOG propulsion system starboard-side, and auxiliary systems. For this overall analysis, both CODOG propulsion systems considered the model of the subdiagram configuration B (Fig. 9.7). For the auxiliary systems' block, it was considered the previous model presented in Fig. 9.9. The same failure and repair data presented in Table 9.2 were used in this analysis.

Analogously to the analysis of the subdiagrams, the analysis for the overall CODOG propulsion system level was also obtained through the simulations, with Reliasoft BlockSim software, the reliability curve, the operational availability, and the number of expected failures as main results as presented in Figs. 9.16–9.18, respectively. In this analysis, 1000 simulations with punctual results every 3 h were also considered for the convergence of results for 1 year of operation (8760 h).

These results of the RAM analysis provide useful information that supports decision-making in organizations. In this example of application, the overall CODOG propulsion system-level analysis was able to simulate and estimate the operational availability over time based on the reliability block diagrams and failure and repair data. As shown in Fig. 9.17, the average availability is around 80% for 1 year of operation. This performance shall be evaluated based on the operational context in which the engineering system is inserted and, depending on the organization, it may be sufficient and acceptable or not.

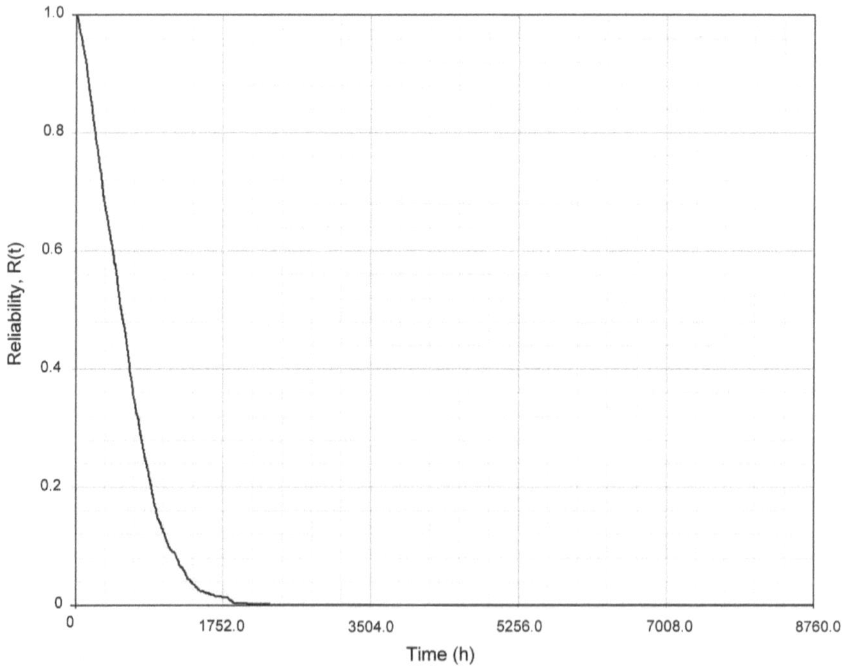

FIG. 9.16 Reliability versus time for the overall CODOG propulsion system level.

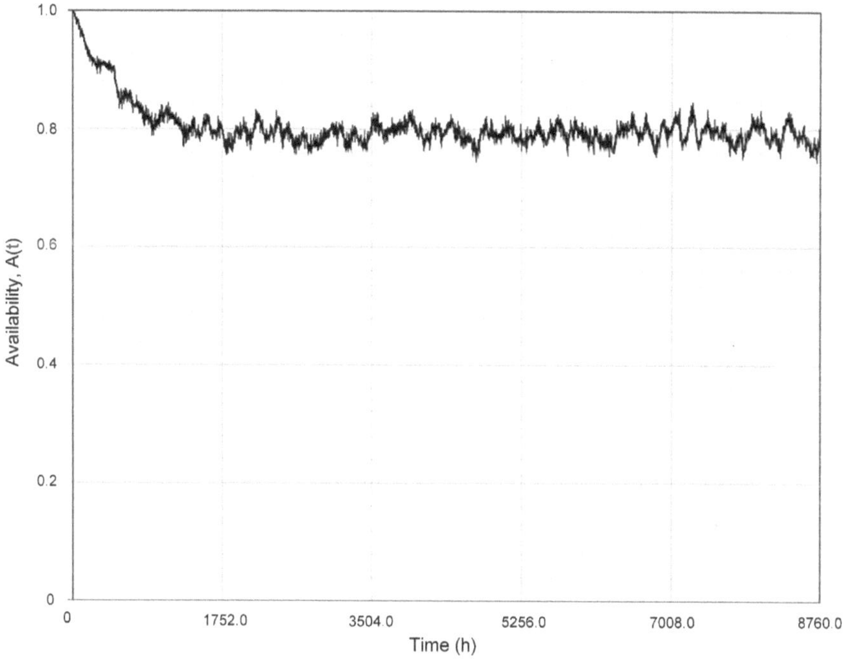

FIG. 9.17 Availability versus time for the overall CODOG propulsion system level.

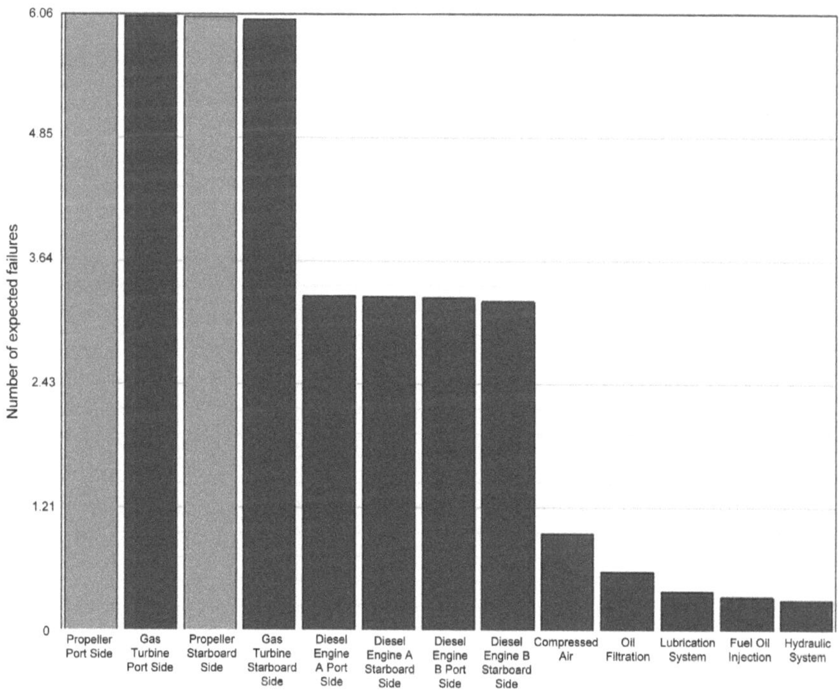

FIG. 9.18 Number of expected failures for the overall CODOG system level.

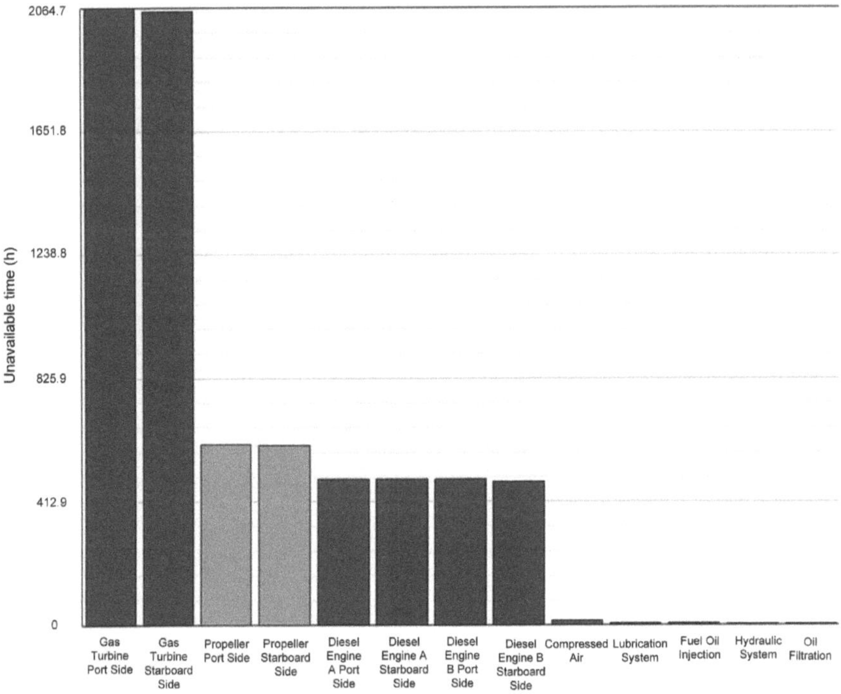

FIG. 9.19 Unavailable time for the overall CODOG propulsion system level.

If the organization decides to improve the performance of its engineering system, the results of the number of failures expected in the RAM analysis may indicate which item has the highest number of failures and, consequently, the least reliability. Although it is a good indicator, it is interesting that the organization also analyze the equipment that causes more downtime, as presented in Fig. 9.19. In other words, in addition to the information about the poor reliability, it is also interesting to consider maintainability for decision-making. For instance, there may be situations in which the equipment is unreliable, but its maintenance is simple and fast, with little impact on the unavailability.

Accordingly, RAM analysis can support performance improvement decisions in the organization as it provides information about the performance of engineering systems. Its results can support the identification of investment opportunities in the portfolio of physical assets, the analysis of engineering systems that deserve more attention, the adequacy of maintenance plans according to reliability estimates, and also the dimensioning of maintenance spare parts stocks. These possibilities corroborate the reason why RAM analysis is widely used and recommended in organizations with modern engineering systems.

9.2.3 HMFD of the CODOG propulsion system

Continuing the first case study, this section demonstrates the implementation of the HMFD framework (presented in Chapter 7) and some results obtained from its application. Considering the system's preliminary study, previously presented in Section 9.2.1, and the general considerations for the HMFD stage, presented in Section 9.1.3, the first step in implementing and applying the framework is data acquisition. In this case, as the vessel considered for this case example has a SCADA system, several variables of interest and descriptors can be collected directly from the database of such a system. Figs. 9.20 and 9.21 show some temperature and vibration measurements of the port-side propulsion system speed reducer gearbox collected over time during a certain period of operation with the data samples collected every hour.

In this application example, the vessel had undergone a general overhaul recently. In this case, the hour meter related to the operational campaign has been restarted, which means that the value of the hours that appear on the abscissa axis in the graphs indicates how long the vessel has been in operation after the overhaul.

Despite presenting an extremely small number of expected failures and, consequently, a high reliability, the speed reducer gearbox is an

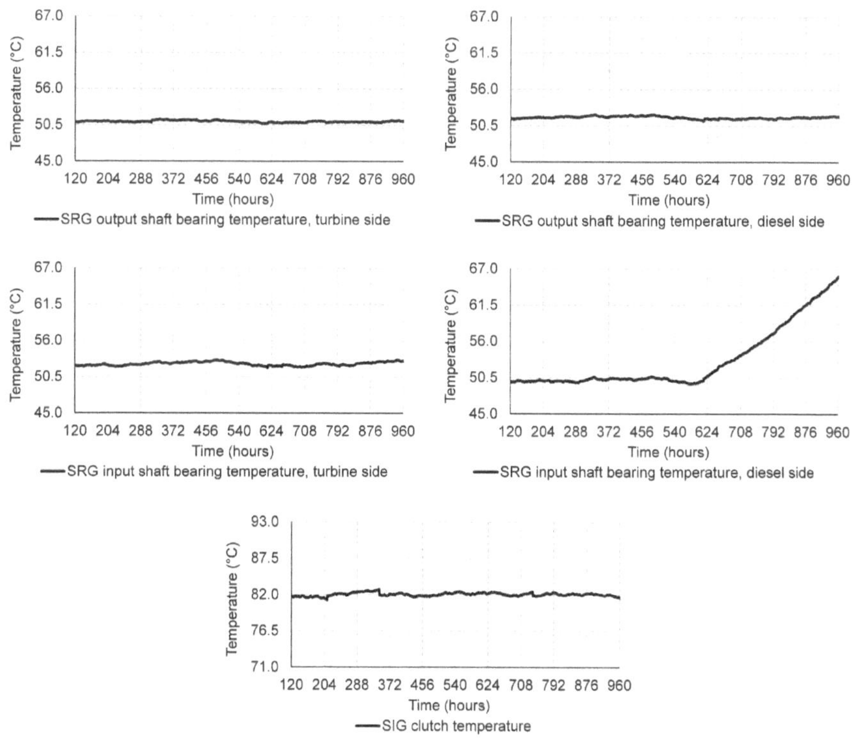

FIG. 9.20 Temperature measurements of the CODOG propulsion system.

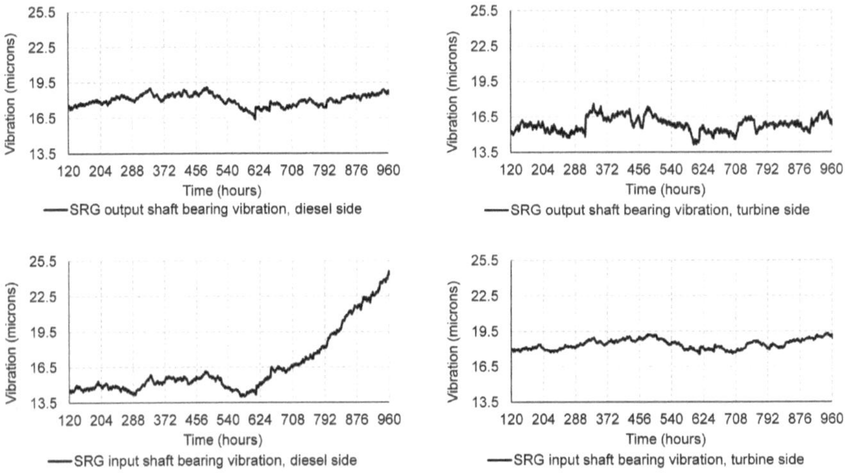

FIG. 9.21 Vibration measurements of the CODOG propulsion system.

extremely important component for the vessel since in all operational configurations, the performance of the ship's performance is directly associated with the performance of this component. Thus, monitoring the health of the SRG and detecting and diagnosing, any fault in its initial phase is a major task for the maintenance team.

During the period of time considered in the graphs of Figs. 9.20 and 9.21, a system's malfunction started to be noticed. This malfunction is related to a rolling-contact fatigue (RCF) in the input shaft diesel side bearing, i.e., Failure Mode 20 of Table 9.1, of the port-side propulsion system. Considering the time in which the ship is in operation after the overhaul, this problem would not be expected. A probable explanation for the origin of this fault (the disturbance, as presented in Chapter 5) would be the execution of an imperfect repair (or imperfect maintenance, as presented in Chapter 4), which, by not bringing the system to an as good as new condition, leads to the early appearance of a fault.

Once the data acquisition process is consolidated and the descriptors are defined, the next step in the HMFD framework is the fault detection process. In this case, as mentioned earlier, the detection method chosen was an adaptive variant of the PCA (Michalski, 2021). This variant uses, in addition to the analyzed data, a set of reference data for each analyzed cluster. It is also necessary to define an analysis window that tracks the data monitored every hour (thus defining the number of observations contained both in the reference sample and in the group of data analyzed). In this example, a 120-h analysis window was considered and the reference data was collected during the first 120 h of system operation after returning from the overhaul.

As mentioned, the T^2-statistic (Mujica et al., 2011) is applied to the PCA output eigenvector matrix as a measure for evaluating the results and defining the fault detection. For each cluster, the detection of an anomaly is defined from the moment that the first statistical moment (average) of the distribution obtained from the T^2-statistic of an analyzed sample data exceeds a preestablished limit value. This threshold is the result of the sum of the average plus three times the standard deviation of the distribution obtained with the application of this same process (PCA followed by T^2-statistic) to a data sample collected when the equipment is in a healthy operating condition, that is, without faults. In this case, a sample of data collected between the 120th and 240th hours of the system operation was considered.

Fig. 9.22 shows the time evolution of the expected value (average value) of the T^2-statistic ($E[T^2]$) for the four clusters associated with the considered failure mode (in this case, clusters C16, C20, C21, and C27, presented in Table 9.2). Also shown in each graph is the limit value to determine the detection, the time when the malfunction of the system starts, and the moment when the FDD of the considered failure mode is performed.

It is important to highlight that the time at which the malfunction effectively starts is generally not available or even known information, since it is very difficult to be obtained. Nevertheless, in the examples of the application of this book, such information is considered for didactic purposes only, assisting the reader to understand the FDD process as a whole.

Likewise, in the proposed applications the fault diagnosis is performed instantly after its detection. As previously mentioned, the Bayesian network was the method selected to perform the diagnosis of failures based

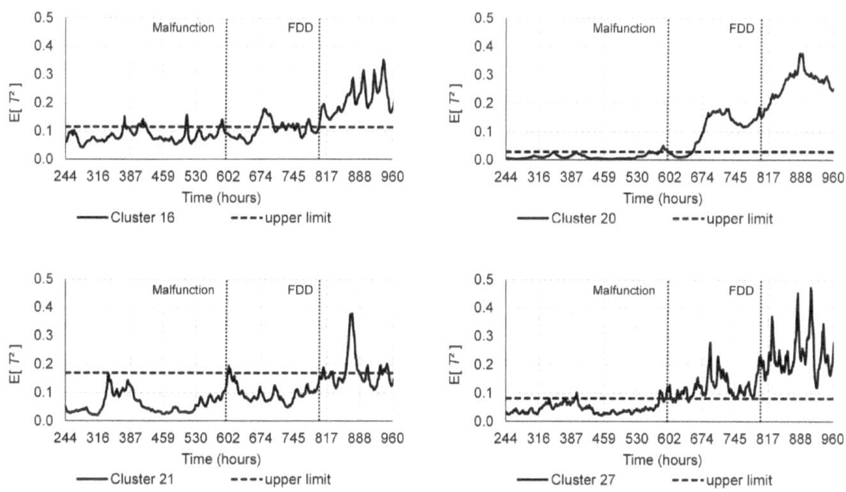

FIG. 9.22 Clusters' analysis of the CODOG propulsion system.

on the results of the detection. The Bayesian network graph representation for this example of application is presented in Fig. 9.23. The correlation between failure modes and clusters is made considering the measurements related to each failure mode (as presented in Table 9.1) and the clusters in which these measurements are grouped, i.e., the effects of a particular failure mode (its symptoms) can be seen in any cluster that has at least one of the measurements associated with this failure mode.

From the Bayesian network, it is possible to obtain the *a posteriori* probability of each failure mode, being known the a priori probabilities of each failure mode (the failure rate, presented in Table 9.2) and the status of each cluster (true or false) over time. Fig. 9.24 shows the result for the failure mode considered in this example (FM20), combined with the status of each related cluster (C16, C20, C21, and C27) over time for a given period of observation of the system.

Analyzing the graph in Fig. 9.24, some aspects can be highlighted. First, it is noted that even before the malfunctioning of the system starts, eventually, the result of the detection process indicated a positive value for the analyzed clusters. Such behavior is associated with temporary disturbances in the systems or even noise in the collected data. For this reason and in order to avoid false alarms, the FDD is only considered effective when the result obtained becomes continuous over time. In the example shown, this behavior occurs after the 800th hour of operation of the system. In other words, it took around 200 h of operation (a little more than 8 days) between the time when the fault started to affect the system behavior effectively and the moment when it was detected and diagnosed.

Another aspect that can be highlighted is the fact that only after three of the four clusters associated with the failure mode presented a positive result for the fault detection (C16, C20, and C27), the fault diagnosis occurs. Finally, it is also interesting to note that the evidence that the failure mode was in progress in the system (the *a posteriori* probability) presented a very high value (practically 100%), which for the subsequent fault analysis and decision-making is extremely positive, since with such a result there is practically no doubt that the failure mode is in progress, allowing any more assertive decisions to be taken.

Once the FDD is performed and a positive result for a given failure mode is obtained, the next step is to perform the fault analysis and calculate the RUL. Considering that the system continues to operate after the FDD, the calculation of the RUL can be done periodically from the moment of detection. This result is shown in Fig. 9.25, where the hours of operation of the system are on the abscissa axis and the RUL, in hours (considering a logarithmic scale), on the ordinate axis.

As the FDD method is multivariable in this case, the calculation of the RUL shall consider all variables affected by the failure mode. Thus, the value shown in the graph in Fig. 9.25 is the lowest value found for

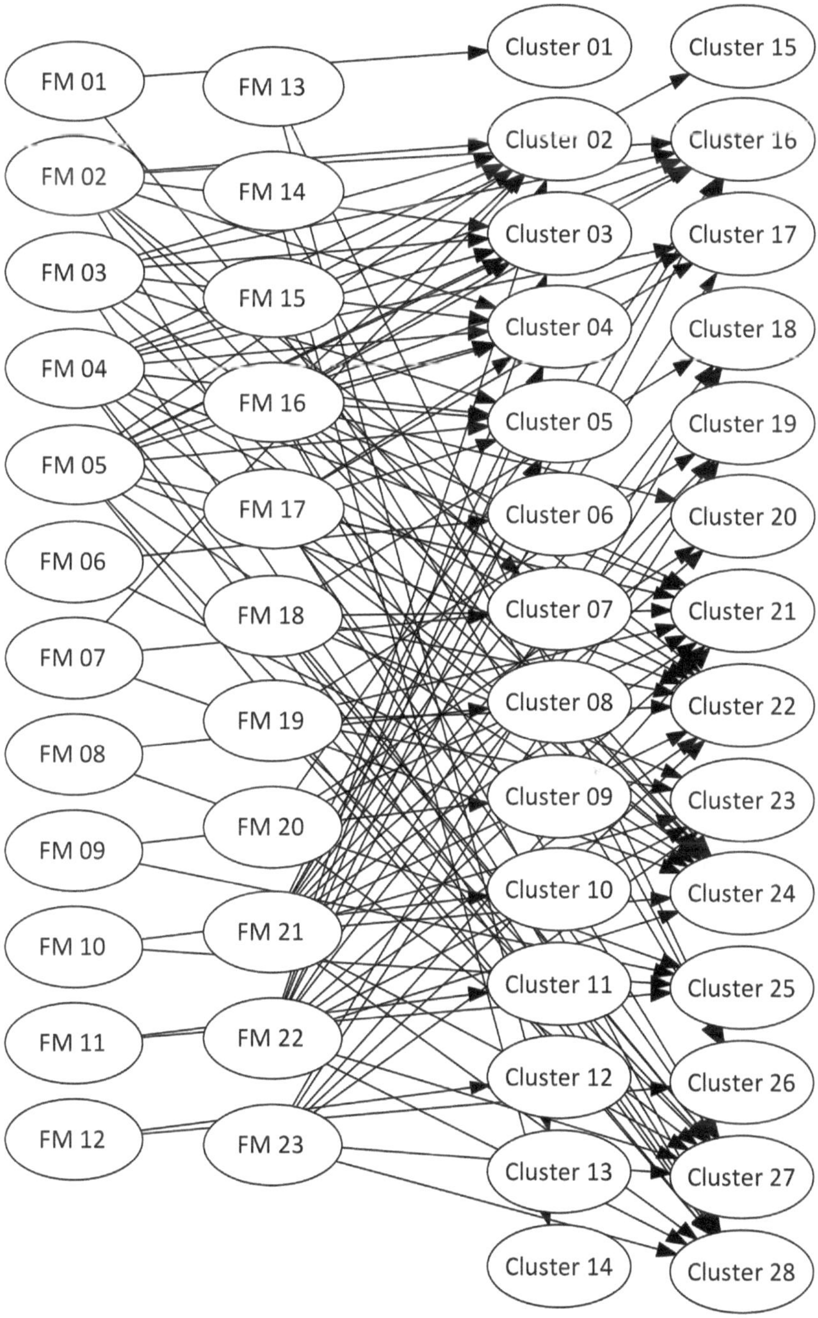

FIG. 9.23 Bayesian network graph representation of the CODOG propulsion system.

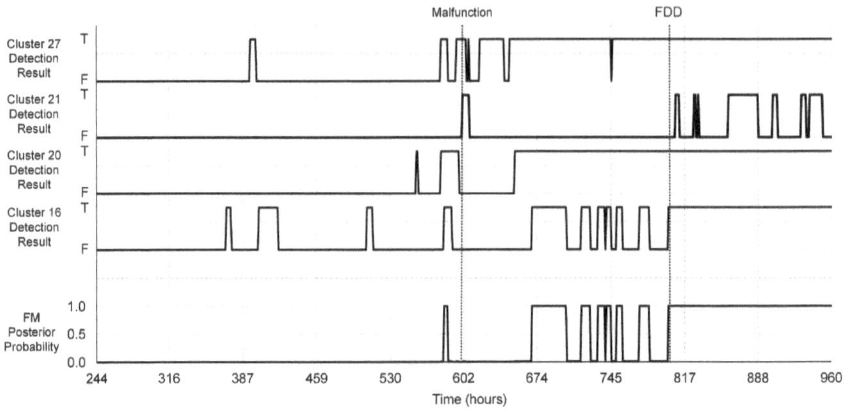

FIG. 9.24 FDD analysis of the CODOG propulsion system.

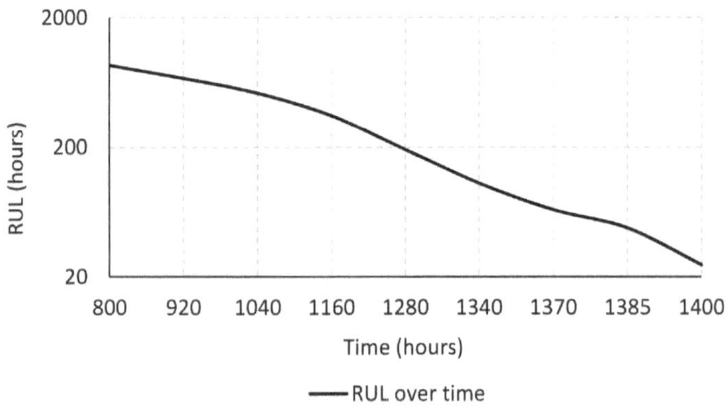

FIG. 9.25 RUL analysis of the CODOG propulsion system.

the set of measurements considered. That is, the worst-case scenario is considered, in which it is sufficient that just one of the variables considered reaches the value established as a limit to maintain the integrity of the system. These values can be the same as those used to define the alerts and alarms of the SCADA system. That is, in this case, the RUL indicates the time for the SCADA system and its protection measures to be activated if the equipment is not turned off and the fault continues to progress.

Note that as soon as the fault is diagnosed, a RUL of around 856 h is calculated. However, as new measurements are considered, the result of the RUL becomes more accurate. Following the system until the effective moment of the failure, with approximately 600 h of operation after the FDD, the calculation of the RUL indicates a result of 24 h.

In practice, it is important to work with a safety margin on such results due to the uncertainties of the process. Likewise, the detection of a fault that is still incipient is of great importance since it allows decisions to be made under less pressure. For example, in the case of a ship such as the one in this example of application, a failure such as the one diagnosed would lead to the loss of one of the propulsion systems, which severely affects the operational capacity of the ship. However, based on the result of calculating the RUL over time, it is possible to schedule the return of the vessel to a port in order to carry out the necessary repairs in the most convenient way.

In the next section are presented some asset management considerations that complement the analyses made considering the results obtained with the implementation and application of the proposed HMFD framework.

9.2.4 Asset management considerations

As seen in Chapter 8 where the so-called 39 asset management subjects were presented, procurement and supply chain management for modern engineering systems must be carefully planned and diligently carried out, particularly where ready access to suppliers and parts may be uncertain. CODOG naval plants certainly fit into this category. Fig. 9.26 gives an idea of the main aspects involved in the efficient maintenance planning of assets such as gas turbines and diesel engines for naval duty.

Broadly defining, naval logistics is the science and art of generating and sustaining material adequacy and availability by means of the integration and synchronization of engineering, maintenance, and supply functions.

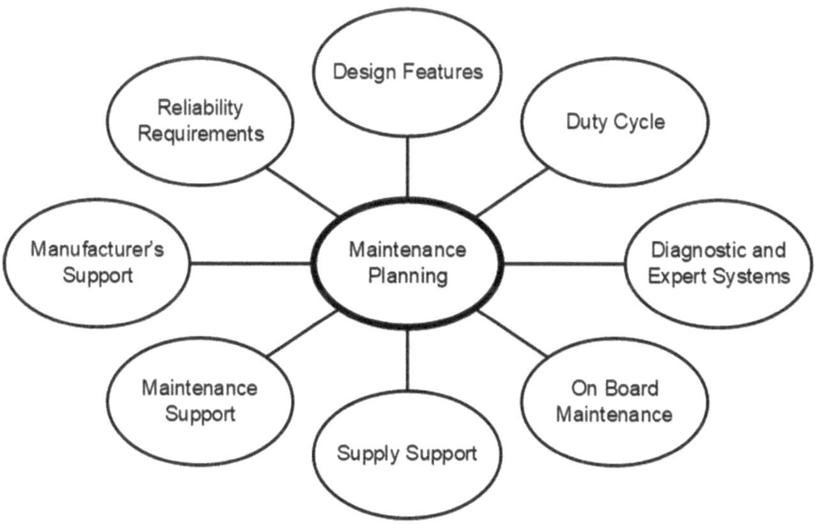

FIG. 9.26 Maintenance planning aspects.

One of the basic principles of naval logistics is to provide the needed capability for seamless support across the asset life cycle (Royal Australian Navy, 2016). This life cycle is illustrated in Chapter 8.

In general, naval ships equipped with CODOG power plants will be submitted to the following periods of maintenance (United States Government Accountability Office, 2020):

Normal Maintenance Period (NMP)—maintenance period normally scheduled at 3-year intervals destined for depot-level maintenance and modernization, i.e., work that cannot be accomplished by ship's crew or exceeds fleet capabilities;
Continuous maintenance—maintenance work conducted out of the NMPs, typically scheduled to be 2–6 weeks in duration; and.
Voyage repair—corrective maintenance actions necessary for the ship to deploy or continue its deployment.

Considering onboard maintenance as Level-I, external maintenance support can be carried out in two levels: Level-II that may be called intermediate maintenance and depot level or Level III. Adequate expertise at all three levels and, as is well known by practitioners, suitable manufacture's support plays an important role in keeping CODOG machinery in full operational capability. Besides, an effective supply chain is certainly fundamental for the maintenance of such assets.

The well-known idea put forward by Shingo (Shingo, 1988) that stock is not a necessary evil but an absolute evil, sometimes quoted as "stock is trouble", applies in one way or another to all kinds of stocks, including naval depots' inventories. This evident perception, however, was conceived in an industrial pull production context where it is less complex to plan for aligning supply with demand. Naval systems, on the contrary, are more likely to require a combination of pull production and what can be termed as modified pull production (Peltz et al., 2014).

Level-II maintenance can operate on a pull system with a one-piece flow. When the required repair actions exceed the capability of this level, however, the asset is sent for Level-III maintenance. In this latter level, bearing in mind funding and procurement needs, careful planning in every fiscal year must be carried out in such a way as to allow depots' maintenance activities to determine the detailed production schedules in order to balance production efficiency and customer needs, and still consider the possibility of occasional shortages of some items. If carried out properly, such a process will produce results close to a pull system. It should be noted that the majority of depot-level ship maintenance takes place during NMPs.

Due to its crucial importance in assuring that CODOG systems will be appropriately maintained along with their life cycles and customers' needs timely attended to, the effectiveness of the supply chain management process should be carefully monitored using a specific combination of indicators relating to customer service and inventory management. Chapter 8 can provide guidance in selecting such indicators.

9.3 Analysis of a FPSO power generation system

Similar to the previous example, this section presents a power generation system of Floating Production Storage and Offloading (FPSO) vessels as the second example of application. This case study also follows the three stages of the analysis presented in Section 9.1. Therefore, Sections 9.3.1–9.3.3 address the system's study, the RAM analysis, and the HMFD for the FPSO power generation system, respectively.

9.3.1 Study of the FPSO generation system

A FPSO unit is a hull-shaped floating production system that receives fluids (crude oil, water, and a host of other things) from a subsea reservoir through risers and separates fluids into crude oil, natural gas, water, and impurities within the topside production facilities onboard. Crude oil stored in the storage tanks of the FPSO is offloaded onto shuttle tankers to go to the market or for further refining onshore. The produced gas is exported, re-injected, or combusted. The FPSO is moored on the production location and, when an existing oilfield is depleted, can be moved to other production locations when necessary.

The FPSO vessels are particularly effective in remote or deepwater locations where seabed pipelines are not cost-effective. They also eliminate the need to lay expensive long-distance pipelines from the oil well to an onshore terminal. The FPSO system has a set of critical subsystems whose reliability and availability affect its overall performance. One of those critical subsystems is the process plant power system that provides electrical power to all pieces of equipment of that plant. The loss of electrical power causes the plant to shut down and the oil well must also be shut down. This event is very critical due to economic losses besides any other maintenance costs.

Fig. 9.27 presents a diagrammatic representation of a gas turbine unit, the principal component of the generating units of the plant power system. The process plant power generation system is designed according to the size of the plant, and in this way, there are FPSOs with two, three, four, or more gas turbines driving electrical generators. The number of turbines used in the power system is also dependent on the nominal power of that device.

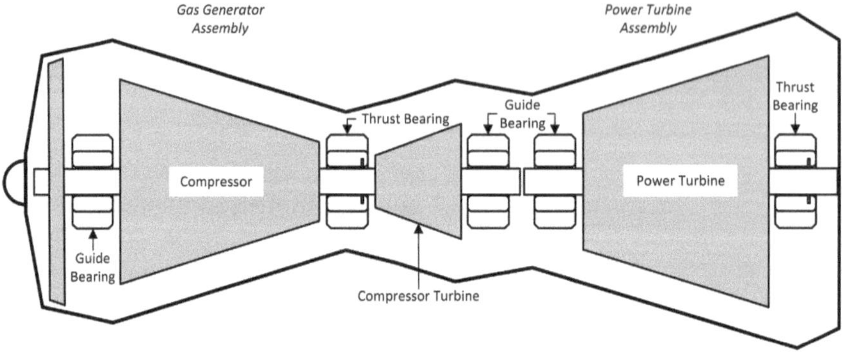

FIG. 9.27 Representation of the FPSO gas turbine unit.

After this brief description of the FPSO power generation system, Figs. 9.28–9.30 present the functional trees for the delimited scope of this second example. First, Fig. 9.28 represents the functional tree for the FPSO unity with its main systems in the first level of the hierarchical structure representation. As the scope of this example of application focused on the production plant power system (1.3) of the FPSO unit, Fig. 9.29 presents the subsystems and equipment of the power system; Fig. 9.30, the components of the gas turbine.

From these functional trees, the next activity in the study of the engineering systems identifies the potential FM and their detectability for the items of interest in the HMFD example of application. Thus, Table 9.6 presents a simplified analysis of the potential failure mode and effects for the gas turbine's components.

Similar to the first example, from the identified potential FM of Table 9.6, a detectability analysis is elaborated and presented in Table 9.7. These analyses combined are of importance for the provision of necessary information later in the HMFD application.

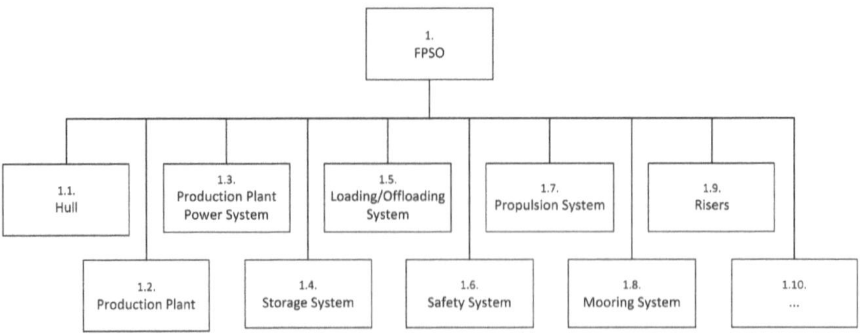

FIG. 9.28 Systems of the FPSO unit.

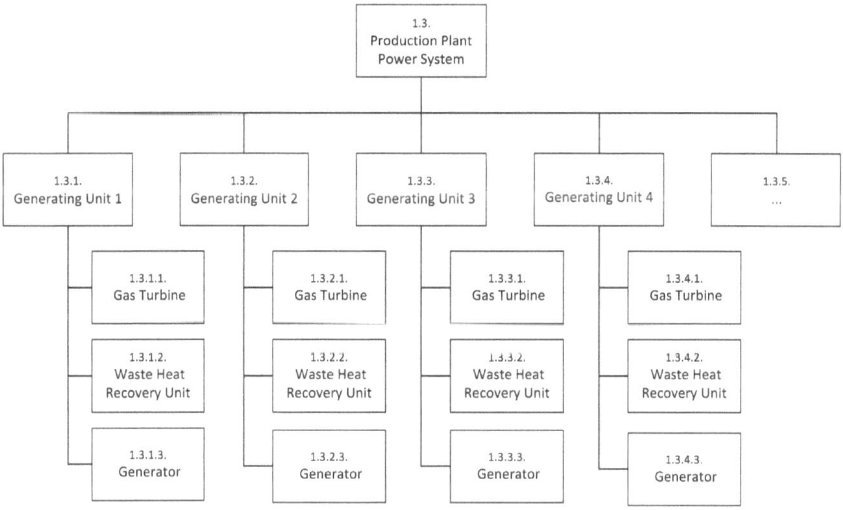

FIG. 9.29 Subsystems and equipment units of the FPSO power system.

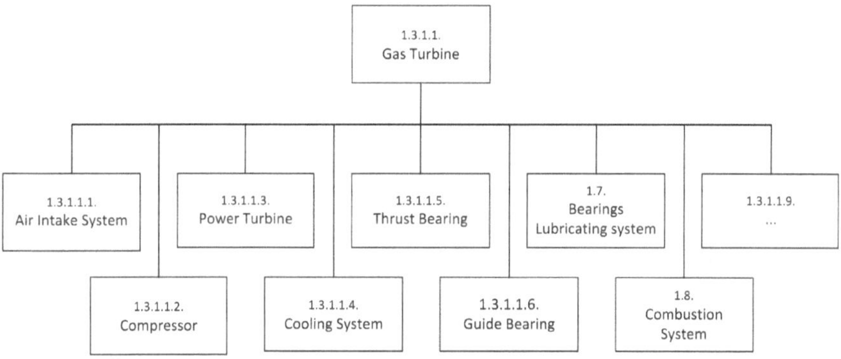

FIG. 9.30 Components of the gas turbine of one generating unit.

Finally, the study of the system gathered the failure and repair data of the equipment of interest for the RAM analysis. It is worth mentioning that these data were not obtained from a specific FPSO unit due to the nature of this example of application. Therefore, Table 9.8 presents the failure and repair data assumed in the FPSO power generation system example of application. It is worth mentioning that, in this application, it was considered that the temporal unit is measured in hours; therefore, λ is expressed in h^{-1}.

With the results of the activities in the study of engineering systems, it was possible to carry out the next two stages of the FPSO power generation example of application: RAM analysis in Section 9.3.2 and HMFD in Section 9.3.3.

TABLE 9.6 Analysis of potential failure modes and effects for the FPSO generation unit gas turbine.

ID item	Item	Function	Potential failure modes	FM #	Potential failure effects
1.3.1.1.1.	Air intake system	Provides clean atmospheric air to the compressor	Insufficient airflow	FM01	Output power reduction
1.3.1.1.2.	Compressor	Supplies pressurized air to the combustion system	Insufficient airflow	FM02	Output power reduction
			Compressor rubbing	FM03	Excessive vibration; potential generation failure
1.3.1.1.3.	Power turbine	Transforms the kinetic energy of flue gases into mechanical energy	Turbine blade rubbing	FM04	Excessive vibration; potential generation failure
1.3.1.1.4.	Cooling system	Provides cooling for lubricating oil	Insufficient lubricating oil cooling	FM05	Overheating; increased wear; potential locking of mechanical parts
1.3.1.1.5.	Thrust bearing	Supports the rotating shaft and its radial and axial loads	Do not support shaft radial loads	FM06	Vibration; potential propulsion impairment
1.3.1.1.6.	Guide bearing	Supports the rotating shaft and its radial loads	Do not support shaft radial or axial loads	FM07	Excessive vibration; potential generation failure
1.3.1.1.7.	Bearing lubricating system	Provides lubrication oil continuously to the turbine bearings	Provide insufficient bearing lubrication	FM08	Excessive vibration; potential generation failure
1.3.1.1.8.	Combustion system	Burns fuel	Insufficient fuel flow	FM09	Power loss; potential generation failure

TABLE 9.7 Failure mode detectability analysis for the FPSO generation unit gas turbine.

FM #	Failure rate (h^{-1})	Measurements	Clusters
FM01	0.00000281	Compressor inlet pressure (air)	C01
		Exhaust temperature (power turbine outlet)	
		Power turbine inlet temperature	
FM02	0.00000623	Compressor outlet pressure (air)	C02
		Exhaust temperature (power turbine outlet)	
		Power turbine inlet temperature	
FM03	0.00000729	Turbine rotation speed	C03
		Compressor case vibration	
FM04	0.00000011	Turbine rotation speed	C04
		Turbine case vibration	
FM05	0.00002587	Cooling water temperature	C05
		Cooling water pressure	
		Turbine case temperature	
FM06	0.00000176	Thrust bearing temperature	C06
		Turbine case vibration	
		Thrust bearing vibration	
FM07	0.00000176	Guide bearing temperature	C07
		Turbine case vibration	
		Guide bearing vibration	
FM08	0.00004523	Turbine case vibration	C08
		Thrust bearing temperature	
		Thrust bearing vibration	
		Guide bearing temperature	
		Guide bearing vibration	
		Lubricating oil inlet pressure	
FM09	0.00000685	Turbine rotation speed	C09
		Fuel flow	

TABLE 9.8 Failure and repair data for the equipment units of the power system.

Item	Reliability ($R(t)$)		Maintainability ($M(t)$)	
	Distribution	**Parameter**	**Distribution**	**Parameter**
1.3.1.1. Gas turbine	Exponential	$\lambda = 0.0000500$	Exponential	$\lambda = 0.0040$
1.3.1.2. Waste heat recovery unit	Exponential	$\lambda = 0.0000400$	Exponential	$\lambda = 0.0020$
1.3.1.3. Generator	Exponential	$\lambda = 0.00000025$	Exponential	$\lambda = 0.0010$

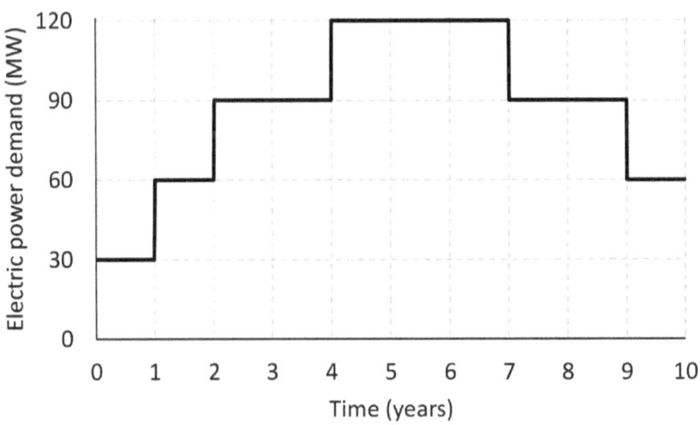

FIG. 9.31 FPSO's electric power demand over 10 years of operation.

9.3.2 RAM analysis of the FPSO generation system

The RAM analysis of the FPSO's power generation system was performed using a GSPN model, as previously mentioned in Section 9.1.2. GSPN has proved to be an important tool for the analysis of system reliability and availability (Ramos, 2018; Ramos & de Souza, 2017). To carry out such an analysis, it was considered that the system consists of four independent generating units, capable of providing 30 MW of power each. It was also considered that the FPSO's electric power demand changes throughout its operational life, as shown in Fig. 9.31.

By analyzing Fig. 9.31, it is noted that, over the first 10 years of the FPSO's operational life, only in three of them the electric power demand reaches a peak of 120 MW. In the other 7 years, not all four generating units need to operate to meet electrical demand, which means that units that are not in operation work as a backup for those that are.

Each generating unit is composed of three subsystems: turbine, Waste Heat Recovery Unit (WHRU), and generator. Each of these subsystems can be represented by the Petri net shown in Fig. 9.32. In this model, there

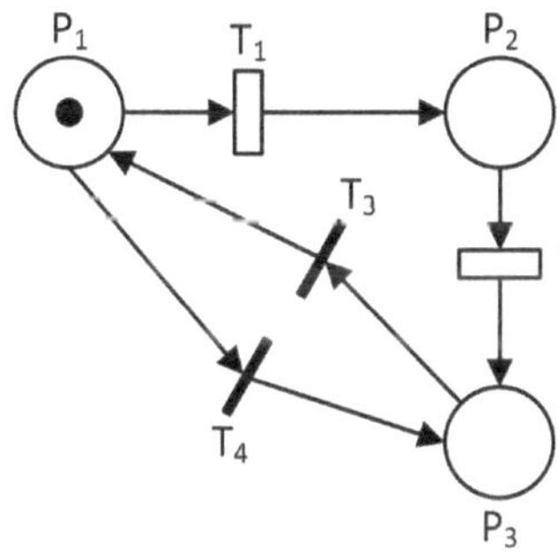

FIG. 9.32 Petri net model of a subsystem of the FPSO's power generation system.

are three places (P_1, P_2, and P_3), four transitions (T_1, T_2, T_3, and T_4), and a token, positioned at P_1. The three places represent the three possible states of the subsystem: in operation (P_1), in failure (P_2), and standby (P_3). The transitions represent the events that cause the subsystem to change state: the occurrence of a failure (T_1), its repair (T_2), the need to enter into operation (T_3), and the need to enter in standby (T_4). Finally, the token represents the current state of the subsystem.

It is important to note, from Fig. 9.32, that T_1 and T_2 are timed transitions whose firing delays are based on probabilistic distributions that represent, respectively, the time that the subsystem takes to present a failure and the repair time. In the model, both distributions were considered exponential and the failure and repair rates from Table 9.8 were used as parameters. Transitions T_3 and T_4, on the contrary, are immediate, being triggered as soon as predetermined rules are reached. In this example, these rules refer to the activation or shutdown of the subsystem so that only the necessary electric power is generated.

Once the Petri net model for a single subsystem is defined, it is necessary to replicate it so that the behavior of the power generation system as a whole is properly simulated. The final model of the Petri net is shown in Fig. 9.33.

It is important to note that, in order to obtain the reliability and availability estimates as a function of time, one must consider the position of all tokens present in the figure model at each moment. This is because, when a failure occurs or when the electric power demand changes, the tokens rearrange themselves, through the firing of transitions, to represent the reconfiguration of the real system. For example, in the first year of operation, when the demand is 30 MW, only one generating unit needs to be in operation and the other three must be on standby, as shown in Fig. 9.33. If, in this case, there is a failure in the turbine of the first generating unit (i.e., T_1 fires), its token goes to P_2 and the WHRU and generator tokens of the first unit go to standby (since there is no way they can operate without

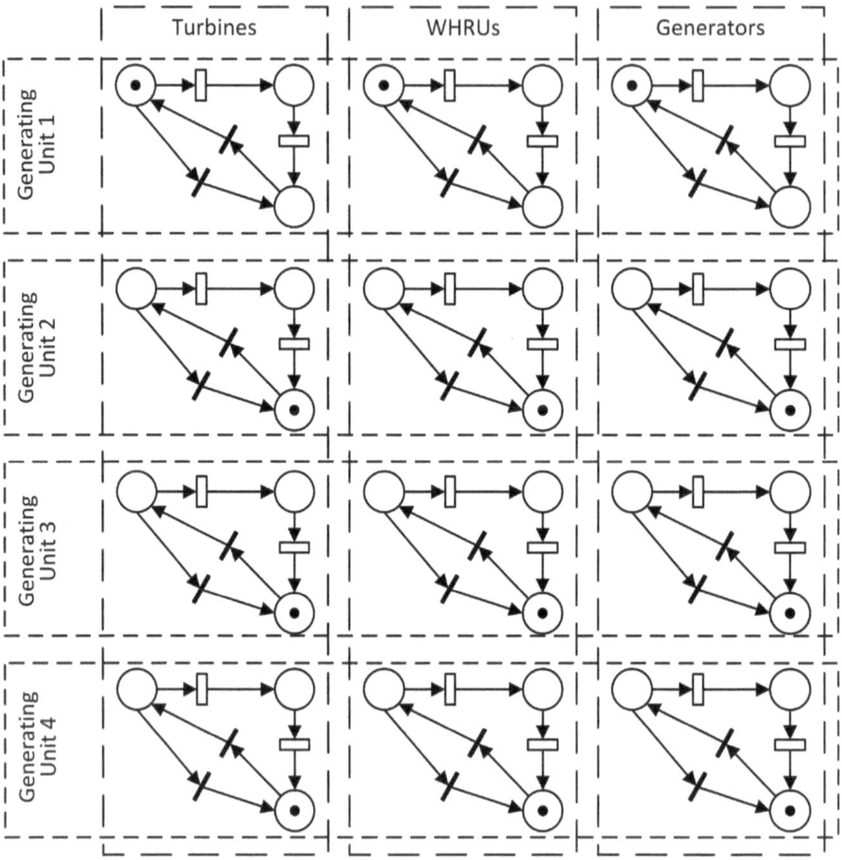

FIG. 9.33 Petri net model of the FPSO's power generation system.

their turbine). Simultaneously, the generating unit 2 goes into operation to be able to generate the electrical power that the first unit failed to generate, that is, the tokens of its three subsystems go from P_3 to P_1. After the failed turbine is repaired, its token goes to P_2.

As one may realize by now, the tokens change places whenever a transition is triggered, i.e., whenever an event that forces a change in the system configuration occurs. Therefore, for the reliability and availability of the system to be properly estimated, the model must always verify, each time a reconfiguration takes place, if the demand has been met, that is, if the system is capable of fulfilling its function.

Once the GSPN model has been developed, it must be simulated just as described in Chapter 4 for the results to be analyzed. Through a simulation that considered ten years of operation of the FPSO, the estimates of reliability and availability for its power generation system were obtained as presented, respectively, in Figs. 9.34 and 9.35.

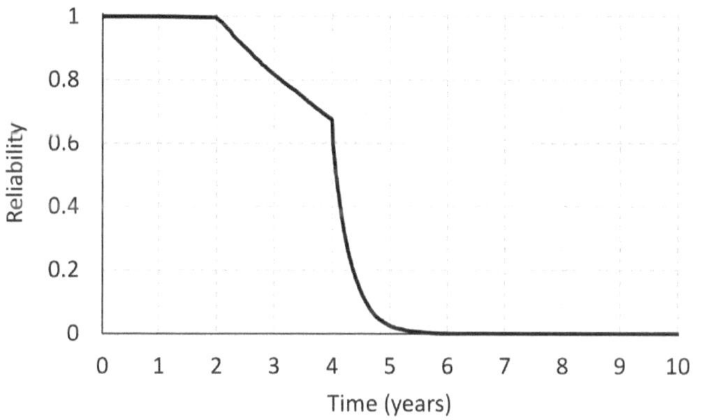

FIG. 9.34 Estimated reliability for the FPSO's power generation system.

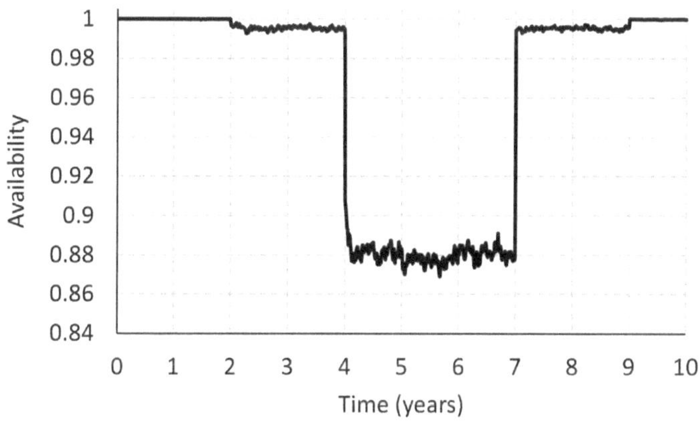

FIG. 9.35 Estimated availability for the FPSO's power generation system.

As shown in Fig. 9.34 that the system's reliability remains close to 100% in the first 2 years of operation. This is because the energy demand in this interval is small when compared to the total generation capacity of the system, making it unlikely that the system will not be able to meet the demand at any point in this time interval. The reliability value starts to drop considerably between the 2nd and 4th year of operation, when demand reaches 75% of total capacity, and drastically reduces when demand reaches full capacity.

Similarly, Fig. 9.35 shows that the lowest estimated value for availability, around 88%, is reached in the time interval corresponding to the highest demand. Such results can contribute positively to the decision-making regarding the FPSO operation. When analyzing the reliability and availability estimates, the system managers, i.e., the decision-makers, can decide to change the FPSO production output in order to flatten the energy demand curve, aiming at a better performance result.

9.3.3 HMFD of the FPSO generation system

Following the same steps as the first application example, this section presents the implementation and application of the HMFD framework presented in Chapter 7 for the FPSO generation system.

As the reader already knows, the first step in implementing and applying the mentioned framework is data acquisition. And, as in the case of the CODOG system, an FPSO has a SCADA system from which it is possible to collect relevant information about the operational conditions of the system and its state over time. Figs. 9.36–9.38 show some temperature, vibration, and pressure data, respectively, over a given time interval. In this example, the FPSO has been in operation for just over a year, that is, right after the demand for electricity generation has gone from 30 MW to 60 MW (as presented in Fig. 9.35).

In this case, a few days before the system completes 12,000 h of operation, a malfunction associated with a fault in the lubrication system of the turbine bearings (FM08 of Table 9.7) started. The natural wear and tear of the system components can be associated, in this example, with the source of the failure.

As in the previous application example, fault detection is the step that proceeds from data acquisition. The same PCA-based method is applied to the monitored data of the FPSO, and a 120-h analysis window is also used in this case. The reference data were collected at the beginning of the first year of operation of the unit, after its commissioning.

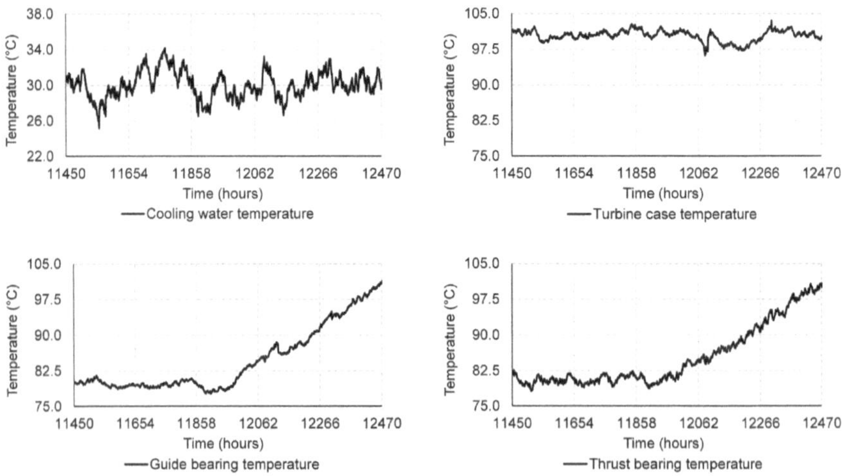

FIG. 9.36 Temperature measurements of the FPSO generation system.

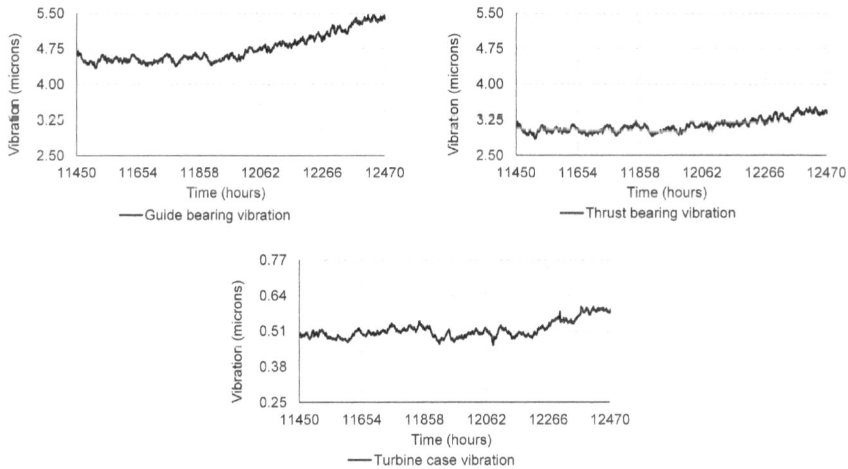

FIG. 9.37 Vibration measurements of the FPSO generation system.

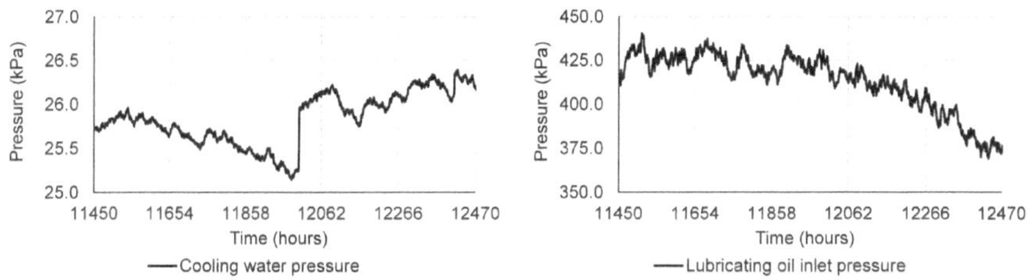

FIG. 9.38 Pressure measurements of the FPSO generation system.

Considering the clusters related to the analyzed failure mode, i.e., the clusters that may be affected by the presence of the fault in the system (clusters C05, C07, C08, and C09, as shown in Table 9.7 in this case), Fig. 9.39 presents the expected value (average value) of the T^2-statistic ($E[T^2]$) over time. It is noted in this case that only two clusters are, in fact, affected by the presence of the failure mode in the system, exceeding the established threshold for the variation of the T^2-statistic average value during the observed period. In this example, the threshold is also established by adding three times the value of the standard deviation to the average value of the T^2-statistic distribution resulting from the PCA application to a data set collected with the system under a health condition in the initial phase of operation of the system.

In this application example, the Bayesian network is also used as a fault diagnosis method and its graph representation is presented in Fig. 9.40.

Being known the failure rates (Table 9.7) of the system and the behavior of the clusters over time, it is possible to estimate *a posteriori* probability of each failure mode from the Bayesian network. Fig. 9.41 shows the result

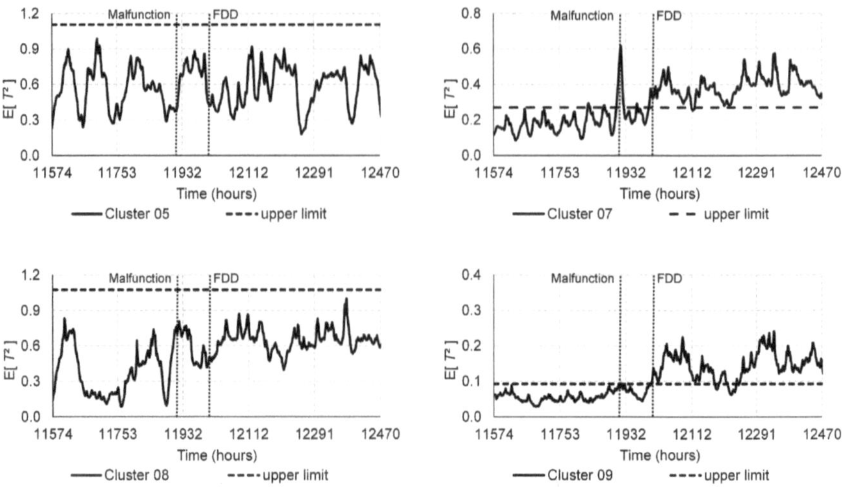

FIG. 9.39 Clusters' analysis of the FPSO generation system.

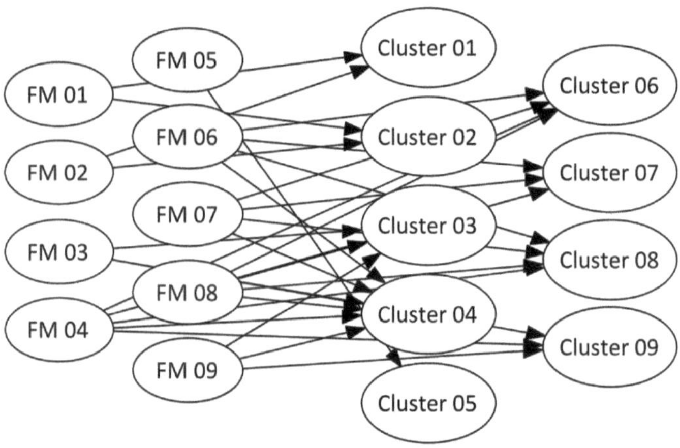

FIG. 9.40 Bayesian network graph of the FPSO generation system.

for the considered failure mode in this example (FM08), combined with the evolution over time of each related cluster (C05, C07, C08, and C09). In this case, the fault was diagnosed only 72 h after it started to affect the system's behavior. The *a posteriori* probability of about 40% for the failure mode is a more than solid indication that decisions can be made based on this result.

From the fault diagnosis, the RUL estimate is performed. The result is shown in Fig. 9.42. The hours of operation of the system are on the abscissa axis and the RUL, in hours (considering a logarithmic scale), on the ordinate axis.

FIG. 9.41 FDD analysis of the FPSO generation system.

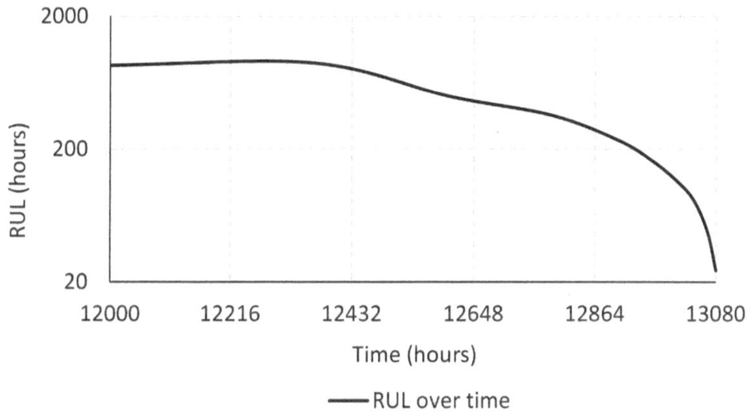

FIG. 9.42 RUL analysis of the FPSO generation system.

The same approach for calculating the RUL applied in the first application example is replicated in this case, where the minimum value between the results obtained for the several variables affected by the failure mode is considered. Note that the behavior of the RUL graph shows an accentuation in the reduction of the RUL value as the equipment remains in operation. This behavior will demand more agility from specialists and decision-makers as the equipment continues to operate.

Another interesting aspect is the effect of an eventual failure in the system considering how the electric power demand changes throughout its operational life and the moment when the failure may happen. For example, strategically the issue is much simpler when the energy demand is still low, as in the considered example of application. Between the 4th and 7th years of operation, however, the situation becomes more complex, requiring a more precise approach, raising the importance of calculating the RUL and the entire FDD process.

9.4 Final remarks

The practical approach of this chapter was designed for the reader to connect the ideas of the previous chapters through two examples of application. Both contexts of modern engineering systems showed the complexity in performing reliability analysis and health monitoring and fault diagnosis for asset management. At the same time, these examples have shown in practice the potential of the results that can be obtained with the different topics covered in this book.

It is evident that not all techniques would be demonstrated in this chapter due to the scope of this book. For this reason, the authors invite the reader to practice and extend the knowledge of this book to different contexts of application. During this practical approach, the reader will be able to reflect on the challenges for the implementation of the techniques and better adapt the application according to expected results and availability of information.

All the techniques presented not only in both examples of application but throughout the book aimed to support decision-making in organizations that have complex engineering systems. In general, this is the final contribution of all these approaches. Although some techniques can contribute at other stages of the life cycle of physical assets such as design and disposal, it can be said that the maintenance stage is the main beneficiary of the content covered in this chapter.

The results of RAM analysis and HMFD can directly support several decisions in maintenance management. Based on the simulations of RAM analysis, maintenance management can have estimates of the general availability of the systems and the reliability of each subsystem and unit of equipment. This can contribute to planning maintenance stops and prioritizing maintenance work orders. Besides, it is possible to estimate the number of failures expected for these subsystems and equipment units and their downtime, contributing to maintenance planning and dimensioning of spare parts stocks.

On the contrary, the application of the HFMD can provide the current condition of the engineering systems based on the monitoring of parameters associated with the observation of the potential FM. This is a highly recommended strategy for organizations of modern engineering systems, as it makes it possible to identify potential failures in advance and to plan interventions in an optimized way. In addition, HMFD contributes to the risk management of engineering systems as an active monitoring protection measure.

Both examples of the application addressed modern engineering systems of asset-intensive industries. In addition to the naval and oil and gas industries, other industries could have been used as a context such as electricity generation and distribution, water supply and treatment, transportation, mining, chemicals, manufacturing, and aviation. The chosen examples should not limit the reader's choice in which contexts to apply these analysis techniques. On the contrary, any organization that has modern engineering systems can benefit from these techniques.

Finally, it is worth mentioning that the reader may face several challenges in order to implement the techniques covered in this book in practice. For instance, it became clear the demand for knowledge and data about the engineering systems for a RAM analysis. Moreover, Chapter 7 covered other aspects that may be barriers to the implementation of HMFD in organizations. Accordingly, there may be a need for a culture change in organizations that support the implementation of these techniques for better decisions in achieving organizational goals.

References

ISO. (2016). *ISO 14224—Petroleum, petrochemical and natural gas industries—Collection and exchange of reliability and maintenance data for equipment*. International Organization for Standardization. https://www.iso.org/standard/64076.html.

Melani, A. H. A., Michalski, M. A. D. C., da Silva, R. F., & de Souza, G. F. M. (2021). A framework to automate fault detection and diagnosis based on moving window principal component analysis and bayesian network. *Reliability Engineering & System Safety, 215*. https://doi.org/10.1016/j.ress.2021.107837, 107837.

Michalski, M. A. D. C. (2021). *Rotating machinery fault identification using model-based and data-based techniques integration*. University of São Paulo.

Mujica, L. E., Rodellar, J., Fernández, A., & Güemes, A. (2011). Q-statistic and T2-statistic PCA-based measures for damage assessment in structures. *Structural Health Monitoring, 10*(5), 539–553. https://doi.org/10.1177/1475921710388972.

OREDA. (2009). *Offshore & onshore reliability data* (5th ed.).

Peltz, E., Brauner, M. K., Keating, E. G., Saltzman, E., Tremblay, D., & Boren, P. (2014). DoD depot-level reparable supply chain management-process effectiveness and opportunities for improvement. In *Research Reports* RAND Corporation.

Quanterion Solutions Incorporated. (2016). Nonelectronic parts reliability data publication (NPRD-2016). In *Reliability databook series* Quanterion Solutions Incorporated.

Ramos, A. G. (2018). *Análise de disponibilidade de um sistema de tratamento de gás em instalações offshore utilizando redes de Petri estocásticas*. University of São Paulo. https://doi.org/10.11606/D.3.2018.tde-13072018-134254.

Ramos, A. G., & de Souza, G. F. M. (2017). Availability assessment of an offshore gas treatment system using stochastic petri nets. In *Proceedings of the 27th international ocean and polar engineering conference* (pp. 1087–1092). https://onepetro.org/ISOPEIOPEC/proceedings-abstract/ISOPE17/All-ISOPE17/ISOPE-I-17-197/17426.

Royal Australian Navy. (2016). Australian maritime logistics doctrine. In *Australian Maritime Doctrine* (1st ed.). Department of Defence CANBERRA ACT 2600. https://www.navy.

gov.au/sites/default/files/documents/Australian_Maritime_Logistics_Doctrine%20%2821Dec16_Amendment%29_web.pdf.

Shingo, S. (1988). *Non-stock production: The Shingo system of continuous improvement* (1st ed.). Productivity Press.

United States Government Accountability Office. (2020). *Navy ship maintenance—Actions needed to address maintenance delays for surface ships based overseas (GAO-20-86)*. United States Government Accountability Office.

CHAPTER

10

Concluding remarks

It's what you learn after you know it all that counts.

John Wooden (1910–2010)

The design of modern engineering systems is a challenging task for engineers. The fast development of new technologies not only associated with electromechanical devices and sensors but also with computer processing and storage capacity has provided new ways to improve system operational performance. The application of these new technologies requires more sophisticated techniques to assess the reliability of those modern engineering systems. Those new technologies also allow monitoring the degradation of system operational performance and defining the root cause of that degradation improving maintenance capability.

In order to take advantage of all the possibilities offered by these new technologies, the reliability engineer must constantly improve its knowledge but also needs to gain experience through the management of modern engineering systems. The combination of knowledge and design experience will consolidate the engineer's capacity to face the new challenges associated with the design of future engineering systems.

10.1 Book contributions

Engineering is the field, discipline, or practice that relates to the development, acquisition, and application of technical, scientific, and mathematical knowledge aiming at designing devices that will improve the quality of human life. The increase in the volume and complexity of knowledge developed by the fundamental sciences, such as mathematics, physics, chemistry, biology, and others, allows the engineers to design modern engineering systems that are capable of executing functions in many fields, such as medicine, telecommunications, transportation, and power generation.

https://doi.org/10.1016/B978-0-12-823521-8.00009-8

The modern engineering systems besides presenting an operational performance must be capable of not only improving the quality of human life but also avoiding damages to the environment during its operation. Environmental damage mitigation is a broad term that encompasses the minimization of CO_2 emissions, the use of recycled materials, and even the minimization of impacts on communities surrounding the system.

The design of modern engineering systems can be based on design methodologies that will consider the system breakdown into subsystems and devices aiming at defining the functional relationships and interfaces between them in order to specify their design requirements and constraints. Among the various design requirements are those related to operational performance such as reliability and availability.

The system reliability is directly associated with the functional architecture and dependency between subsystems and devices. The system reliability is dependent on the reliability of each device and the possible use of redundant devices for critical and/or safety functions. The system availability depends on reliability and maintenance planning.

This book covers aspects associated with studies of reliability and availability of engineering systems. Initially, the book presents the concepts of reliability, risk analysis, and maintenance aiming at relating them to the design of modern engineering systems. System reliability modeling is discussed aiming at improving its performance and avoiding the occurrence of critical failure that may endanger the system safety or threaten human life and the environment. The relationship between system reliability and the consequences of critical failures is defined through the concept of risk. Risk assessment and analysis techniques are briefly presented including qualitative and quantitative methods. Aiming at defining the maintenance plan for engineering systems' reliability-centered maintenance and risk-based maintenance philosophies are presented addressing the main requirements associated with their application. To evaluate the performance of maintained systems, the concept of availability is presented.

In view of the increasing complexity of modern engineering characterized by the use of electromechanical devices associated with control systems, including software, their design will be made according to more sophisticated methodologies that consider the interaction between subsystems to define system performance. Some of the most modern design methodologies with the ability to consider requirements based on reliability and operational risk are presented. The traditional systems' reliability analysis methods are not capable of dealing with some specificities of modern engineering systems' operational aspects such as functional dependency between components, the capacity of reconfiguration in case of critical component degradation or failure, and even difficulties in defining a comprehensive maintenance planning capable of eliminating the degradation of the most critical components. Those typical features of modern engineering systems may affect device failure rate modeling and the definition of systems' operational states. To deal with those characteristics, other systems' reliability modeling methods, such as those presented in Chapter 4, must be used aiming at considering the dynamic variation of system operational states.

By the end of the 1990s and the beginning of the 2000s, there was a great advance in the technology of sensors and monitoring systems. Through the use of a monitoring database, reliability analysis techniques were developed to estimate device degradation and remaining useful life. These developments have given new perspectives to the application of the predictive maintenance practice.

At the beginning of the 2010s, the recent development in wireless telecommunications technology and the use of more sophisticated sensing devices coupled to Supervisory Control and Data Acquisition systems as part of modern engineering system architecture allowed the implementation of health monitoring systems to monitor system performance degradation. Those health monitoring systems are based on the application of artificial intelligence algorithms to detect the occurrence of critical component degradation and estimate their remaining useful life. This analysis is supported by the availability of a great amount of data stored in the cloud database. The methods used to detect operational abnormalities in components of modern engineering systems are presented and discussed in Chapter 5. The chapter does not intend to provide a comprehensive analysis of supervised and unsupervised artificial intelligence machine learning algorithms but aims at presenting their main characteristics and applicability to fault detection in engineering systems. The fault detection process is important to improve maintenance planning since it automates this activity, reducing dependence on human judgment.

Once a fault is detected in an engineering system, a diagnosis method may be used to define the most likely component whose degradation is causing the fault. Some of those techniques are presented in Chapter 6 aiming at providing the reader with information about the most suitable methods to perform fault diagnosis. The fault diagnosis process supports maintenance activities since it abbreviates the time required to define the origin of a system fault.

Based on the information presented in Chapters 5 and 6, the book presents in Chapter 7 the fundamentals of health monitoring and fault detection system development. This development is based on the selection of the most suitable fault detection and diagnosis process considering the system operational context and performance requirements. A framework is presented and is dependent on the definition of the most critical failure modes for system performance. For those failure modes, the analyst shall select the monitoring parameters that are affected by the degradation associated with a given failure mode. After that selection, the most suitable fault detection and diagnosis methods must be selected considering the expected degradation patterns.

The health monitoring and fault detection system is an important part of an asset management system. The asset management system aims at keeping the operational performance and safety throughout the operational life with minimum maintenance costs. The fundamentals of the asset management are presented in Chapter 8 and provide guidance on managing modern engineering systems based on ISO 55000 series of standards. The chapter focuses on maintenance planning incorporating all methods previously presented along with the book. The organization of maintenance management according to asset management philosophy improves the effectiveness of maintenance activity and consequently improves system availability and safety.

The application of health monitoring and fault detection methods developed in this book is presented in Chapter 9. To demonstrate the feasibility of its application, two cases are presented. The first one involves the analysis of a ship-combined diesel or gas propulsion system and the second one analyzes the electrical power generation system for an offshore processing plant. Both studies indicate the advantages of the proposed method for fault detection and diagnosis and show that preventive maintenance practice is improved with the application of the method. The method may be used as a part of an asset management system.

10.2 Future developments

Although the book presented the relationship between reliability and risk analysis concepts with the development of a health monitoring system to support asset management, there are still some challenges that must be faced during the design of modern engineering systems.

The Industry 4.0 concept brings the integration of artificial intelligence, cloud computing, and engineering allowing engineers to create new systems and innovate smart devices. These new designs can increase the functional dependency complexity among system devices and components and may need the improvement of reliability and availability simulation methods.

Modeling and simulation support the design of systems allowing testing of solutions and assessing system performance during design steps avoiding the execution of a great number of reliability tests. This is particularly interesting where system testing on real equipment is not cost-effective. The grand challenge concerns the development of reliability modeling and simulation techniques to allow system analysts to concentrate on critical system parameters/behaviors and evaluate their potential outcomes.

As the complexity and the scale of a system increase, so does the problem of developing an accurate and validated model. When dealing with complex systems involving humans in the loop, the major weakness is the inexistence of a comprehensive model of human behavior or cognitive performance. The ability to build and use complex models has been enhanced by the use of systems engineering techniques such as hierarchical decomposition, object-oriented modeling and programming, model-based system engineering, and agent-based modeling.

Coupled with the problem of system modeling as for reliability analysis, there is the problem of redundancy allocation. The goal of redundancy allocation is to increase system reliability and minimize system manufacturing and operational costs. Depending on the system configuration and functional dependency between devices, the reliability allocation problem must be solved considering advanced algorithms, such as genetic algorithms. That allocation problem may also consider the system maintenance aspects including repair rate and imperfect maintenance. The problem of obtaining optimal design of the engineering systems, including reliability optimization, will continue to play an important role in the design process and will support the enhancement of system performance in a specific operational environment.

Based on the concepts of Industry 4.0, the system design process may consider the development of model-based algorithms to support the development of health monitoring and diagnosis algorithms. The model must represent the effects of critical component failure modes on system process parameters that can be monitored and registered by the Supervisory Control and Data Acquisition system. The digital model representation, named digital twin, is developed for each large piece of equipment that is part of an industrial plant, such as compressors, gas and vapor turbines, boilers, mixing tanks, condensers, and others. The integration of each individual model in a modern complex model, representing the process plant, is also a challenge faced by reliability and maintenance planning engineers. The key challenge of any model and simulation activity is how accurately it represents the system under investigation. The model-based development will be an important tool in the reliability and availability analysis of modern engineering systems.

Another challenge faced by reliability engineers is data gathering to support reliability and availability analysis during system design. The use of innovative components and devices in the system design may impose challenges to reliability engineers since failure and reliability data may not be available for those devices. Devices with technology readiness levels between 6 and 7 may be used on the system design and its reliability on the system operational environment might not be evaluated. The reliability engineer may be capable of designing tests that may be used to furnish failure data that may support device reliability analysis. Data accuracy plays an important role in any reliability-based engineering system design and the tests designed to support reliability evaluation must be based on physics of failure analysis aiming at defining test sequence and environment, including external loading, that shall reproduce the device operational condition.

Moreover, reliability analysis of devices and engineering systems may have difficulty increased by the diversity of system operational conditions. The reliability modeling for various operational conditions may be a laborious task since the criticality of components' failures may vary for each operational condition and the problem of reliability allocation has increased complexity. The reliability analyst may develop a model that supports the optimum reliability allocation and preserves the systems' operational safety in many environments.

Finally, it is important to remind that reliability and availability analysis problems are variable challenges in view of the constant development of new modern engineering systems aiming at attending to the new needs of human beings. Nowadays, important drives for the development of modern engineering systems are the United Nations Sustainable Development Goals aiming at climate change mitigation, industry innovation and infrastructure, affordable and clean energy, and among others. Reliability and operational safety will certainly be design requirements and will require new developments in reliability, risk analysis, and maintenance planning.

Index

Note: Page numbers followed by *f* indicate figures and *t* indicate tables.

E

F

G

H

Kolibri 360 GmbH
Ferdinand-Jühlke-Straße 7
99095 Erfurt, Deutschland
produktsicherheit@kolibri360.de